PHOTOACOUSTIC
AND
THERMAL WAVE
PHENOMENA
IN
SEMICONDUCTORS

PHOTOACOUSTIC AND THERMAL WAVE PHENOMENA IN SEMICONDUCTORS

Edited by

Andreas Mandelis

Photoacoustic and Photothermal Sciences Laboratory
Department of Mechanical Engineering
University of Toronto
Toronto, Canada

North-Holland
New York • Amsterdam • London

Elsevier Science Publishing Co., Inc.
52 Vanderbilt Avenue, New York, New York 10017

Sole distributors outside the United States and Canada:
Elsevier Science Publishers B.V.
P.O. Box 211, 1000 AE Amsterdam, The Netherlands

Library of Congress Cataloging in Publication Data

Photoacoustic and thermal wave phenomena in
semiconductors

Includes bibliographies and index.
1. Semiconductors – Optical properties. 2. Semiconductors – Thermal properties. 3. Optoacoustic
spectroscopy. 4. Imaging systems. I. Mandelis, Andreas.
QC611.6.06P48   1987   621.3815'2   87–19948
ISBN 0–444–01226–5

Current printing (last digit):
10 9 8 7 6 5 4 3 2 1

Contents

I

THERMAL WAVE MICROSCOPY OF SEMICONDUCTORS

II

THERMAL WAVE IMAGING AND CHARACTERIZATION OF SEMICONDUCTOR MATERIALS AND DEVICES

III

NOVEL PHOTOTHERMAL WAVE TECHNIQUES, INSTRUMENTATION AND SEMICONDUCTOR APPLICATIONS

IV

ELECTRONIC TRANSPORT AND NONRADIATIVE PROCESSES IN SEMICONDUCTORS

V

PHOTOTHERMAL WAVE SPECTROSCOPIES OF SEMICONDUCTORS

List of Contributors

Numbers in parentheses indicate the pages on which the author's contribution begins.

A. CLAUDE BOCCARA (237,287), Laboratoire d'Optique Physique- ER 5 CNRS, Ecole Supérieure de Physique et de Chimie, 10, rue Vauquelin, 75231, Paris Cedex 05, FRANCE

LI CHEN (27), Institute of Acoustics, Nanjing University, Nanjing, PEOPLE'S REPUBLIC OF CHINA

HANS COUFAL (147), IBM Almaden Research Center, 650 Harry Road, San Jose, California 95120-6099, USA

LAWRENCE D. FAVRO (69), Department of Physics and Astronomy, Wayne State University, Detroit, Michigan 48202, USA

DANIELLE FOURNIER (237,287), Laboratoire d'Optique Physique - ER 5 CNRS, Ecole Supérieure de Physique et de Chimie, 10, rue Vauquelin, 75231, Paris Cedex 05, FRANCE

TETSUO IKARI (397,425), Department of Physics, Kurume University, 1635 Mii, Kurume, Fukuoka, JAPAN

MASANOBU KASAI (3), Department of Industrial Chemistry, Faculty of Engineering, The University of Tokyo, Hongo, Bunkyo-Ku, Tokyo, JAPAN

GORDON S. KINO (201), Ginzton Laboratory, Stanford University, Stanford, California 94305, USA

YUTAKA KOGA (397,425), Department of Physics, Kurume University, 1635 Mii, Kurume, Fukuoka, JAPAN

PAO-KUANG KUO (69), Department of Physics and Astronomy, Wayne State University, Detroit, Michigan 48202, USA

xiii

ANDREAS MANDELIS (147,323,353), Photoacoustic and Photothermal Sciences Laboratory, Department of Mechanical Engineering, University of Toronto, Toronto M5S 1A4, CANADA

NOBUO MIKOSHIBA (53), Research Institute of Electrical Communication, Tohoku University, Katahira, Sendai, JAPAN

LUIZ C.M. MIRANDA (135,311), Laboratorio Associado de Sensores e Materiais, Instituto di Pesquisas Espaciais, 12200 S.J. Campos, SP, BRAZIL

ALLAN ROSENCWAIG (97), Therma-Wave, Inc., Fremont, California 94539, USA

BARRIE S.H. ROYCE (257), Applied Physics and Materials Laboratory, Princeton University, Princeton, New Jersey 08544, USA

TSUGUO SAWADA (3), Department of Industrial Chemistry, Faculty of Engineering, The University of Tokyo, Hongo, Bunkyo-Ku, Tokyo, JAPAN

SHIGERU SHIGETOMI (397,425), Department of Physics, Kurume University, 1635 Mii, Kurume, Fukuoka, JAPAN

RICHARD G. STEARNS (201), Ginzton Laboratory, Stanford University, Stanford, California 94305, USA

ANDREW C. TAM (173), IBM Almaden Research Center, 650 Harry Road, San José, California 95120-6099, USA

KEIJI TANAKA (441), Department of Applied Physics, Faculty of Engineering, Hokkaido University, Sapporo 060, JAPAN

ROBERT L. THOMAS (69), Department of Physics and Astronomy, Wayne State University, Detroit, Michigan 48202, USA

KAZUO TSUBOUCHI (53), Research Institute of Electrical Communication, Tohoku University, Katahira, Sendai, JAPAN

HELION VARGAS (135,311), Instituto de Fisica, Universidade Estadual de Campinas, 13100 Campinas, SP, BRAZIL

ROBERT WAGNER (323), Photoacoustic and Photothermal Sciences Laboratory, Department of Mechanical Engineering, University of Toronto, Toronto M5S 1A4, CANADA

SHU-YI ZHANG (27), Institute of Acoustics, Nanjing University, Nanjing, PEOPLE'S REPUBLIC OF CHINA

Preface

Over the past ten years and since its modern re-emergence, the field of Photothermal Wave Sciences has witnessed a nearly explosive expansion in many directions, a result of the general applicability and adaptability of the field to many traditional and otherwise areas of research.

Similar to other application directions, the recent very rapid growth of large amounts of seemingly diverse and scattered material in the scientific literature, concerning the use of Photoacoustic and other emerging Photothermal Wave techniques in the investigation of properties of semiconductors, is the foremost factor which determined the necessity and inevitability of this volume. The book is intended to fill the gap generated by the lack of any other major in-depth publication in growing areas of many diversified photoacoustic/ photothermal wave applications to semiconductors by encompassing well established as well as imaginative new approaches to research by acknowledged leaders and pioneering workers in the field.

Having in mind the small population ratio of photoacoustic and other photothermal wave semiconductor workers to the broad semiconductor community at large, it became apparent that the needs of the latter would be served much better through its systematic introduction to, and familiarization with, major available photothermal wave techniques and detailed coverage of their applications to many familiar (and quite a few novel!) research and development problems.

Our principal goal in this book is to advance among the large community of semiconductor workers the understanding and awareness of the existence and versatility of thermal wave techniques, as well as the appreciation of the power and potential of these techniques in semiconductor research and characterization.

This treatise is organized into five parts whose highlights are as follows:

Part I presents detailed reviews of established and emerging thermal wave microscopies with emphasis on the potential of these techniques to identify macroscopic and microscopic subsurface defect structures in semiconductor materials and devices.

Part II explores the imaging capabilities of thermal wave techniques, both theoretically and experimentally, with focus on the measurement of semiconductor parameters such as ion implant profiles, electronic surface states and thermal diffusivities at the wafer or processed device level, which are difficult or impossible to measure non-destructively and/or in a non-contact manner by other non-thermal wave methods.

In Part III we introduce novel photothermal wave techniques which appear to be very promising for the study of semiconductors. Several of these techniques such as Pulsed Photothermal Radiometry and Fourier Transform Photoacoustic Spectroscopy in the far infrared spectral region (which is compatible with the energetics of shallow defects) have hardly been used with semiconductor systems, however, it is our hope that exposure of workers in the semiconductor field to the capabilities and potential of these novel techniques drawn from other condensed phase studies will be a catalyst toward developing such applications in the future.

In Part IV we discuss the abilities of photothermal wave techniques to monitor semiconductor phenomena at the electronic level. We show that electronic transport processes and nonradiative de-excitation mechanisms can be successfully probed *in-situ* in photoexcited bulk materials such as silicon crystals, as well as in surface active devices such as compound semiconductor photoelectrochemical cells.

Part V deals entirely with reviews of progress made in semiconductor spectroscopy through photothermal techniques. Important new measurements of materials properties have been achieved and new spectroscopic ranges have been attained thanks to the sensitivity of such techniques as Mirage Effect (or Photothermal Deflection) spectroscopy. The impact of the techniques is directly linked with basic Solid State Physics and applications in microelectronic and optoelectronic technologies. This part focusses on compound crystalline semiconductors, layered semiconductors and amorphous thin films.

The book is of a research level, however, could also be used as reference material for an upper level graduate course on the subject of photothermal phenomena in semiconductors.

I wish to thank all the contributors to this volume for their enthusiasm and dedication to the project. Special thanks are due to Miss Julie Fuda for her skillful and patient secretarial assistance in word-processing the text. The partial support of Elsevier Science Publishing Co., Inc., through a grant used as "seed" money to defray publication costs is gratefully acknowledged. I also wish to thank Professor R.D. Venter, Chairman of the Department of Mechanical Engineering of the University of Toronto, for kindly putting at my disposal the considerable word-processing, laser-printing and print-shop facilities of the Department, which were used for processing and printing the entire book. Last, but not least, I am grateful to my family for their immense understanding and patience during the many stages of writing and editing of the manuscript. I dedicate this book to my parents Alexandros and Nora.

Andreas Mandelis
Toronto, February 1987

PHOTOACOUSTIC
AND
THERMAL WAVE
PHENOMENA
IN
SEMICONDUCTORS

I

Thermal Wave Microscopy
of Semiconductors

NON-DESTRUCTIVE INSPECTION OF STACKING FAULTS & DISLOCATIONS OF SEMICONDUCTOR WAFERS BY PHOTOACOUSTIC MICROSCOPY (PAM) AND PHOTOTHERMAL BEAM DEFLECTION (PBD)

Tsuguo Sawada and Masanobu Kasai

Department of Industrial Chemistry
Faculty of Engineering
The University of Tokyo
Hongo, Bunkyo-Ku, Tokyo, JAPAN

I. Introduction

Photoacoustic Microscopy (PAM) has attracted remarkable attention as a unique tool which can carry out nondestructive evaluations, because of its distinctive features that cannot be found in conventional optical or electron microscopes. It can be used for both surface and subsurface characterization.

Photothermal Beam Deflection (PBD) is a new technique which detects Photoacoustic signals. Although PAM needs contact with a sample, the use of the PBD technique results not only in non-destructive but also in non-contact imaging. The PBD technique is therefore expected to become an imaging microscope for use with samples under atmospheric and/or high pressure conditions, as well as under vacuum, at high or low temperatures, and so on; in other words, for use even in hostile environments.

In this Chapter a general overview and some broad applications of piezoelectric PAM and PBD to semiconductors are presented, and the differences between PAM and PBD signals are discussed. Specific PAM applications to semiconductors using the microphone-gas cell method are discussed in Chapter 2.

II. Theoretical Principles

1. PRINCIPLES OF PAM AND PBD

Fig. 1 is a diagram showing the principles of PAM and PBD. When a laser beam is absorbed by a solid sample, a part, or all of, the light energy is

Fig. 1 Diagrams of the physical principles of PAM and PBD.

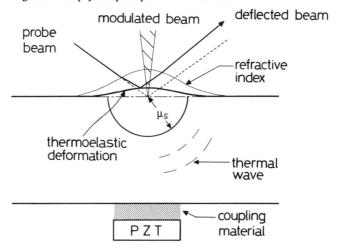

converted into heat. As the exciting laser beam intensity is modulated, the heat is generated repeatedly at the same frequency.

It has been suggested that there are three processes involved in the generation of the PA signal (Rosencwaig, 1980): The absorption of exciting optical energy, followed by the generation and propagation of thermal energy, and finally the generation and propation of elastic energy. It therefore appears that there are three different kinds of PA images, according to the above classifications, that is: 1. Images from local differences in optical properties within the light-reaching region, i.e. within a radius of μ_β, Eq. (1); 2. Images from local differences in thermal properties within the heat-reaching region, i.e. within a radius of μ_s, Eq. (1); and 3. Images from local differences in elastic waves generated at the heat source, and in elastic properties (Young's modulus, Poisson's ratio, the linear expansion coefficient, and local stress and strain) from heat source to the detector. μ_β is the light transmission length and μ_s is the thermal diffusion length, defined as

$$\mu_\beta \equiv \frac{1}{\beta} \; ; \quad \mu_s \equiv \left[\frac{2k}{\omega\rho c} \right]^{\frac{1}{2}} \tag{1}$$

where β is the optical absorption coefficient, k is the thermal conductivity, ρ is the density, and c is the heat capacity of the sample. ω is the angular modulation frequency.

Images generated due to mechanism (1) are the same as those obtained by optical microscopy. Images due to mechanism (2) are unique to PAM and cannot be obtained by either optical or acoustic microscopy. Since μ_s varies with the modulation frequency ($\mu_s \propto f^{-\frac{1}{2}}$), it is possible to carry out depth-profiling by changing modulation frequency as discussed in Chapter 2. The depth-profile and resolution are determined by the thermal diffusion length. This process is called thermal mode signal. Images due to mechanism (3) are similar to those obtained by acoustic microscopy. The depth-profile is determined by the sound wave length, which is generally much greater than the thermal diffusion length. This process is called acoustic mode signal.

In PAM, a modulated laser beam is focused onto a sample and scanned two-dimensionally across its surface, and the thermal wave (elastic wave) arising in the sample can be detected by a piezoelectric transducer attached directly to the sample. In a sample having an optically flat surface, if the probe beam is focused directly on the area irradiated by the exciting beam, it is deflected upon reflection from the surface of the sample. In this case, two processes occur. One is the deflection by the distribution of refractive index in the surrounding gas or liquid as a result of the thermal distribution on the surface of the sample. The other is the deflection by the thermal deformation on the surface of the sample.

Table I

Major PA techniques and measurement capabilities. OBD: Optical Beam Deflection; PTR: Photothermal Radiometry. Other abbreviations are defined in the text.

	Optical	Thermal	Elastic	characteristic	application
Scattered Light	O	×	×		
Microphone	O	O	× ?	nondestructive	solid,powder
PZT	O	O	O	contact vacuum	contact required
OBD	O	O	× ?	noncontact	any sample
PBD	O	O	O	noncontact vacuum	any conditions
PTR	O	O	× ?	noncontact vacuum	any sample

A list of various methods for measuring PA signals is given in Table I, which shows their characteristics and the measured quantities. In Table I a circle indicates measurement ability, a cross indicates no such ability, and a question mark indicates uncertainty. OBD (Optical Beam Deflection) measures deflection of a probe beam parallel to the surface of a sample. As shown in Fig. 2, OBD measures a signal only due to the distribution of refractive index, whereas PBD measures signals due to both refractive index gradient *and* thermal deformation. These two signals will be discussed further below.

2.   THEORETICAL BACKGROUND OF THE PAM AND
     PBD METHODS

Both PAM and PBD measure PA signals induced by heat generation. In order to analyze these signals, the equation for thermal diffusion must first be solved using appropriate initial and boundary conditions.

$$\frac{\partial T}{\partial t} = k\nabla^2 T + \frac{\overline{W}}{c\rho}$$
$$\nabla^2 = \frac{\partial^2}{\partial x^2} + \frac{\partial^2}{\partial y^2} + \frac{\partial^2}{\partial z^2} \tag{2}$$

where T is the temperature and $\overline{W}$ is the heat generation rate per unit time and per unit area.

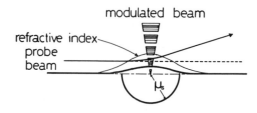

Optical Beam Deflection

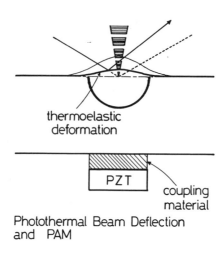

Photothermal Beam Deflection
and PAM

Fig. 2 Physical processes generating Optical Beam Deflection (OBD) and Photothermal Beam Deflection (PBD).

Next the thermoelastic Navier-Stokes equation for the displacement U_q must be solved with appropriate initial and boundary conditions:

$$\nabla^2 U_q + \frac{1}{1-2v}\frac{\partial e}{\partial q} - \frac{2(1+v)}{1-2v}\alpha\frac{\partial Z}{\partial q} + \frac{X_q}{G} = 0 \qquad (3)$$

$$2G = \frac{E}{1+v} \quad ; \quad e = \varepsilon_x + \varepsilon_y + \varepsilon_z \qquad (4)$$

where q is x,y or z; Z denotes differences of temperature; E is Young's modulus; v is Poisson's ratio; and α is the linear thermal expansion coefficient. ε_q and X_q stand for displacement gradient and strain, respectively, in the q-direction.

By use of the elastic displacement potential Φ, Eq. (3) can be transformed as follows:

$$\nabla^2\Phi = \left[\frac{1+\nu}{1-\nu}\right]\alpha Z \tag{5}$$

PAM and PBD methods make use of Eqs. (2) and (3). In what follows, we will discuss some of these methods.

a. The Opsal-Rosencwaig (O-R) Method

Opsal and Rosencwaig proposed a method of thermal-wave depth profiling (Opsal and Rosencwaig, 1982). According to this method a one-dimensional multilayer model of a sample with a non-homogeneous thermal conductivity was considered. The sample was excited by a plane heat source with sinusoidal time dependence, $Q=Q_0\exp(i\omega t)$, at the surface, $x=x_0=0$. Two equations, that of the thermal diffusion and the thermoelastic equation, were solved with two boundary conditions at the surface of the sample. One was the free condition, and the other was the rigid condition. The elastic displacement potential at the n-th layer was found to be

$$\phi_n(x) = \begin{cases} -\dfrac{\gamma_n}{ik}\exp(-ikx)\displaystyle\int_{x_{n-1}}^{x_n} dx'\,T_n(x') & \text{(rigid)} \tag{6a} \\[2em] -\gamma_n\exp(-ikx)\displaystyle\int_{x_{n-1}}^{x_n} dx'x'\,T_n(x') & \text{(free)} \tag{6b} \end{cases}$$

where γ is the thermoelastic constant, k is the elastic wave vector which is determined by Young's modulus, Poisson's ratio and the linear thermal expansion coefficient of the sample. T_n is the distribution of temperature at the n-th layer, and is given by the following equations

$$T_n(x)=A_n\exp(-q_nx)+B_n\exp(q_nx) \tag{7}$$

$$q_n=(1+i)/\mu_n \tag{8}$$

where μ_n is the thermal diffusion length of the sample at the n-th layer.

This method shows that depth profiling is possible by the thermoelastic wave. However, the above theory is too simple to be of value in many cases, where, for instance, a sample has a finite extent. In such cases, this theory is not corroborated by experimental results.

b. Theory of PBD

OBD is the method proposed by Boccara, Fournier, Jackson and Amer (1980), and Boccara, Fournier and Badoz (1980). It detects a signal due to a refractive index gradient. Although theories of OBD have also been proposed by Aamodt and Murphy (1981, 1983), we will not discuss them here, since the signal due to a refractive index distribution is much smaller than that obtained due to a thermal deformation.

Olmstead *et al.* (1983) proposed a theory for the PBD signal due to thermal deformation. This theory was applied to a cylindrical, semi-infinite plate model of a homogeneous sample, and a Gaussian exciting beam. The solutions to Eqs. (2) and (3) above were obtained using Green's functions. The differential displacement can be written according to Olmstead *et al.* (1983).

$$\frac{\partial U_z}{\partial r}(r,0) = f(\beta,R,H,D)$$

where R stands for the thermal properties of a sample, H is for the elastic properties and D is for the measuring conditions, i.e., the power of incident light, the thickness of the sample and the radius of the exciting beam. This theory revealed that the PBD signal was saturated at $\beta\mu_s$ greater than 1.

c. Simulation of PA signals by the Finite Element Method (FEM)

Although the above two theories provided a partial explanation of experimental data, they also left many cases unexplained, mainly because they used very simple models and simple conditions. An important generalization came with the simulation of PA signals by the FEM (Kasai, Ishioka, Kaihara, Fukushima, Sawada and Gohshi; unpublished), which is a numerical calculation using the matrix method. Recently the FEM has come to be used in many fields as the best available method for the solution of the thermal diffusion equation or the thermoelastic equation.

Fig. 3 The cylindrical model of the Finite Element Method (FEM) simulation.

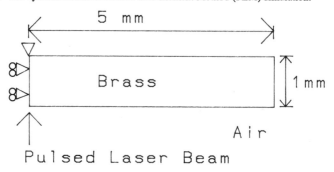

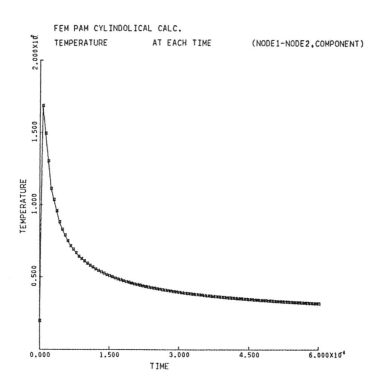

Fig. 4 The transient temperature signal at the center of a sample calculated by FEM

Fig. 3 shows a model sample in FEM calculations, using a pulsed exciting gaussian beam, which has a 6 ns duration, is 100 μm in diameter, and irradiates the center of a cylindrical brass sample. 1% of the energy is assumed to be absorbed by the sample.

Fig. 4 shows the transient temperature at the center of the surface of a sample. The temperature has a maximum at 10 ns and decays exponentially. Fig. 5 shows the distributions of (a) temperature, (b) displacement and (c) stress in radius. Advantages of the FEM method are: 1) this method is applicable to any sample under any conditions; 2) temperature, displacement and stress are obtained for any points on a sample at any time; and 3) using this method, Eqs. (2) and (3) can be reliably solved.

III. Experimental

A block diagram of the apparatus simultaneously used for both PAM and PBD is shown in Fig. 6. An Ar$^+$-ion laser (488 nm, Spectra Physics, Model-164) is used as the pumping beam. The beam of *ca.* 10μm diameter is modulated by an A-O. modulator (Intra Action Corp., AOM-40), and is focused on a

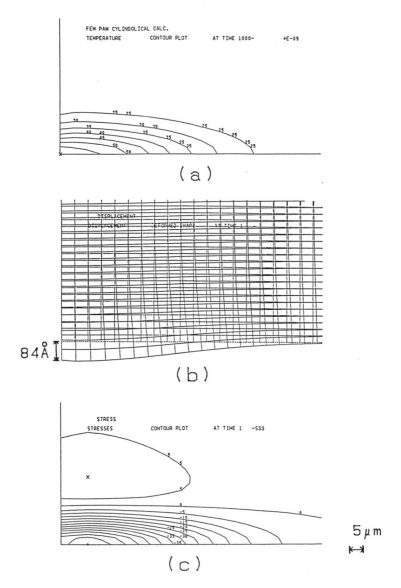

Fig. 5 The distributions of (a) temperature, (b) displacement and (c) radial stress at 1 μsec by FEM.

sample. A He-Ne laser (NEC, GLS-2027) used as a probe beam is focused on the area of the sample irradiated by the pumping beam. The resulting beam deflection is detected with a photodiode (HAMAMATSU S-1223) through a knife edge, which is used to block half portion of the probe beam. There are two methods of measuring beam deflection. One uses a knife edge and a photodiode, and the other uses a Position Sensitive Detector (PSD). Generally speaking, the PSD method has a slower response time. A fast response time is needed for microscopy, and therefore the former method was used with our apparatus. The signal from the photodiode, which is proportional to the amount of displacement of the probe beam, and the signal for the PZT joined to the sample using silver paste, are amplified by a lock-in amplifier (NF, LI-574 or LI-575). We tried several ways of joining the PZT, with optimal results obtained using silver paste. The output is stored in a personal computer (NEC, PC-9801VM2) equipped with an A/D converter (ADTEK SYSTEM SCIENCE CO., Ltd., R-488-AD). PAM and PBD images are obtained by moving the sample in directions X and Y using a controlled X-Y stage (Chuo Prec. Inc. Co., Ltd., PS20XY). The whole system is controlled by a personal computer. A scattered light image, which reflects the optical image of the sample surface, can be detected simultaneously by a photodiode for comparison with the PAM and PBD images.

Fig. 6 A block diagram of PAM and PBD apparatus.

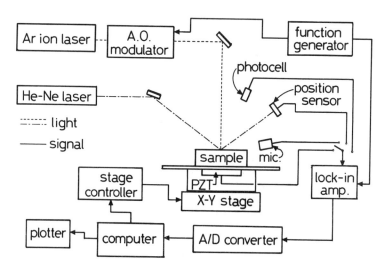

PZT AMP. IMAGE OF 8 ON Al

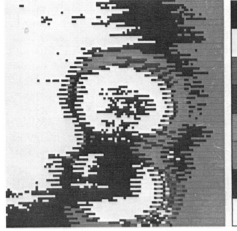

AL8AMP2

MEASURING POINTS
                    100 X 100

MEASURING AREA
            3000 X 3000 (μm)

LASER POWER 500 (mW)

LOCK-IN RANGE
            5000.00(μU)

AMPLITUDE MAX 5451.17(μU)   PHASE MAX   25.45(deg)
           MIN 457.52(μU)   MIN -56.01(deg)

(a)

PZT AMP. IMAGE OF 8 ON Al

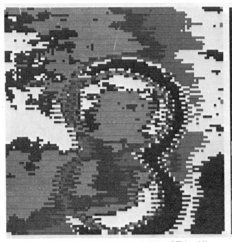

AL8AMP2.ME.ST

MEASURING POINTS
                    100 X 100

MEASURING AREA
            3000 X 3000 (μm)

LASER POWER 500 (mW)

LOCK-IN RANGE
            5000.00(μU)

AMPLITUDE MAX 5021.97(μU)   PHASE MAX   23.56(deg)
           MIN   0.00(μU)   MIN   0.00(deg)

(b)

Fig. 7 PA image of "8" character (a) raw data, (b) processed data.

In our experiments maximum measurement points were 100×100 per frame and maximum scanned area was 2×2 mm. Data stored in the personal computer were processed and imaged on a CRT. Three imaging methods were used; pseudo-color, contrast, and bird's-eye image. The image processing was done as follows: 1) addition; 2) smoothing; 3) median filter; 4) laplacian filter; and 5) equalization of gray level histogram.

Fig. 7 shows the image of a test sample which has the number "8" on its surface. Fig. 7(a) is the raw data and 7(b) is the processed data by equalization of gray level histogram and mediam filter. A comparison of (a) and (b) shows the improvement after image processing.

IV. Results and Discussion

In this section, we will first compare PZT signals with PBD signals via some experimental results (Kasai *et al.*, 1986). We will then go on to demonstrate some applications for semiconductor devices using PZT-PAM and PBD Microscopy (PBDM).

1.   COMPARISON OF PZT AND PBD SIGNALS

Frequency responses of PZT, PBD and OBD signals are shown in Fig. 8. The PBD signal involves two contributions, as described previously and shown in Fig. 2. One is the deflection caused by the distribution of the refractive index in the surrounding gas or liquid; and the other is the deflection caused by thermal deformation on the surface of a sample. OBD detects only the former signal, and PBD detects both signals. Comparing the PBD signal with the OBD signal from the same sample (a GaAs wafer) under the same conditions, the

Fig. 8 The frequency responses of PZT-PAM, OBD and PBDM methods.

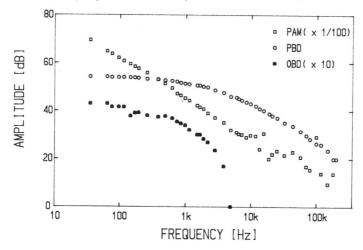

magnitude of the PBD signal is about a hundred times that of the OBD signal. This indicates that in generating the PBD signal, the proportion of the signal caused by thermal deformation is much larger than that caused by the distribution of refractive index. In other words, the PBD signal mainly arises from thermal deformation and has a very small contribution from the distribution of refractive index.

The PZT-PAM signal has an f^{-1} dependency except for some peaks which correspond to resonance frequencies of the PZT or to the parameters of the measurement system. On the other hand, the PBD and OBD signals have an f^{-1} dependency just over 1 kHz, but saturate at frequencies lower than *ca.* 1 kHz. The thermal diffusion length of this sample (GaAs wafer) is *ca.* 100 μm in the saturated frequency regime, and this depth may be the deepest detection limit possible using the PBD method.

In order to compare the depth profile from PZT-PAM with that from PBDM, a model sample, with a subsurface structure as shown in Fig. 9, was measured using one-dimensional scans by both PZT-PAM and PBDM at the same time. In PZT-PAM, the depth profile was determined by the acoustic wavelength of the sample which was greater than the thermal diffusion length. In PBDM, the depth profile was determined by the thermal diffusion length up to about 100 μm. The former is called the acoustic mode, and the latter is called the thermal mode. With PBDM, the PBD signal was much more sensitive than the PZT signal for surface or near surface detection (less than 100 μm). However, although the deep area (more than 100 μm) can be detected by PZT-PAM, it cannot be detected by PBDM. These results agreed with the results from the frequency responses, and suggest that PBDM may be more useful for surface or near surface analysis.

2.   STACKING FAULT IMAGING OF A CZ-Si WAFER BY PZT-PAM

The sample used here was a CZ-Si wafer (Czochralski grown silicon) which contained about 10^{18} atoms/cm^3 of oxygen (Kasai, Shimizu, Sawada and Gohshi, 1985). The surface of the sample was covered with an oxide layer (a few thousand Å). Stacking faults can appear in CZ-Si wafers during thermal oxidation, and these degrade the characteristics of processed devices. It is thus very important to detect the defects, and to estimate their depths non-destructively. At present, a chemical etching method, Transmission Electron Microscopy (TEM), and X-ray topography are the conventional techniques for detection. However, neither of them is rapid and non-destructive, nor can they estimate the depth-profiling.

Fig. 10 shows PA amplitude images of the CZ-Si wafer at about 100, 10 and 1.02 kHz, respectively. We now consider the thermal diffusion length as the amplitude depth sensitivity of PAM. The thermal diffusion length of silicon is about 120 μm at 1.02 kHz. At this frequency a swirl, that is, a crowd of many stacking faults, was observed as an arched image. PA images were also measured at various frequencies lower than about 100 kHz, but this image could not be observed at any frequencies above 1.02 kHz. From these results the position

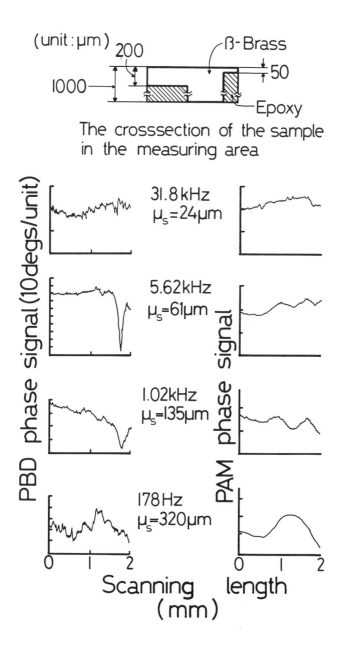

Fig. 9 Comparison between phase signals of PBDM and PAM.

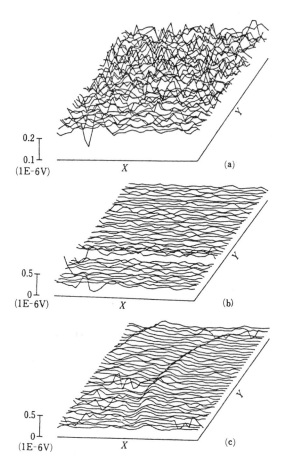

Fig. 10 PAM amplitude images of a CZ-Si wafer, the surface of which is covered with an oxide layer. (a) at 100 kHz, (b) at 10.1 kHz and (c) at 1.02 kHz.

of this swirl could be estimated at about 120 μm below the surface of the CZ-Si wafer, a depth which is reasonable for this sample. No optical image of the swirl was observed in the scattered light image. Consequently the generation of PA signals shown in Fig. 9 was considered to be due to the difference in thermal or elastic properties between a pure Si wafer and defective one. The thermal conductivity decreases at the location of the stacking faults. In addition, in the case of an elastic property difference, a stress can be produced by a strain in the crystal structure at the stacking faults.

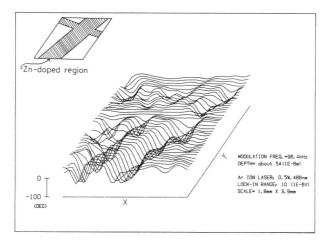

Zn-doped region

MODULATION FREQ. =98.4kHz
DEPTH= about 54 (1E-6m)

Ar ION LASER, 0.5W, 488nm
LOCK-IN RANGE, 10 (1E-6V)
SCALE= 1.8mm X 3.9mm

0

-100
(DEG)

X

Fig. 11 PA image of a Zn-doped GaAs wafer.

3.   Zn-DOPED GaAs WAFER IMAGING BY PZT-PAM

Fig. 11 shows the PA image of a Zn-doped (10^{19} atoms/cm^2) GaAs wafer (Kasai, Sawada and Gohshi, 1985). The crossed Zn-doped region in clearly distinguishable from the pure GaAs wafer. Though no mask (crossed shape of the Zn-doped region) could be observed optically after the mask of organic compounds on the sample had been removed by an etching process, its structure still showed up clearly in the PA image. To elucidate the generation mechanism, the PA image was also measured with a microphone under the same conditions, however, no crossed shape image was obtained. As we mentioned in section 2, the PA image obtained by PZT detection includes three properties of the sample, that is, its optical, thermal and elastic properties. The PA image in Fig. 11 must be an elastic image, because there is no crossed shape in either the optical image by a photocell or the PA image obtained by a microphone. Moreover, the PA signal intensity should increase with decreasing modulation frequency according to f^{-1} or $f^{-3/2}$ dependencies. However, no frequency dependence was observed over the wide range from 20 to 200 kHz in Fig. 11.

Ikoma and Morizuka (1982) have proposed another possible generation mechanism of the PAM image for a p-n junction type semiconductor, according to which the voltage induced by the exciting beam generates acoustic waves in the semiconductor. The present PAM image might therefore be explained as follows: When the electron-hole pair generated in the p-i junction of this Zn-doped GaAs semiconductor by the exciting laser beam diffuses to either the p or i side, voltage is induced in the p-i junction as the result of charge storage. The piezoelectric transducer attached to the sample picks up the elastic wave generated by this modulated voltage, which amounts to a kind of piezoelectric effect on the semiconductor.

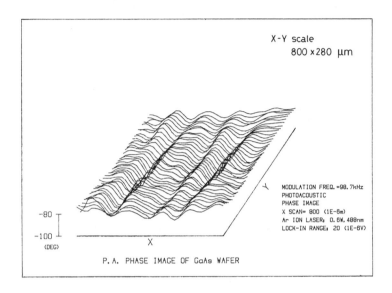

Fig. 12 Typical PA image of n-type (100) GaAs wafer.

4.   DISLOCATION IMAGING OF GaAs WAFERS BY PZT-PAM AND PBDM

a. PZT-PAM Image of Dislocations

Fig. 12 is a typical PAM image of the n-type GaAs wafer (100) used in this experiment (Kasai, Sawada and Gohshi, 1985). The modulation frequency was 98.7 kHz. An X-ray topograph is given in Fig. 13(a). Fig. 13(b) shows the

Fig. 13 (a) X-ray topograph of the GaAs wafer; (b) an illustration of dislocations of the GaAs wafer and the PAM measurement area.

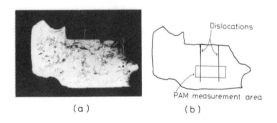

position of the dislocations of the GaAs wafer used and the PAM measurement area. There are two line-like structures in the PAM image in Fig. 12. These are due to dislocations hidden under the GaAs wafer, and this structure corresponds well to that of the X-ray topograph. This could not be observed with a photocell because there was no optical difference.

The PAM measurements were carried out at various frequencies from 10 kHz to 100 kHz, and these two line-like structures could be seen at any frequency. This indicates that the dislocations are perpendicular to the (100) surface of the GaAs wafer. At frequencies less than about 10 kHz, in addition to the above two line-like structures, various other structures were also observed. These might be due to some other defects which are smaller than the dislocations, but they have not yet been fully investigated.

In this experiment, the time required for the measurement of the PAM image was about 20 min. An X-ray topograph requires much longer for the same sample area. Another advantage of PAM is that the image can be measured nondestructively in contact with the atmosphere.

b. PBDM Image of Dislocations

The X-ray topograph of a GaAs wafer (100) and the scanned area are shown in Fig. 14 (Sawada *et al.*, 1985). A number of dislocations which look like straight lines lie perpendicular to the (100) surface of the wafer. Fig. 15 shows a typical PBD phase image of dislocations of this GaAs wafer. Fig. 15(b) was measured at 50 kHz with the thermal diffusion length equal to 75 μm. At this depth two line-like structures were observed similar to those on the X-ray topograph, however, when the thermal diffusion length decreased to 53 μm at

Fig. 14 X-ray topograph of a GaAs wafer.

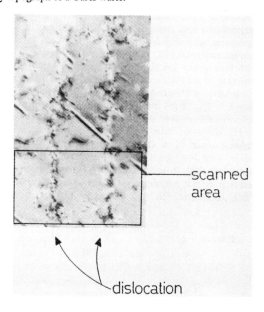

scanned area

dislocation

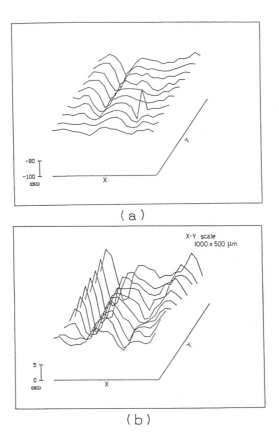

Fig. 15 Typical PBD image of GaAs wafer (a) at 99 kHz and (b) at 50 kHz.

99 kHz, the line on the right was not clearly visible as shown in figure 15(a). This indicates that the dislocation on the right does not occur down to ca. 50 μm below the surface, and is only present underneath the 50 μm layer. The dislocation on the left was observed at all frequencies, and we can therefore conclude that this dislocation extends perpendicularly from the surface to the bulk of the GaAs wafer.

5.   PULSED PAM AND PBDM

When a pulsed laser beam is used as the pumping beam, wide bandwidth measurements, which are unobtainable with CW excitation, can be obtained by means of short pulsed excitation. Furthermore, the pulsed method has an advantage for the theoretical analysis of PA signals, because it provides an impulse response. In this section we describe the basic analysis of pulsed PAM and

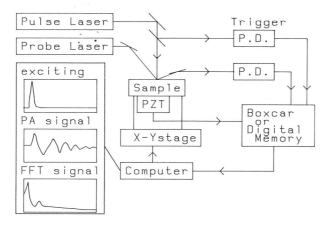

Fig. 16 Block diagram of pulsed PAM and PBD measurements.

PBDM (Kasai, Harata, Sawada and Gohshi; unpublished).

Fig. 16 is a block diagram of a pulsed PAM and PBDM apparatus. A Nd-YAG laser beam (Quanta-Ray) is used as the exciting pulse. The pulse width is about 6 ns, and the incident power is about 1 mJ/pulse. Both the PZT-PA signal and the PBD signal are measured by digital memory (Iwatsu DM-701) or averaged by a Boxcar integrator (NF BX-531), and then processed on a personal computer (NEC-9801VM2).

The main disadvantage of the pulsed method is the destruction of the sample, because the pulsed exciting beam is instantaneously more powerful than a CW beam. The incident power dependences of the PZT-PA signals is shown in Fig. 17 for identification of the incident power threshold which begins to destroy a sample. The sample is a silicon wafer which has an optically flat surface. The pulsed beam diameter is about 500 μm. As shown in Fig. 17, PZT-PA signals show a remarkable increase at or above 0.15 mW average power. It was

Fig. 17 Incident power dependence of PZT-PA signals.

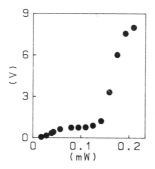

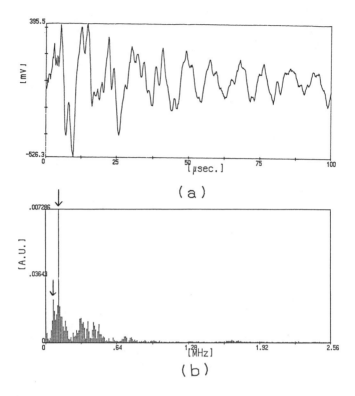

Fig. 18 PZT-PA (a) transient signal and (b) FFT signal

thus concluded that the sample destruction onset occured when the average power reached that threshold, and the power used in subsequent experiments was kept under 0.15 mW.

Fig. 18(a) shows the transient response of a PZT-PA signal. The sample is a GaAs wafer which has an optically flat surface. A few microseconds after firing the laser, a large peak appears and then decays gradually. Fig. 18(b) shows the Fast Fourier Transform (FFT) spectrum of the PZT-PA signal of Fig. 18(a). Two peaks appear, as shown by arrows. The peak on the right at about 100 kHz corresponds to the resonance frequency of the PZT used here, and the left peak at about 60 kHz corresponds to the resonance frequency of the system or to bending modes in the sample. In pulsed PAM, wide band measurements are made possible by the use of the FFT method via single pulse measurements.

Fig. 19(a) shows the transient response of the PBD signal from the sample described above. The arrow indicates the time when the incident pulse is irradiated. The PBD signal has a much faster response than the PZT-PA signal. Fig. 19(b) shows the FFT spectrum of the PBD signal in Fig. 19(a). This spectrum is more narrowband when compared with the FFT PZT-PA spectrum, as expected from the faster PBD response.

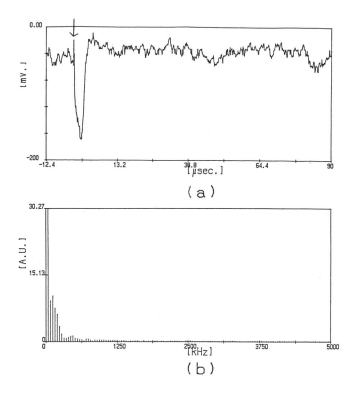

Fig. 19 PBD-PA (a) transient signal and (b) FFT signal

These results indicate that the PZT-PA signal mainly reflects the bulk properties of a sample, whereas the PBD signal mainly reflects elastic properties on, or near, the surface of a sample. This conclusion agrees with the results of the CW method.

V. Conclusions

In this Chapter it has been shown that PAM and PBD are very useful techniques for the non-destructive measurement of semiconductor devices. Information about doped regions and various defects at, or below, the surface of semiconductor devices can be obtained non-destructively using PAM and PBD. This information cannot be obtained non-destructively by conventional methods. Of particular value is the fact that although PAM requires contact with the sample, PBD is both a non-destructive and non-contact method. In addition, a combination of these methods used simultaneously, can yield better estimates of which of the sample's properties is responsible for the observed signals.

At present, the mechanism of the generation of PA signals is not clear, however, this difficulty may be overcome in future work via the comparison of various thermal wave measurements.

VI. References

Aamodt, L.C. and Murphy, J.C., (1981) J. Appl. Phys., **52**, 4903.

Aamodt, L.C. and Murhpy, J.C., (1983) J. Appl. Phys., **54**, 581.

Boccara, A.C., Fournier, D., Jackson, W. and Amer, N.M., (1980) Opt. Lett. **5**, 377.

Boccara, A.C., Fournier, D. and Badoz, J., (1980) Appl. Phys. Lett. **36**, 130.

Ikoma, T. and Morizuka, K. (1982) Oyo-Butsuri **51**, 205 [in Japanese].

Kasai, M., Sawada, T., Gohshi, Y., Watanabe, T., Furuya, K. (1986) Jpn. J. Appl. Phys. Supplement **25-1**, 229.

Kasai, M., Shimizu, H., Sawada, T., Gohshi, Y., (1985) Anal. Sci. **1**, 107.

Kasai, M., Sawada, T., Gohshi, Y., (1985) Jpn. J. Appl. Phys. Supplement **24-1**, 220.

Olmstead, M.A., Amer, N.M., Kohn, S., Fournier, D. and Boccara, A.C. (1983) Appl. Phys. A **32**, 141.

Opsal, J. and Rosencwaig, A., (1982) J. Appl. Phys. **53**, 4240.

Rosencwaig, A., (1980) J. Appl. Phys. **51**, 2210.

Sawada, T., Gohshi, Y., Watanabe, T., Furuya, K. (1985) Jpn. J. Appl. Phys. **24**, L938.

2

PHOTOACOUSTIC MICROSCOPY (PAM) AND DETECTION OF SUBSURFACE FEATURES OF SEMICONDUCTOR DEVICES

Shu-yi Zhang and Li Chen

Institute of Acoustics
Nanjing University
Nanjing,
PEOPLE'S REPUBLIC OF CHINA

I. Introduction

It has been demonstrated that the photoacoustic effect is directly related to the sample's optical characteristics, thermal and mechanical properties, geometric structure, etc. Therefore, surface and subsurface inhomogeneous features of a sample can be investigated by detecting photoacoustic signals. This capability offers a new important technique for microscopic imaging, called photoacoustic microscopy (PAM). Generally, it is also called thermal wave imaging.

About ten years ago, Von Gutfeld and Melcher (1977) initially performed a photoacoustic imaging experiment by pulsed laser at an ultrasonic frequency to detect an aluminum sample with a hole beneath the surface. In the next year, Wong *et al.* (1978) and Wickramasinghe *et al.* (1978) independently performed two PAM experiments in different samples using audio frequency and very high ultrasonic frequency, respectively. Since then, the technique of photoacoustic imaging has received ever increasing attention for its application potential to the nondestructive evaluation of opaque materials and devices, and has been developed rapidly both experimentally and theoretically (Thomas *et al.*, 1986). An important aspect among the applications of PAM is the detection of subsurface features in semiconductor devices. Recently several investigators (McFarlane *et al.* 1980; Rosencwaig and Busse, 1980; Busse and Rosencwaig, 1980; Favro *et al.,* 1980; Busse, 1979; McClelland *et al.,* 1980), nearly simultaneously were successful in detecting inhomogeneous structures, such as diffused and ion-implanted regions, in semiconductors as well as integrated circuits by using PAM. Since 1981, we have also engaged in similar experimental and theoretical studies and derived some valuable results, especially in laminated imaging or depth profiling of subsurface structures and defects in semiconductor devices (Zhang *et al.,* 1982; 1983; 1986; Chen *et al.,* 1985; 1987a; 1987b).

A monograph written by Rosencwaig (1980) has summarized and reviewed the progresses in photoacoustic spectroscopy (PAS) and photoacoustic microscopy (PAM) achieved before 1980, and has pointed out the prospects of the applications on PAM to the semiconductor industry, although PAM was only just emerging then. Other reviews, given by Tam (1983; 1986) and Birnbaum and White (1984), further described photoacoustic techniques and their applications, including the applications of PAM as non-destructive, sometimes even as destructive, testing methods. It has been generally shown that PAM provides some unique capabilities for detection of subsurface structures, such as irregularities, flaws, doping concentrations, etc., that cannot be obtained by other methods. In addition to optical excitation methods, thermal wave imaging can also be readily generated through the absorption of other forms of electromagnetic energy and particle beams. Up to now, thermal wave electron-microscopy has been developed rapidly and successfully (Rosencwaig, 1983; Rosencwaig and Opsal, 1986). These achievements have also promoted the developments of PAM and its applications.

In this chapter, we will put emphasis on the description of the applications of PAM in the detection of subsurface features of semiconductor devices, especially of integrated circuits. In Sections II and III the related theoretical foundations and the experimental instrumentation will be introduced, respectively. Section IV will present various modes of PAM imaging. Finally, the advantages of PAM and its applications will be discussed briefly in Section V.

II. Simplified Photoacoustic Theory - Depth Profiling

1.   INTRODUCTION

There have been quite a number of theoretical models presented for investigating the generation mechanism of the photoacoustic signal in solids for the case of piezoelectric detection. The initial theoretical attempt was a one-dimensional model based on the thermoelastic effect in solids and was presented by White (1963) in order to explain the excitation of acoustic signals in solids by an electron beam or electromagnetic energy. Thereafter, many different theories and various simplified models or calculation methods applicable to condensed matter have been developed (Jackson and Amer, 1980; Patel and Tam, 1981; Opsal and Rosencwaig, 1982; Kino and Stearns, 1985; Rosencwaig and Opsal, 1986). Jackson and Amer (1980) have considered, under some special experimental conditions, the three dimensional stress effects produced by the absorption of optical energy in the sample and presented a three dimensional thermoelastic model. According to this theory, the relationship between the amplitude and the phase of the photoacoustic signal and the modulation frequency can be described explicitly, and the thermal and mechanical properties of the sample can thus be measured quantitatively. Furthermore, the theoretical predictions by Jackson and Amer have been experimentally verified by those authors. For layered samples Opsal and Rosencwaig (1982) developed a one dimensional model for thermal wave depth profiling that provides expressions for the temperature at the surface of the sample and for the thermoelastic response beneath the surface. Detailed descriptions of imaging of electronic processes in semiconductors can be found in other parts of this volume (Chapters 5 and 9).

Most of photoacoustic theories based on thermoelastic models emphasize the contributions of the optical, thermal and elastic parameters of the sample to the acoustic signal generation. Owing to the complexity of the thermoelastic

Fig. 1. One-dimensional thermoelastic model.

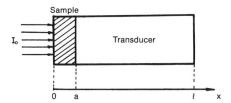

theory of solids, there exist just a few photoacoustic theories which satisfy specific boundary conditions, such as the three dimensional theory presented by Jackson and Amer (1980). These authors further obtained reasonable agreement upon comparison of their experimental findings with the theoretical model. Most of the rigorous theoretical calculations, however, are rather laborious, complex, and difficult to compare with experimental results.

2.   SIMPLIFIED THERMOELASTIC MODEL

In order to qualitatively understand the mechanisms of depth profiling of photoacoustic imaging detected by a piezoelectric transducer, we present a simplified one-dimensional theory as follows (Chen and Zhang, 1987a). Based on the theory by White (1963), and in accordance with some practical experimental conditions, we assume a one dimensional model as shown in Fig. 1. For simplicity, it is assumed that: 1. The sample is opaque and the absorption of light is taking place at the surface; 2. There is good coupling between the sample and the transducer, and acoustic reflection from the interface is negligible; 3. The acoustic wave vector in the transducer is approximately the same as that in the sample; and 4. The sample is thermally thick and the thermal wavelength satisfies the condition $\mu_s \ll a \ll l$, where μ_s is the thermal diffusion length, (Rosencwaig and Gersho, 1976), and a is the sample thickness (Fig. 1).

The energy E_t absorbed at the surface of sample is

$$E_t = \tfrac{1}{2}I_o\beta(1 + e^{i\omega t}) \tag{1}$$

where I_o is the incident laser intensity, β is the optical absorption coefficient, and ω is the modulation angular frequency. The distribution of the thermal field in the sample is given by

$$\frac{\partial\theta}{\partial t} - \frac{\kappa}{\rho c}\frac{\partial^2\theta}{\partial x^2} = 0 \tag{2}$$

where κ is the thermal conductivity, ρ is the density, and c is the specific heat of the sample. From Eqs. (1) and (2), the expression for the harmonic component of the temperature in the sample can be written as:

$$\theta(x,t) = \theta_o e^{-\sigma x + i\omega t} \tag{3a}$$

where

$$\theta_o = \frac{I_o\beta}{2\kappa\sigma} ; \quad \sigma = \frac{1+i}{\mu_s} \tag{3b}$$

Eq. (3) shows that the thermal energy penetrates into the sample in a wave-like form. The temperature wave in the sample is damped out exponentially with the 1/e distance equal to $\mu_s = (2\kappa/\rho\omega c)^{\frac{1}{2}}$.

In the presence of the thermal field, the stress T can be written as follows:

$$T = (2\mu + \lambda)\frac{\partial u}{\partial x} - (3\lambda + 2\mu)\alpha_t\theta \tag{4}$$

where, λ and μ are the Lamé constants, u is the particle displacement and α_t is the thermal expansion coefficient. The wave equation in both sample and transducer is

$$\frac{\partial^2 u}{\partial x^2} + k^2 u = -\sigma\theta'_o e^{-\sigma x} \tag{5}$$

where, k is the acoustic wave vector and

$$\theta'_o = (\frac{3\lambda + 2\mu}{\rho c^2})\alpha_t\theta_o \tag{6}$$

If the surface of the sample and the bottom of the transducer are free, the boundary conditions are

$$T(0) = T(l) = 0 \tag{7}$$

Generally, if in Eq. (4), $\exp(-\sigma l) \to 0$, then we can write:

$$\frac{\partial u}{\partial x}\Big|_{x=0} = \theta'_o \ ; \ \frac{\partial u}{\partial x}\Big|_{x=l} = 0 \tag{8}$$

Eq. (5) can be solved using Green's function formalism:

$$\frac{\partial^2 G}{\partial x^2} + k^2 G = \delta(x - x') \tag{9}$$

Due to the inhomogeneous boundary condition, G must be expressed as

$$G(x,x') = G_1(x) + G_2(x,x') \tag{10}$$

$G_1(x)$ satisfies the homogeneous boundary conditions of Eq. (7). Based on the definition of Green's function, we have the solution:

$$u(x) = \int_0^l G(x,x')f(x')dx' = G_1(x)\int_0^l f(x')dx' + \int_0^l G_2(x,x')f(x')dx' \quad (11a)$$

where

$$f(x') = -\sigma\theta'_o e^{-\sigma x'} \quad (11b)$$

Let $G_2(x,x')$ satisfy

$$\frac{\partial G_2(x,x')}{\partial x}\bigg|_{\substack{x=0 \\ x=l}} = 0 \quad (12)$$

so that:

$$\frac{\partial u}{\partial x}\bigg|_{\substack{x=0 \\ x=l}} = -\theta_o \frac{dG_1(x)}{dx}\bigg|_{\substack{x=0 \\ x=l}} \quad (13)$$

Now, we have the following equations

$$\frac{d^2 G_1}{dx^2} + k^2 G_1 = 0, \quad (14)$$

with

$$\frac{dG_1}{dx}\bigg|_{x=0} = -1, \quad \frac{dG_1}{dx}\bigg|_{x=l} = 0 \quad (15)$$

and

$$\frac{\partial^2 G_2}{\partial x^2} + k^2 G_2 = \delta(x - x'), \quad (16)$$

with

$$\frac{\partial G_2}{\partial x}\Big|_{x=0} = 0, \qquad \frac{\partial G_2}{\partial x}\Big|_{x=l} = 0 \qquad (17)$$

The solutions of Eqs. (14) and (16) with the boundary conditions Eqs. (15) and (17) are, respectively

$$G_1 = \frac{\cos[k(l-x)]}{k\sin(kl)} \qquad (18)$$

$$G_2(x,x') = 2l \sum_{n=0}^{\infty} \frac{1}{(kl)^2 - (n\pi)^2} \cos(\frac{n\pi}{l} x') \cos(\frac{n\pi}{l} x) \qquad (19)$$

Thus, the particle displacement u(x) can be calculated from Eqs. (11), (18) and (19).

3.   DEPTH PROFILING

Using the dielectric displacement D = 0, the piezoelectric equations can be written as

$$E = -h\frac{\partial \mathbf{u}}{\partial x} \qquad (20)$$

Where E is the electric field and h is the piezoelectric coefficient. Therefore, the output voltage of the transducer is the integral of $V(x')$ with respect to x', where

$$V(x') = \int_0^l E(x,x')dx = \theta'_o(\frac{1-\cos kl}{k\sin kl} + 4l\sum_{n=0}^{\infty}\frac{\cos[(2n+1)x'/l]}{(kl)^2 - (2n+1)^2\pi^2})e^{-\sigma x'} \qquad (21)$$

x' is the coordinate of the thermal source at the near-surface. It must be pointed out that only the thermal sources within about one thermal wavelength beneath the sample surface contribute to the integral of Eq. (21). Therefore, the term $\cos[(2n+1)x'/l]$ is approximately a constant in relation to exp $(-\sigma x')$. Moreover, as n increases, $(2n+1)^2\pi^2$ increases rapidly. Therefore, it is reasonable to assume that $\cos[(2n+1)x'/l] \approx 1$. Now, the real part of Eq. (21) can be written as

$$V_{Re}(x',t) = V_o e^{-x'/\mu_s}\cos(\omega t - x'/\mu_s) \qquad (22)$$

where

$$V_o = -\sigma\theta_o' l \left[\frac{1 - \cos kl}{kl \sin kl} + 4l \sum_{n=o}^{\infty} \frac{1}{(kl)^2 - (2n+1)^2 \pi^2} \right] \qquad (23)$$

Eq. (22) represents the output voltage of the detector contributed by the different thermal sources at various depths x' in the sample. It can be seen that the energy of the thermal wave is converted to sound not only at the surface where the laser energy is absorbed, but also along the way of propagation through the sample. The phases and amplitudes of the acoustic signal generated from various points x' are different. Since the thermal diffusion length is very small compared to the acoustic wavelength, these acoustic waves carrying information from different depths in the sample can be considered as generated at the same point as depicted in Fig. 2. The output electric signal is the superposition of the contributions of all these waves. The signal generated at the surface of the sample is, of course, stronger than those from the subsurface structures. In order to perform depth profiling, we must apply some method to strengthen the signal generated from subsurface features of interest, and suppress other signals which are less desirable. For this purpose, the detected signal is mixed with a reference signal

$$V_r = 2\cos(\omega t + \phi_o) \qquad (24)$$

where ϕ_o is the phase shift of the reference signal. The D.C. component of the mixed signal is

Fig. 2. Acoustic waveforms generated at various depths x'.

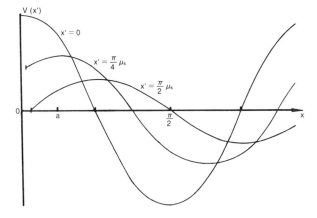

$$\bar{V}(x') = V_o e^{-x'/\mu_s} \cos(\phi_o + \frac{x'}{\mu_s}) \tag{25}$$

This equation expresses the "element" of the output voltage contributed by the thermal source at point x'. The total output voltage is the integral of Eq. (25). Thus, experimentally we can realize depth profiling by adjusting the phase shift. From Eq. (25) we find that when

$$\phi_o = \frac{3\pi}{4} - \frac{x'}{\mu_s} \tag{26}$$

$V(x')$ is maximum. Thus, if we want to display the subsurface features at the depth x', we can set the phase shift ϕ_o to satisfy Eq. (26). Then the signal generated at x' is enhanced and is the largest contribution to the output voltage. In this manner, photoacoustic imaging chiefly displays the subsurface features at the depth x'. Fig. 3 shows the signal amplitude distribution versus x' for different phase shifts. Due to the fact that it is difficult to prepare an artificial multilayered sample with thickness less than one thermal wavelength and characteristics in accordance with our assumptions, the present simplified theory has not been proved quantitatively by experiment. At the time of the writing of this chapter, we are hopeful that the presented model may be helpful in comprehending the principle of the acoustic generation by the thermoelastic effect, although there are substantial simplifications. More detailed theoretical and experimental studies of these problems are in progress.

Fig. 3. Distribution of the output signal $V(x')$ from a lock-in detection system versus x' for different phase shift settings.

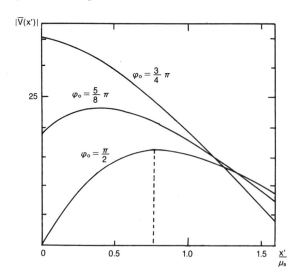

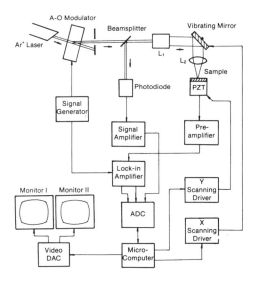

Fig. 4. Block diagram of photoacoustic microscope.

III. Apparatus of Photoacoustic Microscopy

Photoacoustic microscopy (PAM) was developed on the basis of photoacoustic spectroscopy (PAS), which has been described systematically by West *et al.* (1983). The PAM system, however, has different construction as it is mainly used to display the spatial distribution of inhomogeneous characteristics of materials especially subsurface features of opaque samples. As an example, we show in Fig. 4 the schematic diagram of our PAM system (Zhang and Chen, 1986). Although there are many different types of PAM systems, the essential apparatus consists of three parts.

1. PUMPING OPTICAL SOURCE

The pumping optical source usually includes a pump laser, an optical intensity modulator if a CW laser is used, and an optical focus system, which frequently contains optical beam scanning components.

a. Laser Sources

Almost all of the current lasers have been utilized as the optical source, however, the Ar^+-ion laser is widely used in PAM because of its moderate power and visible spectral lines. The Ar^+-ion laser is a highly stable CW emission source, with which some conventional optical components in the remaining parts of the optical system can be used without special selection. On the other hand, almost all of the semiconductor materials are opaque to the Ar^+-ion laser lines, thus it can be favorably used for photoacoustic imaging of subsurface structures of semiconductors. A CW Nd:YAG laser has also been employed as pumping source by McFarlane *et al.*, (1980) and Zuccon and Mandelis (1987). Due to the Nd:YAG laser emission line at 1.06 μm, crystalline GaAs is essentially transparent at that wavelength, whereas the amorphous material has an absorption coefficient of 10^4 cm^{-1}. Therefore, amorphous areas induced by implantation can be readily detected by a photoacoustic imaging technique using a Nd:YAG laser (Zuccon and Mandelis, 1987).

b. Optical Intensity Modulation

In order to excite acoustic signals by use of optical energy, optical intensity modulation is always necessary, except when the pumping source is a pulsed laser. Generally, three types of modulators have been applied to photoacoustic imaging:

(1) Mechanical Choppers

Most of the earlier photoacoustic experiments used mechanical choppers with low frequencies (less than 5 KHz). Mechanical choppers have the advantage of simplicity and 100% modulation depth. However, the resolution of PAM is usually dependent on the thermal wavelength, i.e., the modulation frequency. Thus, mechanical choppers cannot be used with some thermal wave imaging systems for which high resolution is required.

(2) Acousto-optic (A-O) Modulators

A-O modulators are usually used in most modern PAM experiments. The important advantages of the A-O modulator are that its modulation frequency is easy to change in a wide spectral range and the desirable modulation wave form can be selected (sine wave or square wave, etc.). The modulation efficiency approaches 90%, which renders the modulator as a very reasonable modulation medium for PAM.

(3) Electro-optic (E-O) Modulators

Although E-O modulators have the advantage of high modulation depth ($\approx$100%) and variable wave forms, due to the fact that the required voltage is fairly high and the crystal cannot bear high optical beam power, E-O modulators have not been used extensively in PAM as yet.

c. Optical Beam Focusing and Scanning

In the case of a simplified optical system, mechanical scanning of the sample stage effected by stepping motors is always used. This scheme is helpful in improving the optical focus size and extending the scanning range. However, the noise and microvibration induced by stepping motors strongly influence the photoacoustic signal and decrease the signal to noise ratio. Sometimes the scanning speed must be very slow in order to improve the signal to noise ratio. In that case the imaging time is undesirably long.

In order to improve the signal to noise ratio and shorten the imaging time, optical beam scanning is usually exploited. Several optical scanning components, such as rotating multiface mirrors, vibrating mirrors, or moving reflectors, can be used. A mixed scanning technique, which consists of both optical scanning in the X-direction and mechanical scanning in the Y-direction, has been previously described (Luukkala and Penttinen, 1979). For an imaging system with high spatial resolution, however, it is difficult to give consideration to both small optical focusing size and appropriate scanning range by use of a general optical system. Therefore, it is necessary to design a special optical system including complex lenses and scanning components. In our arrangement a vibrating mirror is placed between two sets of lenses, one of which expands and the other focuses the optical beam, in order to achieve a sharply focused spot with diameter 2 μm, as well as a one-directional scanning. The advantage of this method is that the scanning range can be properly extended. Therefore, the optical beam focusing and scanning can be compatibly adjusted (Zhang *et al.*, 1982). On the other hand, sample stage scans in the other direction are accomplished by a low speed mechanical method to reduce the optical path system.

2.   PHOTOACOUSTIC SIGNAL GENERATION AND DETECTION

There are several ways with which condensed matter photoacoustic signal generation can be effected. As a result, sample setting and acoustic signal detection system may vary widely. They can be classified into two types:

a. Gas Cell with Microphone

When an opaque solid sample is put in a hermetically closed gas cell, and according to the Rosencwaig-Gersho (R-G) theory (Rosencwaig and Gersho, 1976), the periodically illuminated surface of the sample heats the nearby gas layer to produce an acoustic signal in the cell. A microphone, condenser or electret, is placed in the cell to receive the acoustic signal. In our laboratory, a specially designed electret microphone has a sensitivity as high as 200 mV/Pa and a flat response in the frequency range between 20 Hz and 20 KHz. This microphone has been shown to be convenient to use for investigating the frequency dependence of the photoacoustic signal of condensed phase samples.

b. Piezoelectric Transducer

In order to improve the spatial resolution of thermal wave imaging, the modulation frequency must be increased to the ultrasonic frequency range. The piezoelectric ceramic PZT is normally used instead of the microphone gas cell as a transducer to detect the photoacoustic signal in a wide frequency range from several tens of Hz to several MHz. PZT transducers can be directly attached to the sample by using a little glue or other viscous fluids as the coupling medium. In our experience, this direct contact method offers much higher sensitivity than that of indirect contact, in which a thin plate reflector is inserted between the transducer and the sample to avoid impinging of the optical beam on the transducer and also to protect the transducer. Several specially designed

PZT transducers can be satisfactorily used at operating frequency ranges from audio to ultrasonic frequencies.

Although some other piezoelectric crystals, such as quartz and lithium niobate, can also be used as acoustic detectors, PZT has some advantages over others: It has high mechano-electric coupling coefficients, it is the least expensive and it is easy to fabricate in any required shape. For these reasons it appears to be the best sensor for detecting photoacoustic signals.

In comparison with the microphone gas cell, the piezoelectric transducer does not only have a wide operation frequency range, but also has some other advantages:

(1) PZT transducers can be used in different modes to detect directly the laser-induced thermoelastic waves with much higher sensitivity than, and without the need for a gaseous medium which is necessary for the operation of microphone detectors. Therefore, the detected signal is only dependent on the properties of the sample and is not affected by those of the adjacent gas.

(2) The characteristics of PZT are fairly stable; it can be operated in either high or low temperature and pressure to study phase transitions and other phenomena.

3.   SIGNAL PROCESSING INSTRUMENTATION

In general, the photoacoustic effect in condensed materials is quite weak. In order to provide the appropriate level of photoacoustic signal to a lock-in amplifier, it is necessary to increase the electric output signal of the detector by a low noise preamplifier, which is close to the detector (Patel and Tam, 1981). Due to the fact that modern lock-in amplifiers have two channels with dual outputs, two components of photoacoustic signals, e.g. the magnitude and phase or the in-phase and quadrature signals, can be detected simultaneously. In addition, it is possible to detect selectively the photoacoustic signal with a suitable phase angle by adjusting the phase shift of the reference signal (McClelland, 1980). This method makes photoacoustic imaging techniques easier for depth profiling of samples. The output signals from the lock-in amplifier via an A/D converter are fed to a microcomputer for data processing. Finally, the output signals are offered to monitors or recorders for display purposes. The computer also controls the optical and the mechanical scanners (Fig. 4).

IV. Imaging Methodology of the Photoacoustic Microscope

It is well known that the characteristics of photoacoustically generated thermal waves can be summarized into two aspects: 1. Thermal waves are crucially damped, their penetration depth being about one thermal wavelength beneath the illuminated surface of the sample. In general, the wavelength is inversely proportional to the square root of the modulation frequency. 2. The propagation of thermal waves takes a finite time, which depends on the thermal properties of the sample. This delay time leads to a phase difference between the pumping optical wave and the detected carrier acoustic wave. Thus, it is possible to display inhomogeneous structures at different locations in the sample by measuring the various components of the photoacoustic signal at different frequencies.

1.   AMPLITUDE AND PHASE IMAGING

It has been shown in Section II that the amplitude of the photoacoustic signal is not only directly dependent on the surface optical properties of the sample, but also on the subsurface features. However, the phase angle of the signal is mainly dependent on the subsurface structures and is insensitive to optical surface structures. The phase angle reflects features below the surface much deeper than the amplitude. Therefore, phase angle imaging essentially displays the subsurface features (Busse, 1979). Comparison between the amplitude and the phase angle images makes it possible to distinguish the subsurface from the surface structures of a condensed phase sample.

2.   IN-PHASE AND QUADRATURE IMAGING

For multilayered samples, it is possible to generate separate photoacoustic spectra from different layers by observing spectra in a phase delayed mode (Adams *et al.*, 1976; 1977). As the pumping laser beam irradiates a sample, the photoacoustic signal from the lock-in amplifier output can be maximized by adjusting the phase of the reference signal. In this case, the detected signal is in-phase with the reference and the magnitude of the in-phase component is proportional to the amplitude of the photoacoustic signal. Subsequently, a phase switch by 90° from the previous setting results in quadrature with the reference, which is approximately equal to the product of amplitude and phase angle. While the optical beam is scanned on the surface of a sample, material inhomogeneities not only affect the amplitude of the signal, but also change the phase angle. In general, the in-phase component is fairly insensitive to the phase angle variations. Thus, in-phase imaging mainly reflects the surface optical structures. The quadrature component, however, is more sensitively affected by changes in the phase angle than changes in the amplitude. Therefore, quadrature imaging is chiefly affected by subsurface structures.

3.   DEPTH PROFILING - LAMINATED IMAGING

To date, the development of photoacoustic and thermal wave techniques has attracted great interest in depth profiling of subsurface features of opaque samples, particularly for integrated circuits. (Rosencwaig, 1982; Rosencwaig and Opsal, 1986). There are three main methods for performing such measurements:

a. Modulation Frequency Scan

PAS with microphone gas cell has been used spectroscopically to perform depth profiling analysis of a piece of apple peel (Rosencwaig, 1978). If the sizes of subsurface structures in a sample are within the thermal wavelength range, depth profiling can also be achieved by recording PAM images at different frequencies. As the depth of subsurface structures becomes shallow and much less than the thermal wavelength, however, thermal wave imaging only displays surface optical structures, but not subsurface structures. If the modulation frequency is simultaneously increased to obtain shorter thermal wavelengths on the order of the sizes of subsurface structures, these features can be exhibited to a certain extent depending on the modulation frequency. (Rosencwaig, 1982).

b. Phase Angle Adjustment

Due to the fact that the phase lag of the photoacoustic signal increases with distance into the sample and with increasing modulation frequency, the depth profile can be determined by controlling these two quantities (Adams and Kirkbright, 1977). It is possible in thermal wave microscopy to adjust the phase of the reference signal so as to maximize the contributions to the signal of either the surface topography or the subsurface features. Therefore, nondestructive depth profiling can be performed at a constant modulation frequency by changing phase angle setting (Rosencwaig, 1982).

c. Correlation Techniques for Photoacoustic Laminated Imaging

For better depth profiling resolution, it is necessary to perform PAM experiments under varying conditions, either by changing modulation frequency, of by adjusting phase angle setting as described above. However, the penetration depth of thermal waves at high frequencies is quite small, and therefore only a very thin layer beneath the sample surface can be imaged upon generation of a photoacoustic signal of small amplitude. This situation means that the depth profiling techniques described above can only be applied to detect very thin layers under strong absorption conditions. In order to overcome this limitation, cross correlation, pseudorandom and frequency chirp techniques have been introduced into PAS and PAM. By using pseudorandom binary sequence (PRBS) modulation of the pumping beam intensity, and cross-correlation techniques instead of a single pulse method, a time-resolved photoacoustic signal related to the subsurface structures of the sample has been obtained (Kato *et al.*, 1980; Kirkbright and Miller, 1983; Sugitani and Uejima, 1985). This method has been alleged to improve the signal to noise ratio over a single laser pulse excitation. Some preliminary results showed that three dimensional analysis of surface and subsurface regions can be obtained. (Takaue *et al.*, 1986). Coufal (1984) has used multiplex excitation methods for photoacoustic applications. Mandelis and Dodgson (1986) performed correlation PAS (CPAS) depth profiling of solid (powdered) and liquid samples (Dodgson *et al.*, 1986). An extensive review of these and other new frequency chirp imaging techniques has been given by Mandelis (1986).

V. Applications of PAM to Semiconductor Devices

Utilizing photoacoustic microscopy as described above, we have detected subsurface features of semiconductor materials and devices. Especially, we have obtained depth profiling of some integrated circuits by both changing the modulation frequency and adjusting the phase angle method. Some examples are given below.

1.   DETECTION OF MICROCRACKS IN SILICON WAFERS

We have been able to demonstrate that mechanical flaws such as subsurface microcracks in silicon wafers, which usually cannot be seen by an optical microscope, can be inspected by photoacoustic microscopy. We have observed some microcracks in a silicon chip from the PAM phase image as shown in Fig. 5(a). The amplitude image obtained from the same chip is shown in Fig. 5(b), which does not display the cracks. Because amplitude imaging represents chiefly the surface optical structures, Fig. 5(b) is an indication that the microcracks are underneath the surface (Chen *et al.*, 1985). These results are similar to that obtained by using electron-acoustic microscopy (Rosencwaig, 1982).

2.   DETECTION OF DISBONDED REGIONS IN POWER TRANSISTORS

Power transistors consist of a silicon wafer bonded to a metal base which serves as a heat sink. If the wafer is not adequately bonded, the power transistor is subjected to thermal cycling that causes failure. Since PAM is especially sensitive to variations of thermal properties of samples, it is capable of detecting these unbonded areas in a non-destructive way. Fig. 6 (a through d) shows the results. The thicknesses of the silicon wafers are in the 300-500 μm range. The

Fig. 5. Photoacoustic images of subsurface microcracks in the silicon wafer of a power transistor. (a) phase image and (b) amplitude image.

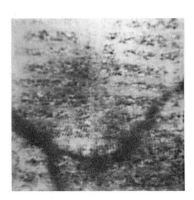 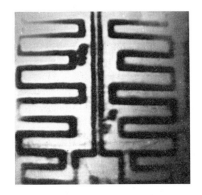

a   b

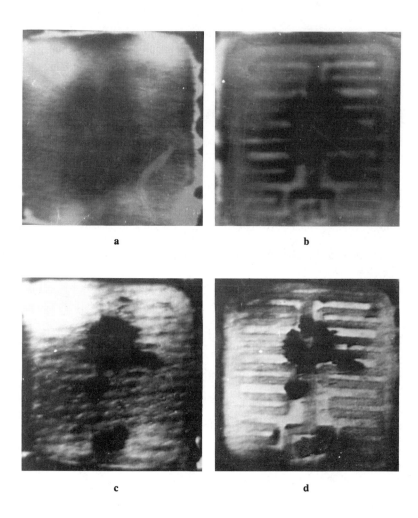

Fig. 6. Photoacoustic images of disbonded region detection in a power transistor; (a): phase image, and (b): amplitude image detected by a PZT transducer; (c) phase image, and (d): amplitude image detected by a gas cell. The bright areas are unbonded.

modulation frequency, 70 Hz, was chosen for proper thermal diffusion length. Figs. 6(a) and (b) are the phase image and the amplitude image, respectively, detected by a PZT transducer. To verify these results, the gas cell with an electret microphone was also used with the same sample, and the results of both phase and amplitude images are shown in Figs. 6(c) and (d), respectively. The bright areas represent unbonded zones. From Fig. 6 we can see that similar results can be obtained by using these two different types of acoustic detectors. These results show the ability for non-destructive evaluation of PAM in testing the existence of disbonded regions in power transistors (Chen *et al.*, 1985).

3. DETECTION OF SURFACE AND SUBSURFACE FEATURES OF INTEGRATED CIRCUITS

The in-phase image of the photoacoustic signal essentially represents, in principle, the surface structures while the quadrature image efficiently reflects the subsurface features. Initial experiments of in-phase and quadrature imaging of integrated circuits have been carried out by Favro *et al.*, (1980). We have also employed the in-phase and quadrature components of the photoacoustic signal to obtain the surface and subsurface images of P-channel Metal-Oxide-Semiconductor (PMOS) integrated circuits, as shown in Fig. 7(a) and (b), respectively.

On the other hand, by changing the modulation frequency, we have also obtained some PAM images of another PMOS integrated circuit. During this experiment, we found that at, or below, 10 kHz, the photoacoustic imaging only reflects the optical surface structures (see Fig. 8(a)), because the thermal wavelength is much longer than the geometrical sizes of the circuit structures. When the frequency is increased to 283 kHz, the subsurface structures can be seen clearly as shown in Fig. 8(b), (Zhang and Chen, 1986).

4. DEPTH PROFILING OF STRUCTURES AND DEFECTS IN SEMICONDUCTOR DEVICES

The depth profiling of samples with multilayer structures (i.e. integrated circuits), can be obtained easily by adjusting the phase angle of the photoacoustic signal as described in Sec. IV of this Chapter. Some of our experimental results using this technique are now discussed in this section.

A set of photoacoustic images depicting the structures of a PMOS integrated circuit obtained 283 kHz is shown in Fig. 9 (a, b, and c). This figure shows that when the phase angle shift is increased gradually, the subsurface structures correspondingly become more apparent. Fig. 9(c) shows the oxidized grids on the silicon wafer underneath the *ca.* 1 μm-thick metallized layer in some regions. For comparison, Fig. 9(d) shows the optical picture of the mask used for oxidation. Good agreement between Figs. 9(c) and (d) can be obtained and the laminated PAM imaging is satisfactory (Zhang and Chen, 1986).

Some photoacoustic images of depth profiling of a MOS FET (field-effect transistor) have also been obtained at a constant frequency (283 kHz) as shown

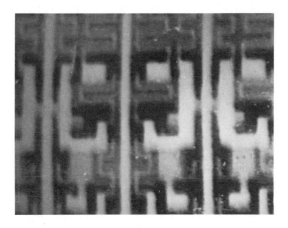

a

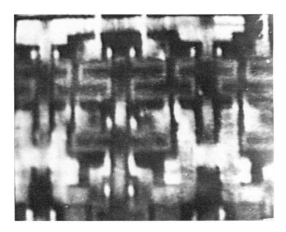

b

Fig. 7. Surface and subsurface structures of a PMOS integrated circuit; (a): in-phase, and (b): quadrature images obtained by PAM.

in Fig. 10 (a, through d). As the phase shift is set for subsurface imaging and is increased gradually, the depth profiling of two defects can be observed. Fig. 10(a) shows a single shallow subsurface defect pointed to by arrow 1. Fig. 10(b), however, displays two defects 1 and 2. Furthermore, Fig. 10(c) only shows the deeper defect 2, while defect 1 disappears. It is interesting to note that defect 2 appears enlarged in this figure. Fig. 10(d) shows only the deeper defect 2 but in smaller size than in Fig. 10(c). Therefore, we can conclude that defect 2 lies deeper in the structure than defect 1. Here, it must be pointed out that although the thickness of the entire structure of the circuits is usually just a

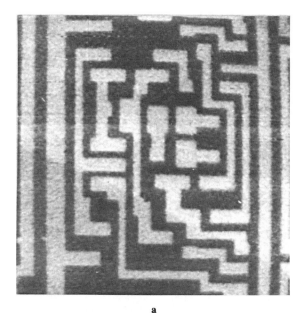

a

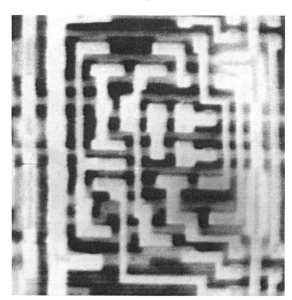

b

Fig. 8. Photoacoustic images of a PMOS integrated circuit. (a) at f = 10 kHz and (b) at f = 283 kHz.

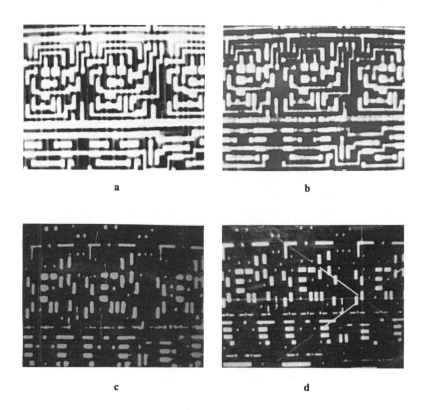

Fig. 9. Laminated images of a PMOS integrated circuit by adjusting the phase angle of the photoacoustic signal; (a) surface imaging; (b) subsurface structures begin to influence the surface picture; (c) laminated imaging of the oxidized grids of the integrated circuit; and (d) an optical photograph of the mask used for oxidation.

few µm, we can perform excellent depth profiling or laminated imaging at frequencies where the thermal wavelength is longer than the thickness by several times. Fig. 10(e) is the optical photograph of this circuit for comparison (Chen and Zhang, 1987b).

VI. Conclusions

Photoacoustic microscopy, due to its unique capabilities for non-destructive examination and subsurface structure detection and depth profiling of opaque materials and devices, has been developed rapidly in different versions. By using PAM, we can "see" inside objects and locate defects and changes in material properties, such as microcracks, delaminations, voids, inclusions, lack of bonding, etc., which are not evident on the outside surface. Therefore, PAM is very attractive for the examination of semiconductor materials and devices, in general, and integrated circuits, in particular, in industrial semiconductor R & D environments. Our experimental results with PAM demonstrate that high depth profiling resolution can be obtained by means of suitable phase angle selection of the photoacoustic signal, even when the thermal wavelength is several times longer than the thickness of the layered sample. Further improvements of PAM's imaging capability should include improving resolution by focusing of the pumping optical beam, higher frequency modulation ability, and higher acoustic detector sensitivity. PAM has the potential to be used for non-destructive quality control and detection of integrated circuits, including large scale integrated circuits (VLSI) and other devices, along the lines of similar applications of electron-acoustic microscopy (EAM) (Rosencwaig and Opsal, 1986).

In comparison with other similar techniques, the advantages of PAM are:

(1) In most opaque materials, the thermal wavelength is about 10 times less than the acoustic wavelength at the same frequency. Thus, in order to obtain the same resolution of subsurface structure imaging, the operating frequency may be 10 times lower than that of acoustic microscopy. As the operation frequencies of PAM can be below or at the MHz range, the resulting electric signals from the transducer can be easily processed by using lock-in demodulation, which conveniently renders depth profiling an easy task.

(2) PAM imaging can be carried out in air, under natural circumstances, and with ease of operation. However, high spatial resolution acoustic imaging must be performed under low temperature or high pressure conditions, whereas electron-acoustic imaging must be performed in a vacuum chamber. Thus, it can be shown that the photoacoustic imaging is much more convenient in practice.

(3) PAM offers a new effective tool for non-destructive and non-invasive detection, because the pumping source is only a laser beam with medium power. By comparison, the mechanisms for the generation of acoustic waves in EAM are more complicated and more difficult to investigate in both experiment and theory (Balk, 1985), due to the fact that electron-beam carried electric charges give rise to coupling into the dielectric and magnetic properties of the sample. For this reason, the EAM method may not be considered as truly non-invasive, however, owing to diffraction limitations in focussing the optical beam diameter, the spatial resolution of PAM is lower than that of EAM.

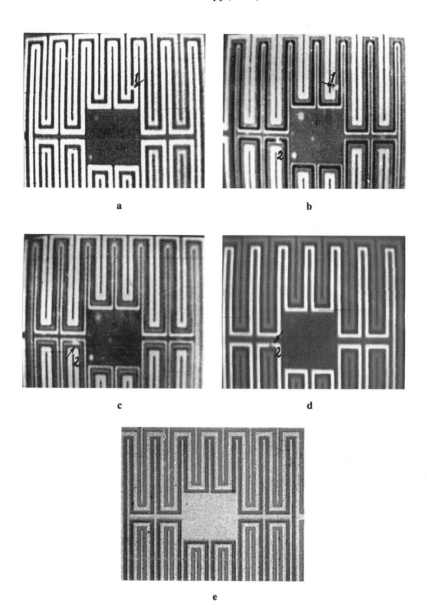

Fig. 10. Photoacoustic depth profiling images of a PMOS FET by adjusting the phase shift. (a) surface aluminum electrodes. Only a shallow defect 1 is displayed; (b) two defects 1 and 2 can be seen and complex subsurface structures of the circuit appear; (c) defect 1 disappears but defect 2 is enlarged; (d) defect 2 becomes small and almost disappears, and the simpler deep structures are displayed; (e) optical photograph of the sample circuit.

Finally, we must point out that, in spite of greater efforts still needed toward developing PAM more precisely in terms of both instrumentation and theoretical calculations, the success gained to-date in its applications to microelectronic and other semiconductor materials and devices, as well as the attained theoretical progress, convince us that in the future this technique will rapidly grow as an important technological non-destructive tool.

VII. References

Adams, M.J., Beadle, B.C., King, A.A., and Kirkbright, G.F. (1976). Analyst **101**, 553.

Adams, M.J., and Kirkbright, G.F. (1977). Analyst **102**, 281.

Balk, L.J., (1985). *4th Int. Topical Meeting on Photoacoustic, Thermal and Related Sciences. Technical Digest,* Ecole Polytech. Montréal, TuC1.

Birnbaum, G., and White, G.S. (1984) *Laser techniques in NDE* in *Research Techniques in Non-destructive Testing,* (R.S. Sharpe, ed.) Academic Press, New York, **7**, 259.

Busse, G. (1979). Appl. Phys. Lett. **35**, 759.

Busse, G., and Rosencwaig, A. (1980). Appl. Phys. Lett. **36**, 815.

Chen, L., Zhang, S.Y., and Wei, L.H. (1985). *Proc. Int. Workshop on Acoust. NDE,* Nanjing, China, E-1.

Chen, L., and Zhang, S.Y. (1987a) Acta Acustica **12**, (in the press).

Chen, L., and Zhang, S.Y. (1987b) Chin. Phys. Lett. **4**, (in the press).

Coufal, H. (1984). J. Photoacoust. **1**, 413.

Dodgson, J.T., Mandelis, A. and Andreetta, C. (1986). Can. J. Phys. **64**, 1074.

Favro, L.D., Kuo, P.K., Pouch, J.J., and Thomas, R.L. (1980). Appl. Phys. Lett. **36**, 953.

Jackson, W., and Amer, N.M. (1980). J. Appl. Phys. **51**, 3343.

Kato, K., Ishino, S., and Sugitani, Y. (1980). Chem. Lett. **103**, 783.

Kino, G.S., and Stearns, R.G. (1985). Appl. Phys. Lett. **47**, 926.

Kirkbright, G.F., and Miller, R.M. (1983). Anal. Chem. **55**, 502.

Luukkala, M., and Penttinen, A. (1979). Electron. Lett. **15**, 325.

Mandelis, A. (1986). IEEE Trans. UFFC, **UFFC-33**, 590

Mandelis, A. and Dodgson, J.T. (1986). J. Phys. C: Solid State Phys. **19**, 2329.

McClelland, J.F. (1980). *Proc. IEEE Ultrason. Symp. 1980,* Boston, MA. 610.

McClelland, J.F., Kniseley, R.N., and Schmit, J.L. (1980). *Proc. Symp. on Scanned Image Microscopy,* (E.A. Ash, Ed.) Academic Press, Orlando, FLA, 353.

McFarlane, R.A., Hess, L.D., and Olson, G.L. (1980). *IEEE Ultrason. Symp. Proc.* 628.

Opsal, J., and Rosencwaig, A. (1982). J. Appl. Phys. **53,** 4240.

Patel, C.K.N., and Tam, A.C. (1981). Rev. Mod. Phys. **53,** 517.

Rosencwaig, A., and Gersho, A. (1976). J. Appl. Phys. **47,** 64.

Rosencwaig, A. (1978). In *Advances in Electronics and Electron Physics,* (L. Marton, ed.) **Vol. 46,** Academic Press, N.Y.

Rosencwaig, A. (1980). *Photoacoustics and Photoacoustic Spectroscopy.* John Wiley and Sons, N.Y.

Rosencwaig, A. (March 1982). Solid State Tech. 91.

Rosencwaig, A. (1983). J. Phys. (Paris) **44,** C6-437.

Rosencwaig, A. and Busse, G. (1980). Appl. Phys. Lett. **36,** 725.

Rosencwaig, A., and Opsal, J. (1986). IEEE Trans. UFFC, **UFFC-33,** 516.

Sugitani, Y., and Uejima, A. (1985). Chem. Lett. (Chem. Soc. Jpn.) **12,** 397.

Takaue, R., Tobimatsu, H., Matsunaga, M. and Hosokawa, K. (1986). J. Appl. Phys. **59,** 3975.

Tam, A.C. (1983). In *Ultrasensitive Laser Spectroscopy.* (D.S. Kliger, Ed.), Academic Press, 1.

Tam, A.C. (1986). Rev. Mod. Phys. **58,** 381.

Thomas, R.L., Favro, L.D., and Kuo, P.K. (1986). Can. J. Phys. **64.,** 1234.

Von Gutfeld, R.J., and Melcher, R.L. (1977). Appl. Phys. Lett. **30,** 257.

West, G.A. (1983). Rev. Sci. Instrum. **54,** 797.

White, R.M. (1963). J. Appl. Phys. **34,** 3359.

Wickramasinghe, H.K., Bray, R.C., Jipson, V., Quate, C.F., and Salcedo, J.R. (1978). Appl. Phys. Lett. **33,** 923.

Wong, Y.H., Thomas, R.L., and Hawkins, G.F. (1980). Appl. Phys. Lett. **32,** 538.

Zhang, S.Y., Yu, C., Miao, Y.Z., Tång, Z.Y., and Gao, D.T. (1982) in *Acoustical Imaging,* Plenum Press. N.Y. **12,** 60.

Zhang, S.Y., Gao, D.T., and Tang, Z.Y. (1983). *Digest Int. Conf. on Lasers.* Guangzhow, China, 452.

Zhang, S.Y., and Chen, L. (1986). Can. J. Phys. **64,** 1316.

Zuccon, J.F. and Mandelis, A. (1987). IEEE Trans. UFFC (In the press).

3

NON-DESTRUCTIVE AND NONCONTACT OBSERVATION OF MICRODEFECTS IN GaAs WAFERS WITH A NEW PHOTO-THERMAL-RADIATION (PTR) MICROSCOPE

Nobuo Mikoshiba and Kazuo Tsubouchi

Research Institute of Electrical Communication
Tohoku University
Katahira, Sendai, JAPAN

I. Introduction

High quality GaAs wafers are required for advanced GaAs devices. GaAs wafers, however, at their present state of fabrication include various defects such as dislocations, microdefects and so on. Recently, dislocation-free GaAs wafers have been obtained by In-doping (Jacob, Duseaux, Farges, Van Den Boom and Roksnoer 1983; Miyairi, Inada, Obokata, Nakajima, Katsumata and Fukuda, 1985), and microdefects other than dislocations have been recognized as an origin of inhomogeneity of such wafers.

To date, two-dimensional observation of microdefects has been carried out by transmission electron microscopy (TEM), chemical etching, photoluminescence (PL) topography and IR transmission topography (Brozel, Grant, Ware and Stirland, 1983; Windscheif, Baeumler and Kaufmann, 1985; Katsumata and Fukuda, 1986). The TEM and the chemical etching are destructive measurements. The PL technique is not a direct observation method for nonradiative defects and a sample must be cooled for the detailed measurement of the PL signals. In IR transmission topography, a GaAs sample must be prepared with a thickness on the order of 3~5 mm and both surfaces must be polished. There is no effective method of providing a simple rapid nondestructive means for the observation of microdefects in industrial type GaAs wafers.

We previously demonstrated the application of a piezoelectric photoacoustic (PA) technique to investigate nonradiative processes and defects in semiconductors (Wasa, Tsubouchi and Mikoshiba, 1980 a,b; Nakamura, Tsubouchi and Mikoshiba, 1984). Although this technique is sensitive to the local temperature rise in semiconductors, its general use is limited because of the necessity of contact.

Nordal and Kanstad (1981) have applied photo-thermal radiometry (PTR) to obtain optical absorption spectra without contact. When a specimen is illuminated by amplitude-modulated light, the modulation of surface temperature by heating through absorption results in similar modulation in thermal reradiation. By measuring such variations in radiant emittance, those authors were able to obtain information on an object's spectral and/or thermal properties. Busse and Renk (1983), and Busse and Eyerer (1983) have applied PTR for nondestructive remote inspection of flaws in metals and polymers, respectively. Tam and Sullivan (1983) have shown that pulsed PTR is useful for remote sensing applications. On the other hand, Ermert, Dacol, Melcher and Baumann (1984) developed a subsurface imaging technique based on the emitted IR radiation, which was generated by a scanned and modulated electron beam.

Recently, we found that a new photo-thermal-radiation (PTR) microscope can provide a nondestructive and noncontact observation of microdefects in GaAs wafers at room temperature (Nakamura, Tsubouchi, Mikoshiba and Fukuda, 1985). We measured the PTR spectra (PTR signal vs. excitation energy) and the PTR images, and compared those images with X-ray topographs, PL images and dislocation-density distributions (Mikoshiba, Akutsu, Nakamura, Tsubouchi and Hosokawa 1986).

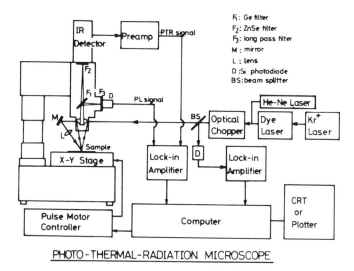

Fig. 1 Schematic diagram of apparatus for the measurement of PTR spectra and PTR images.

II. Experimental Apparatus

Fig. 1 shows a schematic diagram of the apparatus used for the measurements of PTR spectra and PTR images; all measurements were carried out at room temperature (~23°C). A tunable dye laser (HITC, 850~920 nm, ~80 mW) excited with a Kr ion laser (red multiline, 5W) was used as the light source. The laser beam was chopped with an optical chopper and focussed on a spot of about 200 μm diameter at the sample surface. The chopping frequency was about 330 Hz. IR radiation emitted from the heated spot was collected with a reflecting objective mirror and detected with a photoconducting $Hg_{0.8}Cd_{0.2}Te$ IR detector (NERC, MPC11-2-AD1) which was sensitive to IR radiation in the wavelength range from 8 to 13 μm. The PTR signals detected with the IR detector were fed into a lock-in amplifier through a preamplifier with the gain of 60 dB. To normalize the PTR signals by the irradiation-light intensity, we used the reflected light from a beam splitter. We designed the experimental setup so that we could measure PL images at the same time as the PTR measurement. A Kr ion laser (red multiline, 150 mW) was used as the excitation light source. The peak wavelength of PL of GaAs is about 870 nm at room temperature. For the PL measurement, the reflected light from the Ge filter was detected with a Si photodiode through a long pass filter (F3). The wavelength range of the final PL signal was from 830 to 960 nm.

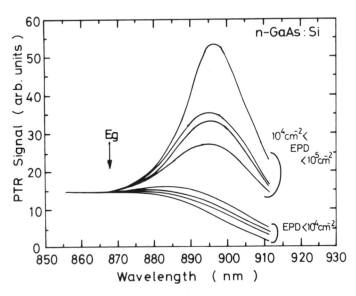

Fig. 2 PTR spectra of n-GaAs wafers with various EPD's, where EPD is the average value of the etch pit density at several points on the GaAs wafer.

III. Results and Discussion

1. PTR SPECTRA AND PTR TOPOGRAPHS OF DISLOCATED n-GaAs WAFERS

Fig. 2 shows the PTR spectra of several dislocated n-GaAs wafers (Si-doped, (100)-oriented, $1 \sim 3 \times 10^{-3}$ Ω cm, $300 \sim 400$ µm thick) with various EPD's, where EPD is the average value of the etch pit density at several points on the GaAs wafer. The bandgap energy of GaAs (1.43 eV at room temperature) is indicated by E_g. The PTR spectra of the GaAs wafer with the higher EPD's have a peak at the wavelength of 895 nm. We suggest that the PTR peak is caused by the increase of local temperature due to the optical absorption and the subsequent nonradiative relaxation at nonradiative defects in GaAs wafers. There is a possibility that the nonradiative defects are microdefects rather than dislocations as discussed later.

Figs. 3 (a) and (b) show the PTR image and the PL image of a dislocated n-GaAs wafer, respectively. The PTR image was obtained by using the excitation wavelength of about 895 nm; the amount of time required to obtain this image was about 90 minutes. The white area in Fig. 3 (a) and (b) indicates the high intensity area of the PTR and PL signals. The increase of the PTR signals in Fig. 3(a) clearly corresponds to the decrease of the PL signals in Fig. 3(b). The PTR signal gives direct information on nonradiative processes at defects in GaAs wafers.

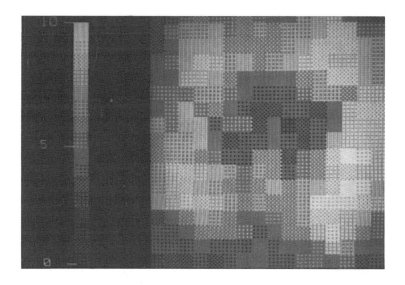

a

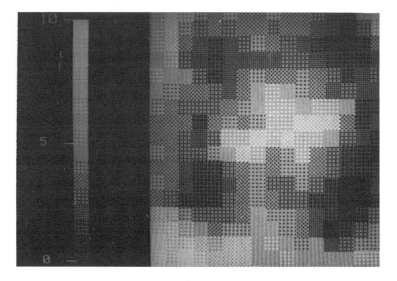

b

Fig. 3 (a) PTR image and (b) PL image of an n-GaAs wafer.

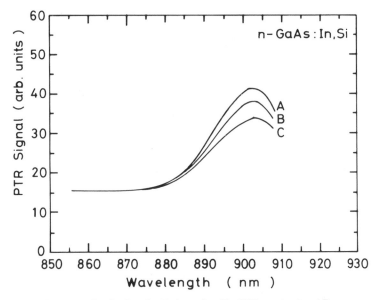

Fig. 4 PTR spectra of an In-doped n-GaAs wafer. The PTR spectra A and B were measured at the dislocated area of the GaAs wafer. The PTR spectrum C was measured at the dislocation-free area.

Fig. 5 X-ray topograph of the In-doped n-GaAs wafer. Points A, B and C indicate the data points of the PTR spectra in Fig. 4.

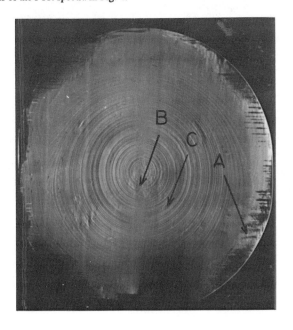

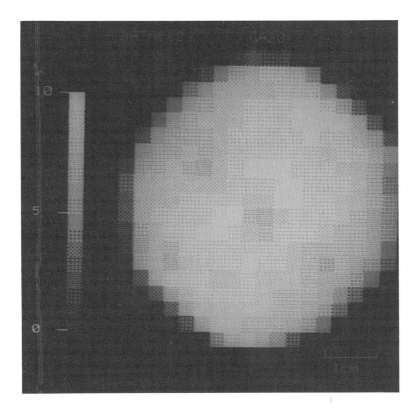

Fig. 6(a) PTR image of an In-doped n-GaAs wafer

2. PTR SPECTRA AND PTR TOPOGRAPHS OF AN In-DOPED n-GaAs WAFER

Fig. 4 shows the PTR spectra at 3 points in an In-doped n-GaAs wafer (Si-doped, $\rho \sim 1 \times 10^{-3}$ Ω cm). The points at which measurements were performed are indicated by A, B and C on the X-ray topograph in Fig. 5. Points A and B are dislocated areas and the point C is dislocation-free area. As shown in Fig. 4, the PTR spectra at the dislocation-free point C as well as at dislocated points A and B have a peak at the wavelength of about 903 nm. It is clear that the PTR peak in Fig. 4 is caused by nonradiative states due to microdefects rather than dislocations. The microdefects cannot be observed in the X-ray topograph.

We found a discrepancy between the wavelength values of the PTR peaks in Figs. 2 and 4. The shift of the PTR peak from 895 nm to 903 nm seems to be caused by the change of band gap energy of GaAs due to In-doping. There is a possibility that the PTR peak (~895 nm) in Fig. 2 and the PTR peak (~903 nm) in Fig. 4 are caused by the same microdefects in the GaAs wafer.

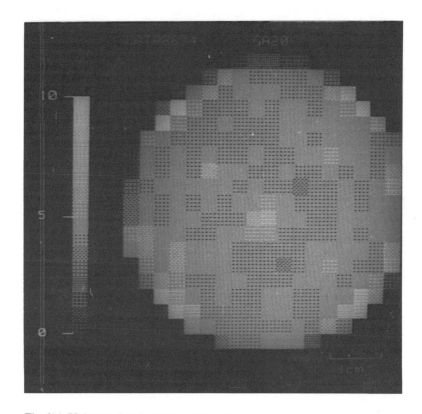

Fig. 6(b) PL image of the In-doped n-GaAs wafer

Figs. 6(a) and (b) show the PTR image and the PL image of the In-doped n-GaAs wafer, respectively. The image in Fig. 6 (a) was obtained by using the excitation wavelength of 903 nm. The increase of the PTR signal in Fig. 6(a) corresponds to the decrease of the PL signal in Fig. 6(b). The PTR image clearly shows microdefect density inhomogeneity which cannot be observed in the X-ray topograph of Fig. 5.

3. PTR IMAGE AND EPD DISTRIBUTION

We measured the PTR image of the GaAs wafer with the excitation wavelength fixed at 895nm, and compared the PTR image with the dislocation-density distribution. We found that the PTR image shows negative correlation with the dislocation-density profile. It was thus deduced that the origin of the PTR signal was the nonradiative microdefects generated predominantly in the low-dislocation-density region.

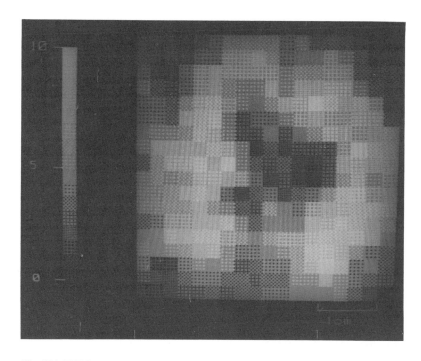

Fig. 7(a) PTR image, of a Si-doped n-type LEC GaAs wafer.

Fig. 7(b) PL image of the Si-doped n-type LEC GaAs wafer

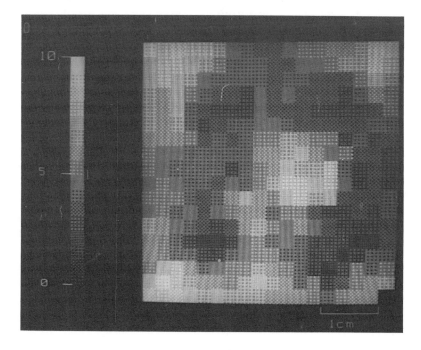

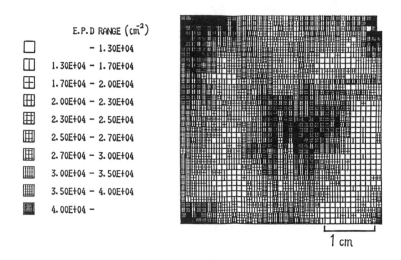

E.P.D RANGE (cm^{-2})

□ $- 1.30E+04$

⊞ $1.30E+04 - 1.70E+04$

⊞ $1.70E+04 - 2.00E+04$

⊞ $2.00E+04 - 2.30E+04$

⊞ $2.30E+04 - 2.50E+04$

⊞ $2.50E+04 - 2.70E+04$

⊞ $2.70E+04 - 3.00E+04$

⊞ $3.00E+04 - 3.50E+04$

⊞ $3.50E+04 - 4.00E+04$

■ $4.00E+04 -$

1 cm

Fig. 7(c) EPD distribution of the Si-doped n-type LEC GaAs wafer

To reveal dislocation-etch-pits and observe their distribution, etching was performed at 370°C for 15 minutes using molten KOH. The distribution of EPD's was evaluated by counting the dislocation-etch-pits with a Wafer Defect Analyzer (Mitsubishi Chemical Industries, GX-11). Figs. 7(a) and 7(b) show a PTR image and a PL image of a dislocated n-type LEC GaAs wafer (Si-doped, (100)-oriented, 310μm thick), respectively. The free carrier concentration of the sample was about $1 \times 10^{18} cm^{-3}$. The PTR image was obtained by using the excitation wavelength of about 895nm. The white areas in Figs. 7(a) and 7(b) indicate high intensity of the PTR and PL signals.

The increase of the PTR signals clearly corresponds to the decrease of the PL signals. The PTR image shows an inhomogeneous distribution of nonradiative defects in the GaAs wafer.

Fig. 7(c) shows the EPD distribution of the GaAs wafer. It should be pointed out that the PTR image in Fig. 7(a) shows negative correlation with the EPD profile in Fig. 7(c).

Figure 8 shows the PTR, PL and EPD profiles along the wafer diameter. The PTR profile is nearly M-shaped, while the PL and EPD profiles are nearly W-shaped.

It has been reported that in various semiconductors some microdefects, such as minute precipitates, vacancy complexes, dislocation loops and so on, are generated in the region where the dislocation density is decreased, because the dislocations act as sinks for the microdefects (Casey 1967; Iizuka 1968, 1971).

Those earlier observations are in agreement with the origin of the PTR signal under the 895 nm-excitation, which is considered to be due to the nonradiative microdefects generated predominantly in the region where the dislocation density is decreased.

As is shown in Fig. 2, the peaks at 895 nm were observed in the PTR spectra of GaAs wafers with higher EPD($1\times10^4\mathrm{cm}^{-2}$<EPD<$1\times10^5\mathrm{cm}^{-2}$), while no PTR peak was observed in the PTR spectra of GaAs wafers with lower EPD (<$1\times10^4\mathrm{cm}^{-2}$). "EPD" indicates the average value of EPD at several points on the GaAs wafer. If we assume that the average value of microdefect density in the GaAs wafer increases with the EPD, the origin of the PTR peak at 895 nm can be interpreted as due to these microdefects. In a GaAs wafer, however, the EPD shows negative correlation with the profile of nonradiative microdefect density as shown in Figs. 7(a) and 7(c). Tajima (1982) investigated deep levels in semi-insulating LEC GaAs by PL measurements at 4.2 K and reported that the PL intensity profile of the 0.80-eV band shows negative correlation with the EPD profile. He concluded that the origin of the 0.80-eV band can be interpreted as due to microdefects generated predominantly in the region where the dislocation density is decreased. Therefore, there is a possibility that the PTR signals under 895 nm-excitation and the PL of the 0.80-eV band are caused by the same microdefects.

Fig. 8 PTR, PL and EPD profiles along the wafer diameter in the Si-doped n-type LEC GaAs wafer.

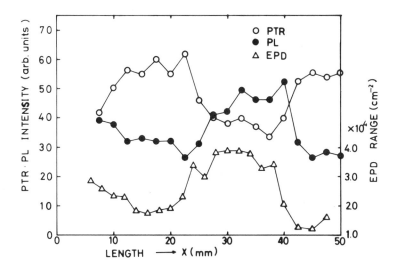

IV. Conclusion

In conclusion, a nondestructive and noncontact method for observation of microdefects in GaAs wafers has been developed with a new Photo-Thermal-Radiation (PTR) microscope. We measured the PTR signal as a function of excitation energy (PTR spectrum) and the spatial distribution of PTR intensity (PTR image) of various GaAs wafers at room temperature. We found that the PTR spectra have a peak due to nonradiative states of microdefects at wavelengths ranging from 895 to 903 nm. We measured the PTR image with the excitation wavelength fixed at 895 nm, and compared it with the dislocation-density distribution. The PTR image shows negative correlation with the dislocation-density profile. The PTR signal apparently originates from the nonradiative microdefects generated predominantly in the low-dislocation-density regions. In IR transmission topography, a GaAs sample must be prepared with a thickness as large as 3~5 mm and both surfaces of the sample should be polished. On the other hand, we were able to evaluate commercially available GaAs wafers (300~400 μm thick) without further preparation, using the new PTR microscope.

V. References

Brozel M.R., Grant I., Ware R.M. and Stirland J.D. (1983). Appl. Phys. Lett., **42**, 610.

Busse G. and Eyerer P. (1983). Appl. Phys. Lett., **43**, 355.

Busse G. and Renk K.F. (1983). Appl. Phys. Lett., **42**, 366.

Casey, Jr.H.C. (1967). J. Electrochem. Soc., **114**, 153.

Ermert H., Dacol F.H., Melcher R.L. and Baumann T. (1984). Appl. Phys. Lett., **44**, 1136.

Iizuka T. (1968). Jpn. J. Appl. Phys., **7**, 490.

Iizuka T. (1971). J. Electrochem. Soc., **118**, 1190.

Jacob G., Duseaux M., Farges J.P., Van den Boom M.M.B. and Roksnoer P.J. (1983). J. Cryst. Growth, **61**, 417.

Katsumata T. and Fukuda T. (1986). Rev. Sci. Instr., **57**, 202.

Mikoshiba N., Akutsu Y., Nakamura H., Tsubouchi K. and Hosokawa M. (1986). *Proc. 1986 Ultrasonics Symp.* (IEEE, New York, 1986).

Miyairi H., Inada T., Obokata T., Nakajima M., Katsumata T. and Fukuda T. (1985). Jpn. J. Appl. Phys., **24**, L729.

Nakamura H., Tsubouchi K. and Mikoshiba N. (1984). Jpn. J. Appl. Phys., **23**, L1.

Nakamura H., Tsubouchi K., Mikoshiba N. and Fukuda T. (1985). Jpn. J. Appl. Phys., **24**, L876.

Nordal P-E. and Kanstad S.O. (1981). Appl. Phys. Lett., **38**, 486.

Tajima M. (1982). Jpn. J. Appl. Phys., **21**, L227.

Tam A.C. and Sullivan B. (1983). Appl. Phys. Lett., **43**, 333.

Wasa K., Tsubouchi K. and Mikoshiba N. (1980a). Jpn. J. Appl. Phys., **19**, L475.

Wasa K., Tsubouchi K. and Mikoshiba N. (1980b). Jpn. J. Appl. Phys., **19**, L653.

Windscheif J., Baeumler M. and Kaufmann U. (1985). Appl. Phys. Lett., **46**, 661.

II

Thermal Wave Imaging and Characterization of Semiconductor Materials and Devices

THERMAL WAVE PROPAGATION AND SCATTERING IN SEMICONDUCTORS

Lawrence D. Favro, Pao-Kuang Kuo, and Robert L. Thomas

Department of Physics and Astronomy
Wayne State University
Detroit, MI 48202 U.S.A.

I. Introduction to Thermal Waves

The flow of heat in a solid depends on its thermal conductivity, κ, its density, ρ, and its specific heat, c. That flow is diffusive in nature and hence is governed by a diffusion equation,

$$\nabla \cdot [\kappa \nabla T(\mathbf{r},t)] - \rho c \frac{\partial T}{\partial t}(\mathbf{r},t) = -f(\mathbf{r},t) \tag{1}$$

Here, the function, $f(\mathbf{r},t)$, represents the heat source and has the dimensions of energy per unit volume per unit time. The term "thermal wave" has come into use in referring to a special class of time-dependent solutions to this equation which occur when the source function $f(\mathbf{r},t)$ is periodic in time. If the source periodicity is characterized by an angular frequency, ω, we can write the source function in complex form as

$$f(\mathbf{r},t) = f(\mathbf{r})e^{-i\omega t} \tag{2}$$

with a corresponding form for $T(\mathbf{r},t)$. The heat diffusion equation, (1), then takes the form

$$\nabla \cdot [\kappa \nabla T(\mathbf{r})] + \kappa q^2 T(\mathbf{r}) = -f(\mathbf{r}) \tag{3}$$

where we have defined a thermal wave number, q, as

$$q = (1+i)(\omega \rho c / 2\kappa)^{\frac{1}{2}} \tag{4}$$

The quantity $\kappa/\rho c$ which appears in the definition of q is called the "thermal diffusivity" of the material. It measures the ability of the material to absorb heat on a transient basis and hence appears in all time-dependent heat diffusion problems. The origin of the term "thermal wave" lies in the form of Eq. (3). This equation is formally equivalent to the Helmholtz equation for wave motion and therefore has wave-like solutions.[*] These solutions, however, represent very heavily damped waves as indicated by the fact that their propagation vector, q, has equal real and imaginary parts. The amplitude of the wave falls to 1/e in a distance called the "thermal diffusion length", μ, given by

$$\mu = (2\kappa/\omega \rho c)^{\frac{1}{2}} \tag{5}$$

---

[*] The recognition of this wave-like behavior goes back at least to Stokes, and probably further. (Carslaw and Jaeger; 1959).

Since this length is inversely proportional to the square root of the frequency, the penetration depth of thermal waves may be varied by varying the frequency. In typical thermal wave microscopes, with common materials (metals, semiconductors, ceramics, etc.) as samples, it may range from the order of microns to a millimeter or so. Values outside this range can be used but are not common. The near-surface specificity caused by this short penetration depth makes thermal wave microscopes complementary to acoustic microscopes, which are ordinarily most effective in imaging deeper features.

The wavelength of a thermal wave is 2π times the thermal diffusion length, so that the amplitude of the wave falls to $\exp(-2\pi)$ (approximately 1/500) in one wavelength. This means that experiments involving the scattering of thermal waves must launch, scatter, and receive the waves in a total propagation distance not much larger than a wavelength. The result of this is that, although thermal waves scatter, diffract, and interfere much like more familiar types of waves, conventional scattering and diffraction formulas, which normally are derived in the far-field limit, cannot ordinarily be applied to thermal wave experiments in an unmodified form. This chapter will be devoted to the description of techniques for calculating formulas which are appropriate for analyzing thermal wave experiments.

II. Reciprocity Theorem for Thermal Wave Scattering

The solution to any thermal wave scattering problem can be imagined to have been obtained by use of a Green's function technique regardless of the actual technique used to solve the problem. This possibility allows one to prove a simple but very useful reciprocity theorem (Kuo and Favro, 1982; Kuo, Inglehart, Favro and Thomas, 1981). The theorem basically says that in any thermal wave scattering experiment, the source and the detector of the thermal waves may always be imagined to have been interchanged without changing the signal received by the detector. Its usefulness stems from the fact that some of the more common detectors (photoacoustic cells, mirage probe beams, etc.) have symmetries which, if they were regarded as sources rather than detectors, would generate thermal waves with a higher degree of symmetry (plane waves, cylindrical waves, etc.) than the actual waves used in the experiment. This results in both calculational and conceptual simplifications in the description of the scattering process.

The proof of the theorem is quite simple. We imagine a sample of arbitrary shape upon which a thermal wave experiment is to be performed. Although it is not necessary to do so, we will imagine the surfaces of the sample to be either insulated or so far from the source that the waves never reach them. Since the sample is not assumed to be homogeneous, any backings, sample holding devices, etc. which the waves do reach can be handled by generalizing the notion of "sample" to include them. The equation satisfied by the temperature in the solid is then given by

$$\nabla \cdot (\kappa \nabla T) + \kappa q^2 T = -f \qquad (6)$$

with the boundary condition

$$\nabla \cdot (\kappa \nabla T)|_{surface} = 0 \qquad (7)$$

and where the thermal conductivity, κ, and the thermal wave propagation vector, q, will ordinarily be functions of position due to inhomogeneities in the sample, the presence of backings, etc. The Green's function for this problem satisfies the same equation with the same boundary condition, but with a delta-function source

$$\nabla \cdot (\kappa \nabla G(\mathbf{r}',\mathbf{r})) + \kappa q^2 G(\mathbf{r}',\mathbf{r}) = -\delta(\mathbf{r}-\mathbf{r}') \qquad (8)$$

If we had not used the simplifying boundary condition (7), the Green's function would have had to satisfy the so-called "adjoint" boundary condition to (7) and would be called the "adjoint" Green's function. The theorem would remain unchanged except that, in addition to interchanging the source and the detector in the inverse problem, one would also have to use the adjoint boundary condition for the temperature in that problem. The application of Green's theorem to Eqs. (6) and (7) results in a simple expression for the temperature

$$T(\mathbf{r}) = \int G(\mathbf{r},\mathbf{r}')f(\mathbf{r}')d^3\mathbf{r}' \qquad (9)$$

The detection system which generates the experimental signal strength can be introduced into the theoretical model by describing the portion of the temperature distribution to which it is sensitive. The detector samples this temperature distribution in some region of the sample (e.g. an area of the surface in a photoacoustic cell, a strip of the surface with a mirage probe beam, etc.) which can be described in general by some weight function, $g(\mathbf{r})$. The signal, S, at the detector is then given by the double integral

$$S = \int T(\mathbf{r})g(\mathbf{r})d^3\mathbf{r} = \iint G(\mathbf{r}',\mathbf{r})f(\mathbf{r}')g(\mathbf{r})d^3\mathbf{r}d^3\mathbf{r}' \qquad (10)$$

On the other hand, if we were to consider the inverse problem, in which $g(\mathbf{r})$ is the source function and $f(\mathbf{r})$ is the detector function, the signal, S_{inv}, would be given by

$$S_{inv} = \iint G(\mathbf{r},\mathbf{r}')f(\mathbf{r}')g(\mathbf{r})d^3\mathbf{r}d^3\mathbf{r}' \qquad (11)$$

which differs from S only by the interchange of $\mathbf{r}$ and $\mathbf{r}'$ in the Green's function. Thus, if the Green's function is symmetric under this interchange, the two signals, S and S_{inv}, will be identical. The symmetry of the Green's function is most easily demonstrated by writing Eq. (8) twice, once with the source point at $\mathbf{r'}$ and once at $\mathbf{r''}$. The application of Green's theorem to this new pair of equations then leads immediately to $G(\mathbf{r'},\mathbf{r''}) = G(\mathbf{r''},\mathbf{r'})$ proving the theorem.

The utility of the theorem stems from the fact that, while the actual source function, $f(\mathbf{r})$, is usually highly localized, the detector function, $g(\mathbf{r})$, very often has a high degree of symmetry, and, when regarded as a source rather than a detector, generates thermal waves with the same degree of symmetry. Thus, instead of having to perform difficult three-dimensional scattering calculations with incident spherical waves (such as would have been generated by the actual source), one can often perform calculations with waves of simpler symmetry. The greatest simplification occurs if the detector has planar symmetry such as is the case with the gas-cell detection scheme. However, even the reduction in the number of degrees of freedom from three to two, which occurs with the cylindrical symmetry of the mirage detection scheme, contributes significantly to the ease of calculating the scattered waves, particularly when numerical computation is involved. Other detection schemes, such as the piezoelectric detection where the acoustic modes of the sample as a whole often determine the function $g(\mathbf{r})$, normally have lower symmetry but occasionally can be simplified by the use of the reciprocity relationship between source and detector.

Another important use of the reciprocity idea occurs when one is not actually trying to calculate the scattered waves, but rather is trying to understand what might happen in a given scattering experiment. An often quoted example is the inability of a gas-cell detector to detect tightly closed vertical cracks in a flat surface. In this case the source in the inverse problem is uniformly distributed over the entire surface of the sample in the cell (the microphone responds equally well to all parts of the cell), and the resulting thermal waves are plane waves propagating away from the surface in the normal direction. Hence, these inverse-problem waves strike vertical cracks edge-on and are not scattered by them. Other more complicated situations can often be analyzed in this same fashion and the insight one gains can be used to guide the design of experiments.

III. Spatial Resolution in Thermal Wave Imaging

1. LIMITING FACTORS FOR SPATIAL RESOLUTION

Many factors can contribute to the resolution of an imaging system. For example, imperfections in lenses or mirrors can limit the resolution of microscopes or telescopes. The coarseness of the detector, for example the grain size of a photographic film, can also be a limiting factor. In scanning microscopes, the illuminated spot size is important. However, all of these limitations can, at least in principle, be reduced by proper experimental design. From a theoretical point of view, a more important consideration is the intrinsic limit placed on the resolution by the underlying physics.

The intrinsic resolution of microscopes is normally discussed in terms of the "Rayleigh criterion", (Jenkins and White, 1976) which limits the resolution to distances of the order of about half of the wavelength of the radiation being used. However, derivations of the formulas leading to this criterion necessarily make use of far-field approximations. These approximations assume that the distance from the source to the point of observation is large compared to the "Fresnel length" (given by the square of the size of the source divided by the wavelength). However, because of their highly damped nature, thermal waves are always observed at distances which are of the order of, or even much smaller than, their wavelength. Thus, far-field approximations, and hence the Rayleigh criterion, are inappropriate when thermal wave scattering is being discussed. Near-field scattering is perhaps most easily understood by studying the approximations involved in the far-field theory.

Consider the Green's function for a generic scattering problem involving an unspecified type of wave with wave number **k**:

$$G(\mathbf{r},\mathbf{r}') = \frac{e^{ik|\mathbf{r}-\mathbf{r}'|}}{|\mathbf{r}-\mathbf{r}'|} \tag{12}$$

Here we may think of **r** as being the point of observation, and **r**' as a typical point of the scatterer. The usual approximation begins by assuming that $|\mathbf{r}'|$ is small compared to $|\mathbf{r}|$. One can then proceed by expanding the phase of the Green's function in powers of $|\mathbf{r}'|/|\mathbf{r}|$, and replacing its denominator by $|\mathbf{r}|$,

$$G(\mathbf{r},\mathbf{r}') = \frac{1}{r} e^{ikr}\left[1-\frac{\mathbf{r}\cdot\mathbf{r}'}{r^2}+O\left[\frac{r'}{r}\right]^2\right] \tag{13}$$

In the far-field approximation (often referred to as the "Fraunhofer limit"), one continues by keeping the zeroth and first order terms in the phase and dropping all higher order terms. The apparent inconsistency in the relative degree of approximation used in the treatment of the denominator of the Green's function and its phase is (correctly) justified by the fact that the dependence of the denominator on **r**' only makes small changes in the *amplitude* of the field, while a relatively small *percentage* variation of the phase can even change the overall *sign* of the field. This first-order Fraunhofer approximation to the phase leads directly to a resolution limit given by the Rayleigh criterion.

The next degree of approximation is the so-called "Fresnel limit" in which the point of observation is not so far, or the size of the source is not so small, that the second order terms in the expansion of the phase are negligible. It is in fact this condition, i.e. that the second order term (of order kr'^2/r) in the phase be comparable with 2π, that defines the Fresnel length

$$r = kr'^2/2\pi = r'^2/\lambda \tag{14}$$

In the Fresnel limit, the Green's function denominator may or may not be approximated to more than the zeroth order, depending on the needs of the particular problem being studied. However, regardless of the degree of approximation being used, both the Fraunhofer and the Fresnel limits are large distance approximations in that they assume that an expansion in powers of $|\mathbf{r'}|/|\mathbf{r}|$ is valid. One can easily imagine situations in which $\mathbf{r}$ is comparable to, or even much smaller than, $\mathbf{r'}$, thus invalidating this assumption. To be sure, this relationship is difficult to arrange in ordinary optics, but in thermal wave scattering these conditions are more often the rule than the exception. This situation requires a totally different kind of approximation, the extreme near-field limit.

2.   THE EXTREME NEAR-FIELD LIMIT

When the quantity, $k|\mathbf{r-r'}|$, which defines the phase of the Green's function (12), is of the order of unity or smaller, it is not the variation of the phase which is principally responsible for image formation (and hence for resolution). It is instead the variation in the denominator of Eq. (12), and hence the amplitude, which causes the field to fall off at the edge of an image. Thus it becomes imperative, not only that the denominator not be set equal to r, but that its variation be treated exactly. This leads to results which are qualitatively different from those of either the Fraunhofer or Fresnel approximation. Since the phase variation is now relatively unimportant in determining the edge of an image, it follows that the wavelength is no longer critical in determining the shape of the edge. It is another quantity with the dimensions of length, namely the source-scatterer distance $|\mathbf{r-r'}|$, which limits the sharpness of the edge and therefore the resolution of the image.

This dependence of the resolution on the depth of the scatterer rather than on the wavelength is illustrated in Fig. 1 (Inglehart, Lin, Favro, Kuo and Thomas, 1983; Inglehart, Grice, Favro, Kuo and Thomas, 1983). In this figure we

Fig. 1 Calculated thermal wave traces of the magnitude of the thermal wave signal from two closely spaced planar subsurface scatterers.

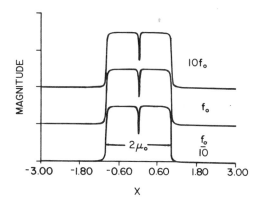

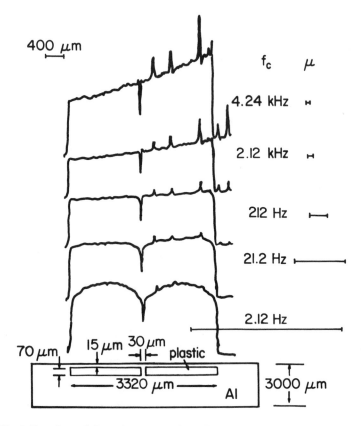

Fig. 2 Experimental thermal wave magnitude line traces corresponding to the situation described by Fig. 1 (see text for a description of the sample).

plot the calculated thermal wave signal scattered from two very shallow insulating layers beneath the surface of a flat sample. The calculated curves are for three different frequencies, corresponding to wavelengths differing by factors of the square root of 10. It is apparent that the changes in the wavelength do not affect the calculated resolution of the small separation between the two insulating layers. The parameters in the calculation were chosen to correspond to experimental measurements on a sample consisting of two very thin plastic strips embedded in an aluminum slab. The plastic strips in the experimental sample were 15 μm beneath the surface and their separation was 30 μm. Fig. 2 (Inglehart *et al.*, 1983; 1983) shows experimental line traces across the two buried strips for a variety of frequencies. The bars on the right indicate the thermal diffusion length (thermal wavelength divided by 2π). It is quite clear that the resolution of 30 μm separation is independent of, and orders of magnitude smaller than, the thermal wavelength. This 30 μm resolution is instead determined by the convolution of the 15 μm depth and the approximate 12 μm size of the heating beam used in the scans, and is in agreement with the extreme near field theory. The curvature of the experimental line scans over the plastic strips for the longest diffusion lengths is believed to be due to multiple scattering between the surface and the strips. The theoretical traces of Fig. 1 were calculated for single scattering only and therefore do not exhibit multiple scattering effects.

It is not always true that extreme near field images will show such sharp edges as those seen above. The sharpness there stems from the fact that the strips were completely flat and at a uniform distance from the surface. McDonald and Wetsel (1984) have performed similar calculations and experiments using scatterers which were rounded on top rather than flat. Line scans over these rounded tops give traces which not only have rounded shapes, but which also exhibit an apparent width which is frequency, and hence wavelength, dependent. This frequency dependence is an indication that the variations in the phase of the signal between the shallower and deeper parts of the rounded top are not completely negligible. Thus, while the extreme near field limit may be characterized by the statement that the $|\mathbf{r}-\mathbf{r}'|^{-1}$ amplitude factor tends to determine the resolution, the phase of the signal may also contribute for some scattering geometries.

IV. Scattering Calculations for Gas-Cell Imaging: Plane Waves

1.   GREEN'S FUNCTION TECHNIQUES

Thermal wave microscopes usually use a focused, amplitude-modulated laser beam as an A.C. heat source on the surface of the sample. In most situations the laser focal spot is small enough to be regarded as a point, so the source represented by the function, $f(\mathbf{r})$, in Eq. (2) can be considered to be a point. In the gas-cell detection scheme, the detector is a microphone which responds to the pressure variations generated in the air above the surface of the sample by surface temperature variations induced by the thermal waves in the sample. This microphone is enclosed in a small, sealed, air-filled cell which is normally much smaller than the acoustic wavelength and therefore isobaric. As such, it responds equally well to the temperature variations of all parts of the enclosed sample surface. Even though the cell is small compared to the acoustic wavelength, it is ordinarily large compared to the thermal wavelength, so that it may be regarded as effectively infinite for purposes of thermal wave calculations. Since the temperature distribution and pressure in the cell are determined by the surface temperature of the solid, and since the microphone responds equally to all parts of the cell, the signal received by the microphone is proportinal to the solid's average A.C. surface temperature. The detector weight function, $g(\mathbf{r})$, corresponding to this surface average can then be pictured as an infinitely thin, uniform sheet just beneath the surface of the solid. In accordance with the reciprocity theorem described in Sec. 2, the A.C. heat source in the equivalent inverse problem would also be a point detector located on the surface of the sample at the position of the source in the direct problem. Thus, the temperature at any given point on the sample surface in the inverse problem corresponds to the signal that would be generated if the heating beam were focused on that spot in the direct problem. Since the direct problem involves the generation and scattering of spherical waves while, as we will see below, the inverse problem involves only the scattering of plane waves, we choose to solve the inverse problem. Therefore the following discussion refers to the inverse problem.

The scattering of thermal waves in some simple geometries can be calculated by Green's function techniques. Perhaps the simplest geometry for which an exact Green's function can be found is a semi-infinite sample with no inhomogeneities to scatter the thermal waves, and with air, or some other gas, above the surface. This corresponds to a flat featureless sample in a gas cell. Since gases have thermal conductivities that are orders of magnitude smaller than those of typical solids, the surface of the solid can be considered to be insulated and the boundary condition (7) is appropriate. A Green's function for this geometry can be obtained by the method of images. One imagines a real heat source at some point $r' = (x',y',z')$ beneath the surface (here the surface is taken to be the x-y plane with the z-axis pointing into the sample), and an *in-phase* image source directly above it at $r'' = (x',y',-z')$. This is to be contrasted with the familiar problem in electrostatics in which a point charge is placed outside a grounded conductor. In that situation, the image is chosen to have the *opposite* sign, thus making the potential, rather than its derivative, zero. This choice of an in-phase image is required by the condition specified by Eq. (7), that the normal derivative be zero at the surface. The Green's function resulting from the two in-phase thermal sources is given by

$$G(\mathbf{r},\mathbf{r}') = \frac{1}{4\pi\kappa}\left[\frac{e^{iq|\mathbf{r}-\mathbf{r}'|}}{|\mathbf{r}-\mathbf{r}'|} + \frac{e^{iq|\mathbf{r}-\mathbf{r}''|}}{|\mathbf{r}-\mathbf{r}''|}\right] \qquad (15)$$

The temperature generated by the uniform surface source in the inverse problem is obtained by putting $g(\mathbf{r}') = C\delta(z')$, where the constant, C, specifies the source strength per unit area. Using Eq. (9) with $f(\mathbf{r}')$ replaced by $g(\mathbf{r}')$ we obtain

$$T(\mathbf{r}) = \frac{C}{4\pi\kappa}\int dx'\, dy'\, dz'\left[\frac{e^{iq|\mathbf{r}-\mathbf{r}''|}}{|\mathbf{r}-\mathbf{r}'|} + \frac{e^{iq|\mathbf{r}-\mathbf{r}''|}}{|\mathbf{r}-\mathbf{r}''|}\right]\delta(z') \qquad (16)$$

The integral over z' simply replaces the delta-function by unity and evaluates the Green's function at $z'=0$. At this point the contributions from the actual source and its image become equal. The remaining integral over the xy plane is most easily performed in polar coordinates and yields,

$$T(r) = \frac{iC}{\kappa q}\, e^{iqz} \qquad (17)$$

This result demonstrates the fact that the surface source in the inverse problem generates plane waves moving in the z-direction. However, other than that, it is relatively uninteresting because the surface temperature (at z=0) is a constant. If we remember that the surface temperature in the *inverse* problem represents the signal that the detector would measure in the *direct* problem, that constant temperature simply tells us that the signal received by the microphone from a featureless sample is independent of the position of the laser spot.

A somewhat more interesting result is obtained if, instead of a semi-infinite solid, one considers a sample with an outside corner as shown in Fig. 3. This could represent the actual corner at the edge of the sample or could represent one side of a deep slot cut into the surface. In this case we imagine a point source at $\mathbf{r}' = (x',y',z')$ with *three* images located at the points $\mathbf{r}_1 = (x',y',-z')$, $\mathbf{r}_2 = (-x',y',z')$, and $\mathbf{r}_3 = (-x',y',-z')$. The Green's function now has four terms and the source in the inverse problem consists of two thin sheets, one on the *xy* plane and one on the *yz* plane. (It is assumed here that both surfaces of the corner are inside the cell and communicate with the micro-phone). The inverse-problem temperature distribution now consists of two plane waves, one propagating away from the *xy* plane, and one propagating away from the *yz* plane,

$$T(r) = \frac{iC}{\kappa q} \left[e^{iqz} + e^{iqx} \right] \tag{18}$$

This temperature distribution does have some physical interest. If we examine the temperature in the vicinity of the corner on, say the *xy* plane, we see that it consists of a constant term coming from the wave propagating away from that plane in the z-direction, and a variable term coming from the wave propagating in the x-direction. This is indicated schematically in the graph of Fig. 4. Since the thermal waves are so heavily damped, the variation in the x-direction dies out within a thermal wavelength of the corner and the temperature is constant thereafter. This result can be interpreted in terms of the *direct* problem in the following way: In the direct problem the laser spot generates spherical waves which die out as they propagate away from it. When the spot is far from the

Fig. 3 Schematic diagram of an outside corner of a sample, with a point source located at $\mathbf{r}' = (x',y',z')$

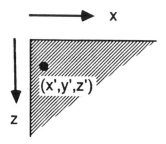

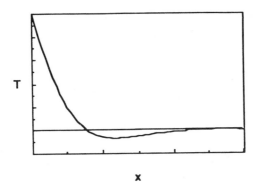

Fig. 4 Schematic temperature variation near the corner of Fig. 3.

corner, the propagation of these waves is undisturbed and the mirophone signal is independent of the position of the spot. As the spot approaches the corner the outgoing waves begin to encounter the vertical surface at the corner, and are scattered (reflected) by it. The surface temperature in the vicinity of the corner is changed by the scattered waves and the signal received at the microphone begins to change. Then, as the spot gets closer and closer to the corner, the signal will undergo oscillations with increasing amplitude as the incident and reflected waves interfere with each other, effectively creating a standing thermal wave. The same thing happens along the vertical surface at x=0. All of this scattering and interference is represented simply by the temperature profile in the *inverse* problem.

While these exact Green's function solutions are of some interest, such solutions can only be obtained for a very limited number of geometries. A more important use for Green's functions is that of a basis for approximation schemes. To illustrate this use, we imagine a sample with some subsurface flaw such as a crack, an inclusion, or a void. The sample shape will be imagined to be one for which we can calculate an exact Green's function, or at least one for which we can calculate a reasonably accurate approximate Green's function. For purposes of definiteness, we will imagine it here to be a semi-infinite solid. The defect will be imagined to be a void upon the surface of which the boundary condition (7) is appropriate. We then imagine that we have written down Eqs. (6) and (8) and solved Eq. (8) for the Green's function for the problem without the defect. At this point we apply Green's theorem to Eqs. (6) and (8) but integrate only over the region of the sample exterior to the void. Since both $T(\mathbf{r})$ and $G(\mathbf{r}',\mathbf{r})$ satisfy the correct boundary condition on the surface of the sample, the surface integrals generated by Green's theorem vanish there, but, since $G(\mathbf{r}',\mathbf{r})$ does not satisfy (7) on the surface of the void, the integral over the surface of the void does not vanish. Thus, Eq. (9) is replaced by,

$$T(\mathbf{r}) = \int G(\mathbf{r},\mathbf{r}')f(\mathbf{r}')d^3\mathbf{r}' - \kappa \oint_{void} [\hat{n}\cdot\nabla G(\mathbf{r},\mathbf{r}')]T(\mathbf{r}')dS' \qquad (19)$$

If we remember that we are supposed to be thinking of the *inverse* problem, the source function, $f(\mathbf{r}')$, should be replaced by a thin surface sheet as was done above. The temperature distribution then becomes

$$T(\mathbf{r}) = \frac{iC}{\kappa q}e^{iqz} - \kappa \oint_{void} [\hat{\mathbf{n}}\cdot\nabla G(\mathbf{r},\mathbf{r}')]T(\mathbf{r}')dS' \qquad (20)$$

The first term on the right represents the incident wave (a plane wave in this case), while the second term represents the wave scattered from the void. The difficulty with this "solution" to the problem is that it is not a solution at all, but rather is an integral equation for $T(\mathbf{r})$. If we are to make use of this equation we must find some way to approximate the surface temperature of the void in the second term on the right hand side. There are many ways of doing this and the accuracy of the resulting temperature distribution depends on the cleverness of the approximation. One approximation which works quite well for large voids whose surface curvature is gradual on the scale of a wavelength, is to say that the incident plane wave simply undergoes specular reflection at the surface of the void. The effect of such a specular reflection would be that the temperature on the void's surface would just be twice that given by the incident plane wave. The approximate temperature distribution would then be given by

$$T(\mathbf{r}) = \frac{iC}{\kappa q}e^{iqz} - \frac{2iC}{q} \oint_{void} [\hat{\mathbf{n}}\cdot\nabla G(\mathbf{r},\mathbf{r}')]e^{iqz'}dS' \qquad (21)$$

where the integral over the void is to be taken only over the unshadowed portion of the surface. As is always the case in this kind of inverse problem, the temperature distribution of Eq. (21) should be evaluated on the sample surface in order to obtain the calculated microphone signal.

Approximation schemes, like the one above, which rely on the cleverness of guesses, while sometimes very successful, are neither very reliable nor very satisfying to use. However there are systematic methods for solving integral equations like Eq. (20) above. One of these, the Born approximation, is described in the next section.

2.   THE BORN APPROXIMATION

The Born approximation is a successive-approximation scheme in which Eq. (20) is formally solved by repeatedly substituting the entire right hand side of the equation into the surface integral term on that side in place of $T(\mathbf{r}')$. The effect is to generate an infinite series of terms, with each succeeding term having one more power of the Green's function and one more integration over the surface of the scatterer. If one keeps only the zeroth and first order terms in the expansion the result is known as the "first Born approximation". The result is an approximate temperature distribution given by

$$T(\mathbf{r}) = \frac{iC}{\kappa q} e^{iqz} - \frac{iC}{q} \oint_{void} [\hat{\mathbf{n}} \cdot \nabla G(\mathbf{r},\mathbf{r}')] e^{iqz'} dS' \qquad (22)$$

Since voids tend to be rather strong scatterers of thermal waves, this first-order approximation is not a particularly good one for voids, although its similarity to Eq. (21) indicates that it may give qualitative correct answers. On the other hand, the first Born approximation is quite well suited to the calculation of thermal wave scattering from closed cracks or thin films of some foreign material in a solid, because these features scatter only weakly. Such thin scatterers can often be characterized by saying that the heat flux across them, that is the negative of the thermal conductivity times the normal component of the temperature gradient, is continuous, but that the temperature, while discontinuous across them, has a discontinuity that is simply proportional to that flux

$$T_{|}-T_{||} = \beta \kappa \hat{\mathbf{n}} \cdot \nabla T \qquad (23)$$

Here we have designated the two sides of the crack by | and | | and have called the propportionality constant β. The unit normal vector is taken as positive in the direction from | to | |. If this condition is used in place of (7) in Green's theorem above, Eq. (19) is replaced by

$$T(\mathbf{r}) = \int G(\mathbf{r},\mathbf{r}')f(\mathbf{r}')d^3\mathbf{r}' - \kappa \int_{crack} [\hat{\mathbf{n}} \cdot \nabla G(\mathbf{r},\mathbf{r}')][T_{|}(\mathbf{r}')-T_{||}(\mathbf{r}')]dS' \qquad (24)$$

or

$$T(\mathbf{r}) = \int G(\mathbf{r},\mathbf{r}')f(\mathbf{r}')d^3\mathbf{r}' + \beta\kappa^2 \int_{crack} [\hat{\mathbf{n}} \cdot \nabla G(\mathbf{r},\mathbf{r}')][\hat{\mathbf{n}} \cdot \nabla T(\mathbf{r}')]dS' \qquad (25)$$

The first Born approximation then yields the equivalent of Eq. (22)

$$T(\mathbf{r}) = \frac{iC}{\kappa q} e^{iqz} + \frac{i\beta\kappa C}{q} \int_{crack} [\hat{\mathbf{n}} \cdot \nabla G(\mathbf{r},\mathbf{r}')][\hat{\mathbf{r}} \cdot \nabla e^{iqz'}]dS' \qquad (26)$$

It was this formula which was used to calculate the curves plotted in Fig. 1 above. Since it only involves one integration over the surface of the scatterer, it is apparent that it represents only the single-scattering contribution to the temperature. The Born approximation is capable of representing multiple scattering but only if higher order terms are retained. The computation necessary to complete such calculations increases rapidly as the order increases because of the multiple integrations which are involved.

3. PARTIAL WAVE EXPANSION

A method called the "partial wave expansion", which is in common use in nuclear physics, can be a useful tool for studying thermal wave scattering from spherically or cylindrically symmetric scatterers. The basis for the method lies in the existence of simple expansions of plane waves (such as are generated by the surface sources in gas-cell inverse problems) in either spherical or cylindrical waves. The expansion in spherical waves (each of which individually satisfies the wave equation) is given by

$$e^{iqz} = \sum_{n=0}^{\infty} (i)^n (2n+1) j_n(qr) P_n(\cos\theta) \tag{27}$$

where $j_n(qr)$ is the so-called "spherical Bessel function" and where $P_n(\cos\theta)$ is a Legendre polynomial. The variables r and θ refer to the radial distance and polar angle (measured from the z-axis) in spherical coordinates. The corresponding expansion in cylindrical waves involves ordinary cylindrical Bessel functions and trigonometric functions, and is expressed in terms of cylindrical, rather than spherical coordinates. Since the cylindrical calculation parallels the spherical one exactly, we will not present it here. The technique will again be illustrated by considering the scattering from a void, but this time a spherically symmetric one. The center of the void will be taken to be at a depth z_0, beneath the surface and its radius will be taken to be a. The, as yet undisturbed, incoming plane wave can be expanded about the center of the void as,

$$e^{iqz} = e^{iqz_0} \sum_{n=0}^{\infty} (i)^n (2n+1) j_n(qr) P_n(\cos\theta) \tag{28}$$

To understand the next step we must recognize that each of the Bessel functions $j_n(qr)$ contains both incoming and outgoing spherical waves. These can be separated by writing the Bessel function as one-half the sum of the incoming (superscript "(2)") and outgoing (superscript "(1)") spherical Hankel functions,

$$e^{iqz} = e^{iqz_0} \sum_{n=0}^{\infty} (i)^n \frac{(2n+1)}{2} [h_n^{(1)}(qr) + h_n^{(2)}(qr)] P_n(\cos\theta) \tag{29}$$

This combination of incoming and outgoing waves is such that, in one direction from the scatterer (the direction of propagation of the original plane wave), the outgoing waves add constructively and the incoming waves add destructively, producing a net outgoing wave. In the opposite direction the reverse is true, so that the combination just reproduces the original plane wave. When this combination of waves interacts with a scatterer, such as the void we are considering, an additional scattered wave is generated and must be added to the plane wave. Since the scattered wave propagates away from the void, the expansion of this additional wave involves only *outgoing* spherical waves. The net effect of the scatterer, then, is to change the expansion coefficients of the outgoing (but not the incoming) spherical waves in Eq. (29). The combined temperature distribution of the incident and scattered waves can then be written as,

$$T(r) = e^{iqz_0} \sum_{n=0}^{\infty} (i)^n \frac{(2n+1)}{2} [b_n h_n^{(1)}(qr) + h_n^{(2)}(qr)] P_n(cos\theta) \tag{30}$$

where we have introduced constants b_n to account for the distortion of the outgoing spherical waves by the void. The temperature distribution can be rewritten as a sum of an unscattered plane wave and an outgoing scattered spherical wave as,

$$T(r) = e^{iqz} + e^{iqz_0} \sum_{n=0}^{\infty} (i)^n \frac{(2n+1)}{2} (b_n-1) h_n^{(1)}(qr) P_n(cos\theta) \tag{31}$$

The constants, b_n, may be evaluated by requiring that the boundary condition (7) be satisfied on the surface of the void at $r=a$. The result is,

$$b_n = -\frac{h_n^{(2)'}(qa)}{h_n^{(1)'}(qa)} \tag{32}$$

or equivalently,

$$b_n-1 = -2\frac{j_n'(qa)}{h_n^{(1)'}(qa)} \tag{33}$$

where the primes indicate derivatives with respect to the arguments of the functions. This can then be substituted into Eq. (31) to get the temperature distribution everywhere. To get the gas-cell signal strength, the resulting expression must be evaluated on the sample surface at $z=0$. If we call the distance of the heating beam from the point directly over the center of the sphere R, the variables r and θ can be evaluated on the surface as (See Fig. 5),

$$r = \sqrt{R^2 + z_0^2} \tag{34}$$

and

$$cos\,\theta = -cos(\pi-\theta) = \frac{-z_0}{\sqrt{R^2 + z_0^2}} \tag{35}$$

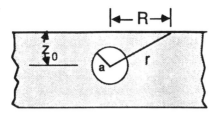

Fig. 5   Geometry for thermal wave scattering from a void.

Combining Eqs. (31), (33), (34), and (35) we obtain the final expression for the gas-cell signal

$$T(r)=1-e^{iqz_0}\sum_{n=0}^{\infty}(-i)^n(2n+1)\frac{j_n'(qa)}{h_n^{(1)'}(qa)}h_n^{(1)}(q\sqrt{R^2+z_0^2})P_n\left[\frac{z_0}{\sqrt{R^2+z_0^2}}\right] \quad (36)$$

If instead of a void, the scatterer were an inclusion, the boundary condition given by Eq. (7) would have to be replaced by two conditions, one specifying the continuity of the heat flux across the surface, and one specifying the continuity of the temperature. This calculation will not be carried out here, but, if the thermal conductivity and wave number inside the inclusion are given by $\bar{\kappa}$ and $\bar{q}$, respectively, then the only change in the calculated result is the replacement of Eq. (33) by

$$b_n-1=-2\frac{\bar{\kappa}\bar{q}j_n'(\bar{q}a)j_n(qa)-\kappa qj_n(\bar{q}a)j_n'(qa)}{\bar{\kappa}\bar{q}j_n'(\bar{q}a)h_n^{(1)}(qa)-\kappa qj_n(\bar{q}a)h_n^{(1)'}(qa)} \quad (37)$$

The limit of this expression as $\bar{\kappa}$ approaches zero is identical to Eq. (33). Plots of the calculated signal for both conducting and insulating spherical inclusions are given in Fig. 6, and the corresponding plots for cylindrical inclusions are given in Fig. 7. These plots are in agreement with experiments done on model samples in the authors' laboratory. A word of warning about the use of these formulas is perhaps in order. As one can see from the derivation, the final expression gives an exact formula for the single scattering of a thermal wave from a spherical void or inclusion. It does not, however, include the effect of the reflection of the scattered wave from the surface of the sample back onto the scatterer to be scattered again. If the surface of the spherical scatterer approaches the surface of the sample, i.e. as the depth of the center, z_0, approaches the radius, a, such multiple scattering effects are expected to become important and the single-scattering result will no longer be accurate.

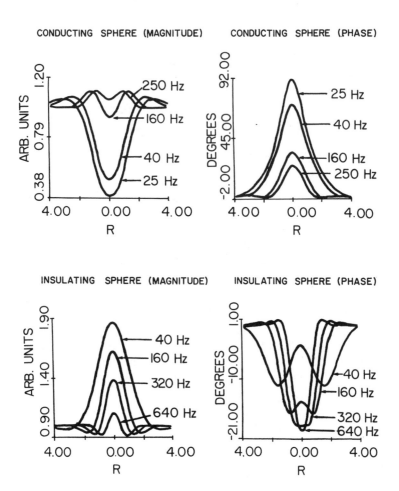

Fig. 6 Magnitude and phase of the thermal wave signal for conducting and insulating spherical inclusions.

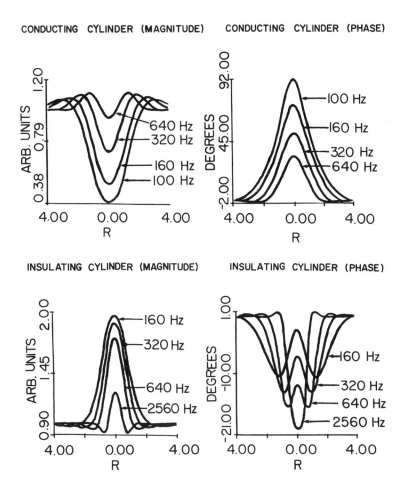

Fig. 7 Magnitude and phase of the thermal wave signal for conducting and insulating cylindrical inclusions.

Opsal (1986) has recently described the use of the partial wave expansion to obtain approximate scattering results for non-spherical scatterers. When one applies boundary conditions on any shape other than a sphere, the terms in the partial wave expansion with different values of the index, n, become coupled. This results in an infinite set of coupled equations for the coefficients b_n. Opsal's approximation consists of truncating the series at some point, beyond which the remaining terms are hopefully negligble, and solving the resulting matrix equations numerically.

V. Scattering Calculations for Mirage-Effect Imaging: Cylindrical Waves

1. UNIFORM SEMICONDUCTING MATERIALS: DETERMINATION OF THERMAL DIFFUSIVITY

The thermal diffusivity of a uniform material (e.g. a semiconductor wafer) can be determined by means of mirage effect thermal wave measurements. In effect, these measurements determine the value of the thermal wavelength (2π times the thermal diffusion length given by Eq. (4)) as a function of frequency. The slope of a plot of that wavelength versus the reciprocal of the square root of the frequency then yields a value for the thermal diffusivity $\alpha = \kappa/\rho c$. In order to explain exactly how this is accomplished, one needs to use a rather detailed theory of the mirage effect, so that theory will be developed first. In the mirage detection scheme, thermal waves in the solid are detected through the effect their temperature distribution has on the temperature gradient, and hence on the index of refraction gradient, in the air above the sample. The gradient of the index of refraction is detected by a second laser beam directed along the surface of the sample in the vicinity of the heated spot which is generating the waves. It is the deflection of this probe beam, similar to the mirage seen over a hot road in the summer, by the index gradient which produces the measured signal. The deflection angle is very small, typically less than a milliradian, so that the path of the probe beam may be considered to be a straight line. The measured signal then corresponds to an average (to be explained in more detail below) of the temperature distribution over a straight line across the surface. Using the same reasoning as in the proof of the reciprocity theorem in Section 2, we can imagine experiments utilizing the mirage detection scheme to have *line* sources. As a result we will be discussing the generation of two-dimensional, i.e., cylindrical waves.

Formulas for the deflection of the mirage probe beam have been developed by several groups (Boccara, Fournier and Badoz, 1980; Jackson, Amer, Boccara and Fournier, 1981; Aamodt and Murphy, 1983; See chapters 10 and 12). These formulas express the angle of deflection in terms of an integral of the fractional change in the index of refraction, n, over the path of the beam. If we denote the vector angular deflection by **M**, and the temperature in the air by $T(\mathbf{r})$, the deflection can be written as,

$$\mathbf{M} = -\int \frac{1}{n} \frac{dn}{dT} \nabla T(\mathbf{r}) \times d\mathbf{r} \tag{38}$$

Since the expected temperature excursions are quite small, it is reasonable to assume that the ratio $(dn/dT)/n$ will be essentially constant over the path. Also, since the angular deflection is small, the direction of the beam path, $d\mathbf{r}$, can be taken to be constant, say in the y-direction, so the deflection can be written to a high degree of accuracy as,

$$\mathbf{M} = -\frac{1}{n} \frac{dn}{dT} \nabla \overline{T}(x,z) \times \hat{\mathbf{y}} \tag{39}$$

where the bar indicates an average over the y-coordinate. It is this average that reduces the problem to a two dimensional one.

Since the temperature that appears in Eq. (39) is the temperature of the air above the sample surface, the thermal wave problem that must be solved is one with two media: one the air above the sample, the parameters of which will be indicated by a subscript, g (for "gas"), and the other the sample itself whose parameters will be indicated by a subscript, s. The gas will be taken to occupy the region $z<0$ and the sample the region $z>0$. The wave equations that must be solved can then be written as

$$\nabla \cdot [\kappa_g \nabla \overline{T}_g(x,z)] + \kappa_g q_g^2 \overline{T}_g(x,z) = 0; \qquad z<0, \text{ (gas)} \tag{40}$$

and

$$\nabla \cdot [\kappa_s \nabla \overline{T}_s(x,z)] + \kappa_s q_s^2 \overline{T}_s(x,z) = -\delta(x)\delta(z); \quad z\geq 0; \text{ (sample)} \tag{41}$$

where we have assumed (temporarily) that the heating beam is a point source of unit strength at $x=0$ and $z=0$. These equations are perhaps most easily solved by writing the temperature, $\overline{T}(x,z)$, as a Fourier transform in x,

$$\overline{T}(x,z) = \int_{-\infty}^{\infty} dk \, e^{ikx} t(k,z) \tag{42}$$

Eqs. (40) and (41) then become,

$$\frac{\partial}{\partial z}[\kappa_g \frac{\partial}{\partial z} t_g(k,z)] + \kappa_g k_g^2 t_g(k,z) = 0 \tag{43}$$

and

$$\frac{\partial}{\partial z}[\kappa_s \frac{\partial}{\partial z} t_s(k,z)] + \kappa_s k_s^2 t_s(k,z) = \frac{-1}{2\pi}\delta(z) \tag{44}$$

where

$$k_s = (q_s^2 - k^2)^{\frac{1}{2}} \tag{45}$$

and

$$k_g = (q_g^2 - k^2)^{\frac{1}{2}} \tag{46}$$

Since the only source term is located at z=0, the thermal waves must propagate away from that plane, i.e., in the negative z-direction for z<0, and in the positive z-direction for z>0. Thus they can be written as

$$t_g(k,z) = C_g e^{-ik_g z} \tag{47}$$

$$t_s(k,z) = C_s e^{ik_s z} \tag{48}$$

The boundary conditions that must be applied to these waves are that the temperature be continuous across the plane, z=0, and that the heat flux have a discontinuity which corresponds to the source strength in Eq. (44) on the same plane,

$$C_g = C_s \tag{49}$$

$$i\kappa_s k_s C_s + i\kappa_g k_g C_g = -\frac{1}{2\pi} \tag{50}$$

These equations can be solved for the constant C_g,

$$C_g = \frac{i}{2\pi(\kappa_s k_s + \kappa_g k_g)} \tag{51}$$

The mirage deflection signal can now be obtained by combining Eqs. (39), (42), (47), (48) and (51),

$$\mathbf{M} = \frac{i}{2\pi}\frac{1}{n}\frac{dn}{dT}\hat{\mathbf{y}}\times\nabla\int_{-\infty}^{\infty}dk\frac{e^{i(kx-k_g z)}}{(\kappa_s k_s + \kappa_g k_g)} \tag{52}$$

This expression correctly describes the signal that would be produced by an infinitesimally narrow probe beam passing in the vicinity of a point heating beam. However, we are interested in producing a theoretical expression that can be compared with data from real experiments in which both of these beams have finite radii. Accordingly, we will assume that the beams have Gaussian profiles with R_1 and R_2 denoting the radii of the heating and probe beams, respectively. The heating beam will be assumed to be centered at the origin, and the probe beam centered a distance, x, horizontally away from the origin and a height, h, above the surface. Since Eq. (52) contains only simple exponential functions of x and z, these Gaussian averages can be performed analytically, yielding final expressions for the normal (z-direction) and tangential (x-direction), deflections of the probe beam,

$$\mathbf{M}_{norm} = \frac{-1}{\pi}\frac{1}{n}\frac{dn}{dT}e^{-\frac{q_g^2 R_2^2}{4}}\int_0^{\infty}dk\frac{k_g\cos(kx)e^{ik_g h}e^{-k^2 R_1^2/4}}{(\kappa_s k_s + \kappa_g k_g)} \tag{53}$$

and

$$\mathbf{M}_{tan} = \frac{-i}{\pi}\frac{1}{n}\frac{dn}{dT}e^{-\frac{q_g^2 R_2^2}{4}}\int_0^{\infty}dk\frac{k\sin(kx)e^{ik_g h}e^{-k^2 R_1^2/4}}{(\kappa_s k_s + \kappa_g k_g)} \tag{54}$$

In these expressions we have used the symmetries of the integrands to express the deflections as integrals from zero to infinity with trigonometric functions in x, rather than from minus infinity to infinity with exponential functions.

It is clear from Eqs. (53) and (54) that the normal deflection is an even function of the horizontal offset, x, between the heating and probe beams, and that the tangential deflection is an odd function of the offset. The technique for measuring the thermal diffusivity involves measuring $\mathbf{M}_{tan}$ as a function of the offset distance, x, in the limit of very small probe height, h (Thomas, Inglehart, Lin, Favro and Kuo, 1985; Thomas, Favro, Kim, Kuo, Reyes and Zhang, 1986; Kuo, Lin, Reyes, Favro, Thomas, Kim, Zhang, Inglehart, Fournier, Boccara and Yacoubi, 1986; Kuo, Sendler, Favro and Thomas, 1986). The dependence of $\mathbf{M}_{tan}$ on the thermal diffusivity of the sample is through the quantity k_s which appears in the denominator of the integrand and which was defined in Eq. (45). Except for the overall phase factor involving the probe beam radius, R_2 (henceforth to be ignored), its dependence on k_g, at least for small values of h, is quite weak. This is true because, while k_s and k_g are ordinarily of about the same order of magnitude, k_s is typically orders of magnitude larger than k_g, thus making the term involving k_g in the denominator negligible. The resulting function is related to, but not quite the same as, the special variety of Bessel functions known as Kelvin functions. Plots of the real part of the function versus the offset, x, for a small value of R_1 are given in Fig. 8. The antisymmetry of the function is quite apparent as is the fact that it represents a very heavily damped wave. The two points at which this plot of the real part of $\mathbf{M}_{tan}$ first goes to zero on either side of the central zero are of particular interest. Since the real part of the function vanishes there, the function is purely imaginary, corresponding to values of the phase of $\pm\pi/2$. Inasmuch as we are dealing with wave propagation, one might naively think that these ninety degree points would correspond to a quarter wavelength distance on each side of the center. This is not so for two reasons. First, the effect of the finite heating beam size, R_1, is such as to effectively add a constant to this distance. Second, cylindrical waves are not exactly periodic, and the first zeros do not occur precisely $\pm\pi/2$ from the origin. A numerical analysis shows that the distance x_0, between the two ninety degree phase points on either side of the origin is given by

$$x_0 = d + \sqrt{1.4}\,\frac{\lambda}{2} = d + \sqrt{1.4\pi\alpha/f} \tag{55}$$

Fig. 8 Plots of the real part of $\mathbf{M}_{tan}$ (Eq. (54)) as a function of the offset distance, x, between the probe and heating beams for three different frequencies.

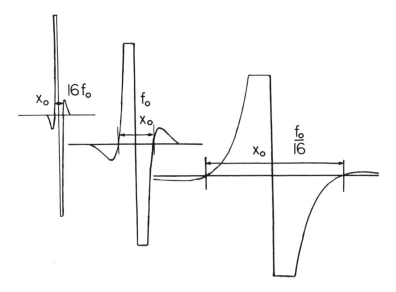

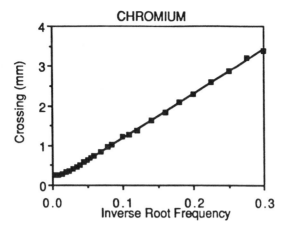

Fig. 9 Experimental (squares) and theoretical (line) plots of x_o (see text) versus the reciprocal of the square root of the frequency for a pure single crystal of chromium.

where d is a distance of the order of the heating beam diameter and where f is the frequency ($=\omega/2\pi$). Thus, a plot of x_0 versus the reciprocal of the square root of the frequency should have a slope given by $(1.4\pi\alpha)^{\frac{1}{2}}$ and an intercept which is dependent on the size of the heating beam. A theoretical and an experimental plot of this type for chromium is shown in Fig. 9. It is important, when making experimental measurements, that the *low-frequency* portion of this plot be used to calculate the slope. This is because we have assumed that the probe beam height, h, is small. "Small" in this case means small compared to the thermal wavelength in air. The thermal wavelength increases with decreasing frequency, decreasing the effective height of the beam and making our assumption more valid. The high-frequency portions of both the theoretical and experimental curves show deviations from linearity associated with the finite beam height and the properties of the gas.

2.   LAYERED SEMICONDUCTING MEDIA: THERMAL CHARACTERIZATION OF OXIDIZED OR COATED MATERIALS

The method used above to measure thermal diffusivity can also be applied to layered semiconductor structures. This will be illustrated here by considering a material with a coating or an oxide on its surface. The theoretical description of the mirage effect for a coated material is very similar to the description of the uncoated material above. We will again let the subscript "g" indicate the gas above the surface, but will use the subscript "s" to indicate the coating material, rather than the bulk material under it. The bulk of the sample will be indicated by the subscript "b". Then, if the coating or oxide has a thickness, a, we can describe the coated material by simply adding one more equation to Eqs. (40) and (41),

$$\nabla \cdot [\kappa_b \nabla \overline{T}_b(x,z)] + \kappa_b q_b^2 \overline{T}_b(x,z) = 0; \qquad z>a, \ (bulk) \qquad (56)$$

with a corresponding addition to Eqs. (43) and (44). In place of Eqs. (47) and (48) we must write

$$t_g(k,z) = C_g e^{-ik_g z}; \qquad z<0, \ (gas) \qquad (57)$$

$$t_s(k,z) = C_s \sinh(\theta_s + ik_s a); \qquad 0<z<a, \ (coating) \qquad (58)$$

$$t_b(k,z) = C_b e^{ik_b(z-a)}; \qquad z>a, \ (bulk) \qquad (59)$$

The form of Eq. (58) was chosen to include waves propagating in both the positive and the negative z-directions. This is because the wave propagating away from the source (Eq. (48)) will now scatter at the oxide-substrate interface at $z=a$. In fact Eq. (58) is sufficient to include all scattered waves as they bounce back and forth between the two surfaces of the oxide or coating, because the two complex constant θ_s and C_s give enough freedom for the function to satisfy the boundary conditions at both surfaces. When these conditions are applied to Eqs. (57-59) one can again solve for the constant, C_g, which is necessary to calculate the probe beam deflection,

$$C_g = \frac{i}{2\pi(\kappa_s k_s \coth \theta_s + \kappa_g k_g)} \qquad (60)$$

where θ_s is determined by,

$$\tanh(\theta_s + ik_s a) = \frac{\kappa_s k_s}{\kappa_b k_b} \qquad (61)$$

The only change in the final expressions, Eqs. (53) and (54), for the mirage deflections in the uncoated cases is the insertion of the factor $\coth \theta_s$ in the first term in the denominator of each. With the resulting modified expressions it is possible to extract information about the substrate as well as the oxide or coating from experimental plots of M_{tan}. In this case, however, one needs more information than just the slope of x_0 versus $f^{-\frac{1}{2}}$. The full procedure is to make a multi-parameter fit of the formulas to the entire curves of both the real and imaginary parts of M_{tan} versus the offset distance, x. Depending on the magnitudes of the parameters involved, one can find any or all of the oxide thicknesses, the diffusivities of the substrate, oxide, and gas, and such experimental parameters as the probe height, etc. If some of these parameters are known from independent measurements, the fitting process is of course much quicker. Also, in some cases it is possible to do the fitting, not on the entire signal, but on an x_0 versus $f^{-\frac{1}{2}}$ plot and obtain a particular piece of information. For instance, if the diffusivities of the gas, the substrate and the oxide, as well

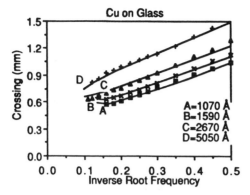

Fig. 10 Experimental (symbols) and theoretical (lines) plots of x_0 (see text) versus the reciprocal of the square root of the frequency for thin (1000 Å to 5000 Å) Cu films on glass substrates

as the beam height, are known, the oxide thickness can be obtained by fitting such curves. An example of this is shown in Fig. 10, where the thicknesses of independently measured 1000 Å to 5000 Å films of copper on glass were determined by this kind of curve fitting. In this regard it should be pointed out that these measurements were made in the low frequency regime, with thermal wavelengths which were orders of magnitude larger than the thicknesses of the films, thus demonstrating that it is not necessary to use ultra-high frequencies to measure thin films.

VI. Conclusions

In this chapter a rigorous theoretical basis has been developed for the quantitative investigation of propagation and scattering of thermal waves in condensed matter, in general, and semiconductor structures, in particular. The formalism has treated imaging resolution factors, approximate scattering calculations from subsurface defects and their applications to uniform and layered structures via photoacoustic gas-cell and photothermal beam deflection (Mirage effect) probes. Good agreement with experiments using mirage effect detection in air has been obtained.

VII. References

Aamodt, L.C. and Murphy, J.C., (1983) J. Appl. Phys. **54**, 581.

Boccara, A.C., Fournier, D., and Badoz, (1980) J., Appl. Phys. Lett. **36**, 130.

Carslaw, H.S. and Jaeger, J.C. (1959) *Conduction of Heat in Solids* (second edition), Clarendon, Oxford.

Inglehart L.J., Grice, K.R., Favro, L.D., Kuo, P.K., and Thomas, R.L., (1983), Appl. Phys. Lett. **43**, 446.

Inglehart, L.J., Lin, M.J., Favro, L.D., Kuo, P.K., and Thomas, R.L., (1983), *Proc. 1983 Ultrasonics Symposium*, (B.R. McAvoy, ed.) IEEE, New York, p. 668.

Jackson, W.B., Amer, N.M., Boccara, A.C., and Fournier, D., (1981) Appl. Opt. **20**, 1333.

Jenkins, F.A. and White, H.E. (1976). *Fundamentals of Optics*, McGraw-Hill (4th ed.), p. 332.

Kuo, P.K., Inglehart, L.J., Favro, L.D., and Thomas, R.L., (1981) *Proc. 1981 Ultrasonics Symposium*, (B.R. McAvoy), IEEE, New York, p. 788.

Kuo, P.K. and Favro, L.D. (1982), Appl. Phys. Lett. **40**, 12.

Kuo, P.K., Lin, M.J., Reyes, C.B., Favro, L.D., Thomas, R.L., Kim, D.S., Zhang, Shu-yi, Inglehart, L.J., Fournier, D., Boccara, A.C., and Yacoubi, N. (1986), Can. J. Phys. **64**, 1165.

Kuo, P.K., Sendler, E.D., Favro, L.D., and Thomas, R.L., (1986), Can. J. Phys. **64**, 1168.

McDonald, F.A. and Wetsel, G.C., (1984) *Proc. 1984 Ultrasonics Symposium*, (B.R. McAvoy, ed.), IEEE, New York, p. 622.

Opsal, J.L., (1986), *Review of Progress in Nondestructive Evaluation*, Vol. **5A**, (D.O. Thompson and D.E. Chimenti, eds.), Plenum, New York, p. 295.

Thomas, R.L., Inglehart, L.J., Lin, M.J., Favro, L.D., and Kuo, P.K., (1985), *Review of Progress in Nondestructive Evaluation*, Vol. **4B** (D.O. Thompson and D.E. Chimenti, eds.) Plenum, New York, p. 859.

Thomas, R.L., Favro, L.D., Kim, D.S., Kuo, P.K., Reyes, C.B., and Zhang, Shu-yi, (1986), *Review of Progress in Nondestructive Evaluation*, Vol. **5B**, (D.O. Thompson and D.E. Chimenti, eds.), Plenum, New York, p. 1379.

THERMAL WAVE CHARACTERIZATION AND INSPECTION OF SEMICONDUCTOR MATERIALS AND DEVICES

Allan Rosencwaig

Therma-Wave, Inc.
Fremont, CA 94539
USA

I. Introduction

During the past few years, thermal wave physics has been successfully applied to studies of semiconductor materials and microelectronic devices. Previously, we described how these materials and devices might be examined with thermal wave imaging (Rosencwaig, 1985). This imaging is performed in a scanning electron microscope with a thermoelastic technique for thermal wave detection. The thermoelastic technique is based on the measurement of ultrasonic signals from the thermoelastic waves generated by the periodic heating from the modulated electron beam (Rosencwaig and Opsal, 1986). Thus, this detection requires a direct coupling between the sample and piezoelectric detector.

Although this technique is a very sensitive detection method for thermal waves, it has limited applicability for integrated circuit process control and inspection because it requires contact to a transducer and thus is potentially contaminating. Furthermore, the results are often strongly dependent on difficult to control acoustic variables such as sample geometry and sample/detector coupling.

To overcome these problems we have developed several noncontact thermal wave techniques that can be used at the high modulation frequencies required for micron-scale resolution. One of these methods generates the thermal waves in a sample with a modulated focused laser pump beam and detects the thermal waves by measuring with a probe laser the small local periodic deformation of the sample surface caused by the periodic local heating induced by the modulated pump laser (Rosencwaig, Opsal and Willenborg, 1983; Opsal, Rosencwaig and Willenborg, 1983). Although this is a sensitive, noncontact and reproducible means for detecting thermal waves in most materials, an even more powerful method has been developed for use with semiconductors (Rosencwaig, Opsal, Smith and Willenborg, 1985; Opsal and Rosencwaig, 1985; Smith, Rosencwaig and Willenborg, 1985). This method is described below.

When a material is excited with an intensity-modulated energy source, or pump, its optical properties can be altered by the absorption of the incident energy. This results in the sample's complex refractive index undergoing periodic variations at the modulation frequency of the pump intensity excitation. The induced changes in the sample's optical refractive index can be detected by measuring the modulated reflectance of an optical probe from the sample surface. By the use of phase-sensitive detection methods, probe beam modulated reflectances, $\Delta R/R$, as low as 10^{-7} can be measured (Rosencwaig *et al.* 1985).

This phenomenon, called modulated photoreflectance, was first observed (Wang, Albers and Bleil, 1967) nearly twenty years ago and developed into a useful spectroscopic technique for investigating the optical properties of various materials (Wang *et al.*, 1967; Gay and Klauder, 1968; Albers, 1969; Nahory and Shay, 1968; Nilsson, 1969; Cerdeira and Cardona, 1969; Aspnes, 1970). These spectroscopic studies employ low intensity pump ($< 1 \text{W/cm}^2$) and probe ($< 1 \text{mW/cm}^2$) optical beams and operate at low pump intensity modulation frequencies (typically < 1kHz). The wavelength of the probe beam is scanned and the modulated reflectance of the sample is recorded as a function of the probe wavelength. In most materials the modulated photoreflectance signal arises from a purely thermoreflectance effect (Nilsson, 1969; Berglund, 1966; Batz, 1966; Batz, 1972). That is, since the optical properties of most materials are dependent to some extent on the sample temperature, and since the temperature of the sample surface undergoes a periodic variation from the heat produced by the absorbed intensity-modulated pump beam, the reflectance of the probe beam experiences a corresponding modulation. In semiconductors, however, there are often more significant contributions to the photoreflectance signal from the free carriers that are generated by the pump beam. These contributions arise from thermal effects associated with free carrier recombination (Nilsson, 1969), from a free carrier Drude effect on the optical refractive index (Bonch-Bruevich, Kovalev, Romanov, Imas and Libenson, 1968), and also from an electroreflectance effect (Seraphin, 1972) whereby the photogenerated carriers modulate the amount of surface band bending, or surface electric field, in the semiconductor. Under the experimental conditions used for conventional photoreflectance, it has been found that the electroreflectance term is the dominant effect on the recorded signal (Nahory and Shay, 1968; Nilsson, 1969; Cerdeira and Cardona, 1969; Aspnes, 1970).

We have developed a somewhat different modulated reflectance technique (Rosencwaig *et al.*, 1985). In this method we employ a highly focused (~ 1 μm spot size) laser beam as the pump with a typical intensity in the 10^4–10^5 W/cm^2 range, and we intensity-modulate the pump laser in the MHz range. In addition, the probe beam is a laser at a fixed wavelength that is usually far removed from any critical points in the sample's band structure. Our interest is not to perform optical spectroscopy as in the conventional photoreflectance method, and thus the probe wavelength is not scanned. Instead we are interested in characterizing the spatial and frequency dependence of the photogenerated thermal waves in non-semiconductors and of the photogenerated thermal and electron-hole plasma waves (Opsal and Rosencwaig, 1985) in semiconductors. By so doing we are able to detect and quantify those intrinsic and defect characteristics in the near-surface region of the sample that affect the local thermal and electrical transport properties. In particular, we and others have used this laser-induced modulated reflectance method for monitoring the dose and uniformity of ion implantation in Si and GaAs wafers (Smith, Rosencwaig and Willenborg, 1985; Smith, Taylor and Schuur, 1985; Smith, Powell and Woodhouse, 1985; Guidotti and van Driel, 1985), for detecting and quantifying the residual damage in silicon wafers that results from chemopolishing and scrubbing processes (Smith, Hahn and Arst, 1986), and for detecting and characterizing radiation damage and contamination in silicon from reactive ion and plasma etch processes (Geraghty and Smith, 1986).

As we will discuss in what follows, our different experimental conditions result in signals from semiconductors that do not have the same origin as the signals observed in conventional photoreflectance experiments. In addition, we have observed an interesting temporal behavior of our modulated reflectance signal in silicon; a phenomenon that has apparently not been detected in the conventional photoreflectance experiments.

Section II of this chapter describes the experimental conditions for our laser-induced modulated reflectance measurements. In Section III we present a phenomenological model for the signal obtained in silicon. In Sections IV, V and VI we describe how the modulated reflectance signal is used to obtain quantitative information about chemopolishing, ion implantation and dry etch processes on silicon wafers. Section VII examines the temporal behavior of this signal and shows how this behavior can be used to obtain information about defect-related electronic surface states in silicon. Section VIII contains conclusions and a summary.

II. Experimental Methodology: The Therma-Probe System

The experiments described in this paper were performed with a commercial system, the Therma-Probe^{TM} system,$^{(*)}$ that utilizes the laser-induced modulated reflectance technique. Fig. 1 depicts the basic optical arrangement in this system. The 488-nm beam of a 35 mW Ar$^+$-ion laser is intensity-modulated in the 1-10 MHz frequency range with an acousto-optic modulator. The pump beam is then directed through a beam expander and focused to an ~ 1 μm diameter spot on the sample with a sample incident power of ~ 10 mW. The 633-nm beam of a 5-mW He-Ne laser, the probe beam, is directed through a beam expander, a polarizing beam splitter and quarterwave plate, reflected off a dichroic mirror and focused collinearly with the Ar$^+$-ion pump beam onto a 1 μm diameter spot on the sample with an incident power of ~ 3 mW. The two

Fig. 1 Schematic depiction of apparatus used for performing laser-induced modulated reflectance measurements.

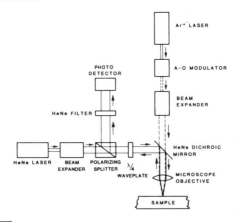

$^{(*)}$ Therma-Probe systems are products of Therma-Wave, Inc., Fremont, CA. Therma-Probe is a TRADEMARK of Therma-Wave, Inc.

laser spots are thus coincident on the sample surface. The 633-nm probe beam undergoes a small modulation in its reflected power as a result of the pump beam-induced variation in the local optical refractive index of the sample at the probe beam wavelength. The retroreflected probe beam passes through the quarterwave plate again, and since its polarization is now rotated 90° with respect to the incoming beam from the He-Ne laser, the retroreflected beam is directed by the polarizing beam splitter to the photodetector, which measures the high frequency modulation in the reflected probe beam power. The photodetector, which is shielded from any modulated pump laser light by suitable blocking filters, is purposely underfilled by the probe beam that strikes its surface. This is done so that the photodetector measures only modulations in the probe reflected power and is insensitive to the small variations in probe beam size or reflected angle that can also occur from thermal lens and thermoelastic deformation effects at the sample surface (Opsal *et al.*, 1983). The probe beam reflectance modulations are measured with a phase-sensitive synchronous detection system capable of measuring reflectance changes at 1-10 MHz frequencies as small as $10^{-7}/\sqrt{Hz}$. As we will see below, most of the samples produce modulated reflectance signals in the 10^{-4}–10^{-2} range, and thus are readily measured with this apparatus.

III. Theory of Laser-Induced Modulated Reflectance in Si

Optical modulation effects in semiconductors are complicated by the fact that several mechanisms can contribute to the change in optical properties of the material under investigation. As pointed out in Chapters 12 and 13 and by numerous investigators (Nahory and Shay, 1968; Nilsson, 1969; Cerdeira and Cardona, 1969; Aspnes, 1970; Bonch-Bruevich *et al.*, 1968; Lompre, Liu, Kurz and Bloembergen, 1984), the photogenerated electron-hole plasma can modify the refractive index in silicon and other semiconductors directly through both a Drude effect and through a modulation of the surface band bending. Also, as we have noted earlier (Opsal and Rosencwaig, 1985), the laser-induced modulated reflectance can have a significant thermal component due to the heating of the lattice by the photogenerated plasma. Furthermore, if the perturbation on the refractive index occurs over very small distances (< 100 Å) then, as pointed out by Aspnes and Frova (1969), the nonuniformity of the index perturbation can itself significantly affect the magnitude and spectral lineshape of the modulated reflectance.

While it can be difficult to identify which of the above photoreflectance mechanisms are most significant in a given set of measurements, we do have some general observations that limit the possibilities, at least for the case of silicon and the experiments we are reporting here. First of all, our modulated reflectance signals are *linear* in the pump beam intensity. Thus, it is unlikely that our signals are due to modulating the surface band bending which in general is a nonlinear effect (Aspnes, 1980). Second, our measurements on ion-implanted silicon wafers show much more sensitivity to the amount of implant than we would expect from a purely thermal effect (Smith, Rosencwaig and

Willenborg, 1985; Smith, Taylor and Schuur, 1985). Therefore, it is most likely that the bulk of our observed modulated reflectance signal is due to the photo-generated plasma and its effect on the refractive index through the Drude effect. Since the density of the plasma can be strongly influenced by any local electrical fields (e.g., due to charge trapping) we will also include the possible effects of having a nonuniform perturbation on the refractive index.

Let us first start with a thermally induced modulation of the refractive index. To see how this refractive index modulation affects an optical probe beam, we consider the reflectance from a half-space. Letting R denote the unperturbed reflectance, then in terms of the complex index of refraction, $n+ik$, we have

$$R = \frac{(n-1)^2+k^2}{(n+1)^2+k^2} \tag{1}$$

and applying a small perturbation $\Delta n+i\Delta k=T_o \ (dn/dT+idk/dT)$ we find for the thermal wave-induced modulated reflectance,

$$\Delta R/R = \mathrm{Re}\left[\frac{4T_o}{(n+ik-1)(n+ik+1)}(dn/dT+idk/dT) \right] \tag{2}$$

In deriving this expression we have tacitly assumed that the refractive index modulation is slowly varying with respect to the optical wavelength in the material. This is generally valid, since thermal wavelengths are typically much longer than optical wavelengths and, therefore, the modulated reflectance essentially measures the surface temperature, T_o. One particularly significant feature of this detection scheme is the high degree of localization it affords. By focusing the pump and probe beams onto the same spot, measurements with better than 1 μm spatial resolution are possible which is especially important in semiconductor applications.

As we have already noted there are in semiconductors electronic effects that also affect a modulated reflectance measurement. An optical pump beam incident on a semiconductor will also generate an electron-hole plasma similar in many respects to the thermal wave (Opsal and Rosencwaig, 1985). Thus, we need to include in the modulated reflectance any significant plasma effects. One of the simplest and one that depends linearly on the plasma density, N, is the optical Drude effect, given by (Lompre *et al.*, 1984)

$$dn/dN = -\lambda^2 e^2/(2\pi nmc^2) \tag{3}$$

$$dk/dN = -(k/n)(dn/dN) \tag{4}$$

which is valid for silicon when relaxation effects in the plasma are negligible.

Assuming a probe wavelength, $\lambda = 633$ nm, we also have $k \ll n$ and Eq. (2) (with T replaced by N) to a good approximation then reduces to

$$\Delta R/R = \frac{-2\lambda^2 e^2}{\pi n (n^2-1)mc^2} N_o \tag{5}$$

To evaluate Eq. (5) we use for the electron's charge $e=4.8\times10^{-10}$ esu, the velocity of light $c=3.0\times10^{10}$ cm/sec, the effective mass $m=0.15m_o$ where the free electron mass is $m_o=9.1\times10^{-28}$ gm, and the index of refraction $n = 3.9$. With these values we have $\Delta R/R \sim -10^{-22}N_o$ so that a plasma density $N_o=10^{18}$/cm^3 implies a modulated reflectance $\Delta R/R=-10^{-4}$ which is of the same order as, but of opposite sign to, the expected thermal wave-induced modulated reflectance (Rosencwaig *et al.*, 1985).

Another optical effect to consider is that due to having a spatially nonuniform perturbation on the refractive index. Following the analysis of Aspnes and Frova (1969) we have for the modulated reflectance:

$$\Delta R/R = \text{Re}\left[\frac{4<N>}{(n+ik-1)(n+ik+1)}(dn/dN+idk/dN) \right] \tag{6}$$

with $<N>$ the weighted average of $N(x)$

$$<N> = -2iK\int dx\, N(x)\exp(2iKx) \tag{7}$$

and where K is the electromagnetic wave-vector in the material,

$$K = (2\pi/\lambda)(n+ik) \tag{8}$$

We first note that Eq. (6) reduces to the same form as Eq. (2) in the limit that the variation in $N(x)$ is slow compared to the spatial variation of the probing optical beam. Next in the other extreme where $N(x)$ goes to zero beyond $x=\delta$ and $|2K\delta| \ll 1$, we have

$$<N> = -2iK\delta N_1 \tag{9}$$

where N_1 is the unweighted average of $N(x)$. For silicon with $k \ll n$ we then obtain for the modulated reflectance

$$\Delta R/R = \frac{16\lambda nke^2}{(n^2-1)^2 mc^2} N_1\delta \tag{10}$$

which we note is positive and opposite in sign to the normal Drude effect of Eq. (5). Using for the extinction coefficient $k = .025$, we have that this latter surface-type effect is equal to the bulk-type effect of Eq. (5) when $N_1\delta=N_o\times10^{-4}$. That is for $\delta=100$ Å this would require $N_1 = 10^{20}/\text{cm}^3$. Such a magnitude is possible if, for example, there are trapping sites at the surface ($\sim 10^{14}/\text{cm}^2$) that can effectively pin some fraction of the plasma at the surface. However, we should point out that the effect becomes much more significant if k differs from the value we have assumed here. In fact, increasing k in a thin layer near the surface by an order of magnitude would dramatically affect the modulated reflectance while leaving the D.C. reflectance essentially unchanged. Thus, near surface lattice damage which may be insignificant in a normal optical reflectivity measurement could be readily observed in modulated reflectance through its effects on k (in addition to any plasma wave propagation effects).

We have hypothesized a simple multiple trapping model as a mechanism for producing the nonuniform spatial effects on the optical properties discussed above (Opsal, Taylor, Smith and Rosencwaig, 1987). Under conditions of intense illumination with above band gap light we create electron-hole pairs which, in addition to diffusing and eventually recombining, can be trapped into available surface states. Trapped electrons will pin holes at the surface and, conversely, trapped holes will pin electrons. Assuming that the intrinsic dangling bond surface states trap electrons which could then be reemitted into bulk states with some characteristic time τ_1, we would expect for the modulated component of the pinned hole density a frequency dependence of the form $(1-i\omega\tau_1)^{-1}$. If, in addition to these intrinsic states, there are defect or damage related states which can trap either electrons or holes with a characteristic emission time τ_2, then we similarly expect their dependence to be of the form $(1-i\omega\tau_2)^{-1}$. Since the total number of carriers pinned at the surface depends on the net charge that has been trapped, we would then have for the modulated component of the total number of pinned carriers

$$N_p = \gamma N_o[(1-i\omega\tau_1)^{-1} + \beta(1-i\omega\tau_2)^{-1}] \qquad (11)$$

where β is the ratio of the amount of charge trapped into defect surface states to the amount trapped into intrinsic surface states. Also in Eq. (11), γ is a constant of proportionality between the total number of pinned carriers and the number of photogenerated carriers left in the bulk, N_o. If the intrinsic and defect surface states trap charge of opposite sign, then β is taken to be negative. Furthermore, since the Drude effect does not depend on the sign of the charge, we also take the absolute value of the real part of Eq. (11) while retaining the imaginary part to preserve phase.

For the modulated reflectance we then have (including these trapping effects),

$$\Delta R/R = [\Delta R/R]_o \, | \, 1-\alpha[(1-i\omega\tau_1)^{-1}+\beta(1-i\omega\tau_2)^{-1}] \, | \qquad (12)$$

where $[\Delta R/R]_o$ is the bulk-type Drude effect given by Eq. (5)

$$[\Delta R/R]_o = \frac{-2\lambda^2 e^2}{\pi n(n^2-1)mc^2}N_o \qquad (13)$$

and α is a constant, $\alpha = \gamma[8\pi n^2 k\delta/\lambda(n^2-1)]$. Also in Eq. (12), as we did in Eq. (11), we take the absolute value of the real part of the quantity in square brackets while keeping the imaginary part as it is. This preserves the charge independence of the Drude effect while maintaining the relative phase difference between the bulk and surface contributions to the modulated reflectance.

There are some interesting predictions of this phenomenological model that we should emphasize. The most apparent is that $\Delta R/R$ can *increase* with increasing modulation frequency from a minimum of $[\Delta R/R]_o | 1-\alpha[1+\beta] |$ to a maximum of $[\Delta R/R]_o$ whenever $\alpha| 1+\beta | \leq 1$. This behavior is consistent with all of our measurements (Opsal *et al.*, 1987) on silicon samples for which the diffusion coefficient is large enough to ensure that the bulk density N_o is not decreasing significantly with increasing frequency. More interesting, however, are the possibilities when β is a time dependent quantity. That is, if the defect surface states are somehow modified in time by the photogenerated plasma either through recombination or trapping processes, then one can expect to observe a time dependence in $\Delta R/R$ that increases, decreases, or does both with time depending on the initial value of β, the emission times, and the modulation frequency. If in a p-type sample the signal decreases with time, one could perhaps expect an increasing signal in an n-type sample since the defect states, if due to the doping, would trap charge of the opposite sign. In general we expect the time dependent effects to disappear at sufficiently high modulation frequencies. These, as well as other effects have been observed and reported (Opsal *et al.*, 1987), and will be discussed in more detail in Sec. VII.

IV. Wafer Chemopolishing Characterization

The final manufacturing process for silicon wafers includes a chemo-mechanical polishing step. This process gives the silicon wafer its mirror finish, but potentially leaves a thin layer of structurally damaged silicon at the wafer surface. There is at present some concern that circuit yield and performance in shallow junction, highly integrated devices, often with gate oxide thickness in the 100 to 250 Ångstrom range, are harmed by such residual damage on incoming wafers. Rather than being completely eliminated by the anneal cycles of the subsequent processing, regions of structural damage on the surface (i.e., polishing damage, scratches, saw marks) have been reported to function as local nucleating sites for stacking faults during oxidation (Fisher and Amick, 1966). Similarly, localized surface flaws in the form of minute pits in the silicon surface are found to provide preferential sites for defect generation during dopant diffusion (Ravi, 1972). Also, recent studies by Kugimiya and Matsumoto (1981) have shown that some types of surface defects such as polishing marks, dimples, etc., survive various device fabrication processes and result in reduced yield.

Previous effects to characterize polished wafer surfaces have employed the techniques of x-ray topography (Yasuami, 1980), infrared photoelasticity (Kotake and Takasu, 1981; Takasu and Matsushita, 1975), preferential etching , x-ray rocking curve (Takasu and Matsushita, 1975; Kundsen, 1963), and x-ray diffuse scattering (Kashiwagura, Harada and Ogino, 1983; Yasuami, Harada and Wakamatsu, 1979; Yasuami and Harada, 1981). However, these techniques have been of limited use due to complexity or lack of sensitivity or reproducibility. Thus, there has been a need for a noncontact, nondestructive surface characterization tool with good sensitivity and flexible rapid mapping capability such as provided by the modulated reflectance technique.

Fig. 2 (a) Modulated reflectance map of a Si wafer exhibiting a highly damaged surface region from the polishing process. This is indicated by the large average $\Delta R/R$ signal (1.53×10^{-3}). Also note the substantial nonuniformity across this wafer, indicated by the large standard deviation ($\sigma = 5.97\%$); (b) Modulated reflectance map of Si wafer exhibiting minimal damage from the polishing process (avg. $\Delta R/R=3.4\times10^{-4}$; $\sigma=0.92\%$).

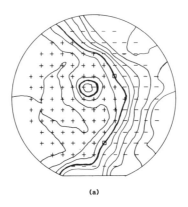

(a)

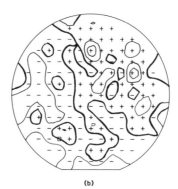

(b)

Full wafer contour mapping is available with the Therma-Probe system described in Section II. In the mapping procedure for each wafer, the modulated reflectance signal is automatically measured at 137 sites uniformly distributed over the wafer and the data displayed as a contour map, two examples of which are shown in Fig. 2. In these maps the +, − and square symbols represent sites having a modulated reflectance signal greater than, less than, or within 1/4% of the average signal on the wafer, respectively. The bold line represents the average modulated reflectance, $\Delta R/R$, value on the wafer, and the contour interval is selected appropriately. Each map datum is the average modulated reflectance signal from a 2.5-mm scan across the wafer surface at that particular location.

Fig. 3 (a) $\Delta R/R$ map of an incoming silicon wafer (avg. $\Delta R/R = 1.01 \times 10^{-3}$; $\sigma = 4.86\%$). (b) Remap of the wafer as in (a), showing good reproducibility of $\Delta R/R$ signal and map pattern (avg. $\Delta R/R = 1.00 \times 10^{-3}$; $\sigma = 4.80\%$). (c) Remap of the wafer as in (a), but with the wafer rotated by 90 degrees in the clockwise direction (avg. $\Delta R/R = 1.01 \times 10^{-3}$; $\sigma = 5.02\%$).

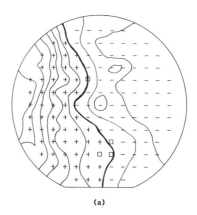

(a)

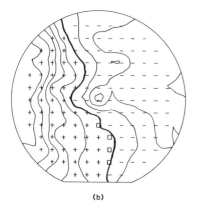

(b)

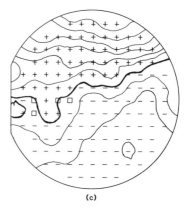

(c)

As an example of the reproducibility of the modulated reflectance data, we show in Fig. 3 the results from three successive maps made from the same wafer. The wafer was unloaded and reloaded into the Therma-Probe system for each map. As shown in Fig. 3a, this incoming wafer has a relatively high average $\Delta R/R$ signal (1.01×10^{-3}) as well as substantial nonuniformity (4.86% standard deviation) in the form of a linear gradient with higher damage on the left side of the wafer. Note that the map pattern is reproduced very well in the remap in Fig. 3b. Furthermore, to explicitly demonstrate that the observed pattern originates entirely within the wafer and is not an artifact of the Therma-Probe system, we show in Fig. 3c a map taken after rotating this same wafer 90 degrees in the clockwise direction. Again the same map pattern (rotated) is obtained. The $\Delta R/R$ signal values from these three maps are within 0.4% of each other. The time required to measure a standard Therma-Probe map such as shown above is approximately 5 min. and no special preparation or processing of the wafer is required.

The range of modulated reflectance signal typically observed from chemopolished silicon wafers is illustrated by the maps in Fig. 2. The map in Fig. 2a is an example of a wafer having a relatively large $\Delta R/R$ signal (1.5×10^{-3}) indicating extensive polishing damage, and also showing a large variation in the damage across the wafer, as demonstrated by the relatively large standard deviation of 5.97%. On the other hand, the low $\Delta R/R$ signal (3.4×10^{-4}) and standard deviation (0.93%) on the wafer mapped in Fig. 2b are typical for a wafer having minimal detectable damage over the entire wafer.

An important result from these experiments is that a modulated reflectance signal of $\sim 3\times10^{-4}$ is consistently measured on wafers having "damage free" surface quality. It appears that this value represents a lower limit for high-quality silicon surfaces. These low $\Delta R/R$ values are consistently measured on wafers processed through a low-damage polishing process, as well as on wafers that have undergone thermal oxidation followed by wet chemical removal of the oxide layer. This finding is consistent with results from studies of pre-anneal and post-anneal ion implanted silicon wafers (Smith *et al.*, 1985; Smith, Rosencwaig, Willenborg, Opsal and Taylor, 1986). This result is also in agreement with theoretical modeling that undamaged silicon will exhibit the lowest modulated reflectance signal (Opsal and Rosencwaig, 1985; Opsal *et al.*, 1987). Thus, the polished incoming wafer in Fig. 2b appears to have a negligible level of polishing damage, while the wafer of Fig. 2a appears to have a high level of polishing damage. As will be discussed in a later section, the $\Delta R/R$ signal from wafers exhibiting high polishing damage is comparable to that obtained from low-dose ion implants.

To characterize the polishing damage on standard incoming silicon wafers, product-grade polished wafers with <100> and <111> orientation were obtained from four different silicon vendors and mapped on a Therma-Probe system. Table I shows the modulated reflectance signal obtained on these wafers. The data in column 2 show that a large variation in signal was found from vendor to vendor. Vendor B wafers consistently show quite low $\Delta R/R$ signals indicating negligible polishing damage. The wafers from vendor C tend to

Table I

Average modulated reflectance, $\Delta R/R$, signal ($\times 10^{-5}$) versus wafer type and polishing method. Standard deviation values in percent indicating variation of $\Delta R/R$ signal across the wafer are given in parenthesis.

Wafer Type	Incoming as is	Repolished only by A	Repolished only by B	Repolished by A then B
A	94-160 (2%-5%)	203 (2.3%)	32.7 (0.82%)	32.6 (0.84%)
B	29-33 (0.3%-0.6%)	147 (2.1%)	32.7 (0.85%)	32.2 (0.89%)
C	40-43 (2%-3%)	226 (2%)	33.3 (0.77%)	32.2 (0.67%)
D	85-107 (2%-12%)	148 (2.6%)	32.8 (0.67%)	31.8 (0.6%)

exhibit somewhat higher levels of damage. On the other hand wafers from vendors A and D consistently show quite high $\Delta R/R$ values, indicating substantial residual polishing damage. In addition, a considerable spread in $\Delta R/R$ signals is found on wafers from vendors A and D, indicating variability in the polishing process. However, $\Delta R/R$ signals from vendors B and C are tightly clustered, indicating better control of the polishing process.

When a set of A, B, C and D wafer samples have been simultaneously repolished by vendor A (column 3, Table I), the $\Delta R/R$ data show the typical large values otherwise only seen on the incoming vendor A wafers. Similarly, when a second set of A, B, C and D wafers was repolished by vendor B the results (column 4) show the typical low values otherwise seen only on the incoming vendor B wafers. Finally, a third set of A, B, C and D wafers was polished first by vendor A, resulting in high $\Delta R/R$ signals, then repolished by vendor B who removed roughly 250 microns of silicon to assure that the surface region was influenced only by the last polishing process. Again the $\Delta R/R$ data from these wafers (column 5) show the low values indicative of the low-damage polishing process of vendor B.

It thus appears that a primary source of variations in the $\Delta R/R$ signal measured on polished silicon wafers is caused by differences in the wafer polishing process. One particular way in which the processes may differ is in the use of a scrubbing step in the final cleaning cycle. To examine the impact of the scrubbing step on the wafer surface quality, a further experiment was performed. The objective was to test whether the Therma-Probe system could detect a change in the surface damage due to a standard automated scrubbing cycle.

A group of wafers was polished together in the same process and then separated into three groups. One group received a 20-30 sec. brush scrubbing cycle and a second group received a 20-30 sec. scrubbing by high-pressure water jet. The third group received no scrubbing. The results, plotted in Fig. 4, show that the scrubbing cycle increases the $\Delta R/R$ signal by about 5×10^{-4}, thus indicating that both the brush and jet scrubbing processes further damage the silicon surface.

We realize that, for the manufacturing of silicon wafers, there is a general belief that some type of mechanical scrubbing is beneficial in removing particulates and stain from the wafer surface. However, with the ever more stringent requirements for damage-free surface regions as discussed in the introduction, this requirement for scrubbing must be scrutinized carefully.

Finally, we are able to make a connection between the damage induced by polishing and the lattice damage induced by ion implantation. In Fig. 5, we plot the $\Delta R/R$ signal data from Table I along with the $\Delta R/R$ signal from maps on five silicon wafers implanted with boron ions at an ion energy of 50 keV. The implanted doses are 1×10^{10} to 9×10^{10} ions/cm^2. It is of course true that the distribution of the damage with depth into the silicon is likely to be different for these types of damage. But because the $\Delta R/R$ signal represents a measure of the total damage beneath the surface to a depth of approximately three microns, such a comparison can be made. We see that the $\Delta R/R$ signal from vendor A and D wafers, for example, indicates an amount of damage equivalent to a B^+ ion implantation dose of 2×10^{10} to 4×10^{10} ions/cm^2.

Fig. 4 Modulated reflectance signal for scrubbed and non-scrubbed Si wafers

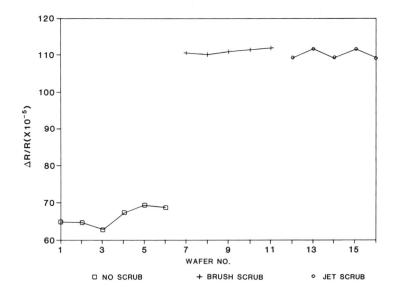

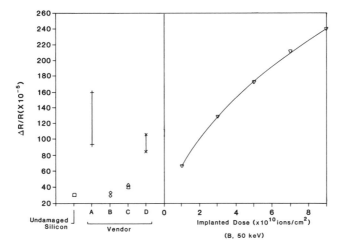

Fig. 5 Display of $\Delta R/R$ signal for wafers which have received damage either from wafer polishing or from ion implantation. Also plotted is the $\Delta R/R$ signal for undamaged silicon.

This study demonstrates that the modulated reflectance technique is an effective means for characterizing polished silicon surfaces. This method is noncontact and nondestructive and provides rapid, flexible mapping capabilities. The high sensitivity of the modulated reflectance technique makes it possible to quantify the damage resulting not only from various polishing processes, but also from the post-polishing scrubbing processes as well.

V. Ion Implant Inspection

It is widely recognized that integrated circuit performance and yield are strongly dependent on the accuracy and uniformity of the dose of dopant ions implanted into the wafer. This is especially true for metal-oxide-silicon (MOS) integrated circuit (IC) technology where there are several important implant steps, the most critical of all being the low-dose implants used to adjust the threshold voltages of the transistors in the IC. Nevertheless, ion implant monitoring has always been a complex and difficult task, as present monitoring techniques such as capacitance voltage (Nicollian and Brews, 1982), optical dosimetry (Golin, Schell and Glaze, 1985), spreading resistance (Masur and Gruber, 1981), and sheet resistance (Keenan, Johnson and Smith, 1985) usually involve substantial processing time before evaluation or lack sensitivity in a critical dose range. The greatest and fundamental difficulty, however, is that real time measurements must presently be taken from test wafers rather than the actual product wafers. This is because the conventional methods of monitoring the implant process require more area than can be made available on the product wafer surface and furthermore are usually destructive to the product wafer.

As we have shown above, the modulated reflectance signal is very sensitive to the presence of lattice damage or disorder in or near the surface region of a Si wafer. Since the ion implantation process can produce a significant amount of lattice disorder, the Therma-Probe inspection system is a sensitive monitor for this process and therefore can be used to measure the dose and uniformity of the implant (Smith, Rosencwaig and Willenborg, 1985; Smith, Taylor and Schuur, 1985; Smith, Powell and Woodhouse, 1985; Smith, Rosencwaig, Willenborg, Opsal and Taylor, 1986).

The dependence of the modulated reflectance signal on ion dose from 1×10^{11} to 1×10^{15} ions/cm^2 is shown in Fig. 6 for 50 keV $^{11}B^+$ ions implanted into silicon wafers. The three sets of data plotted on this graph correspond to three different thicknesses of oxide overlayer ("bare", 360 Å, 660 Å) through which both the implant and the modulated reflectance measurements were made. The wafers were measured directly after implantation with no subsequent anneal. We see in Fig. 6 that the lattice damage induced by the ion beam causes the modulated reflectance signal to increase, above that of the nonimplanted Si, in a pronounced manner. The functional dependence exhibited by the data is smooth, monotonically increasing and gradual such that reasonable sensitivity for dose measurements is maintained over four orders of implanted dose. We have found appreciable ion dose sensitivity to extend below 1×10^{10} ions/cm^2 and up to 1×10^{17} ions/cm^2. Detailed studies of boron, phosphorous, and arsenic implants into both Si and GaAs wafers have been conducted and published (Smith *et al.* 1985; 1985; 1985; 1986). It should be noted that all modulated reflectance values can be converted to actual dose values by a simple calibration procedure.

Ion implantation uniformity is best monitored with a two-dimensional contour map of the wafer. Examples of contour maps on implanted wafers generated with the Therma-Probe system are shown in Figs. 7(a)-7(d). The map in Fig. 7(a) shows the 0.5% contours derived from the modulated reflectance signal

Fig. 6 Modulated reflectance signal vs dose for 50 keV$^{11}B^+$ implanted into silicon wafers through a surface oxide layer of 0, 360, and 660 Å thickness.

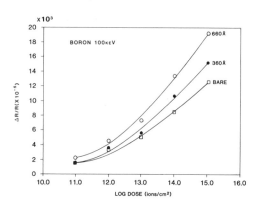

measured at 137 points uniformly distributed across a wafer implanted with 80 keV $^{11}B^+$ through a 640-Å oxide overlayer to a dose of 8×10^{11} ions/cm^2. This wafer shows a smooth gradation of modulated reflectance signal and thus of ion dose from higher values on the right side to lower values on the left. The contour map of another low-dose wafer in Fig. 7(b) shows a pattern of stripes resulting from the frequently encountered problem of inadequate control of the ion beam scan frequencies in the x and y directions, causing a readily detected "scan lockup" pattern of implanted dose.

Using the modulated reflectance signal, one can also detect the presence of ion channeling as illustrated with Fig. 7(c). The distinctive broad band of low signal in the center of the map in Fig. 7(c) arises from the planar channeling along the <110> planes to substantially greater depths in this region (Turner, Current, Smith and Crane, 1985; and references therein). To prove that ion channeling is indeed the cause of the pattern in Fig. 7(c), a second wafer was implanted under identical conditions except that the implant was performed through a 250-Å oxide layer. It is well known that implantation through amorphous dielectric layers can reduce this unwanted ion channeling considerably (Turner *et al.*, 1985). As the map in Fig. 7(d) shows, the pattern attributed to planar channeling is no longer present. The standard deviation of the 137 measurements was reduced from 3.4% in Fig. 7(c) to 0.23% in Fig. 7(d).

In addition to wafer maps, valuable information can be obtained from one-dimensional line scans. This is especially true for patterned, product IC wafers. As a final example of modulated reflectance measurements of implanted ion dose, we show in Figs. 8(a) and 8(b) a scan across a 40-μm distance on a high-density read only memory (ROM) cell that is programmed with a patterned ion implant step (Smith, Taylor and Schuur, 1985; Smith, Rosencwaig, Willenborg, Opsal and Taylor, 1986). Figure 8(a), an artist's depiction of the portion of the circuit that we examined, shows the array of n-channel MOS transistors that comprise the single bits of a ROM device. The gate, field oxide, and source/drain regions of the circuit are pointed out. Also pointed out is the photoresist mask that was used to shield selected transistor gates from the programming implant to form the OFF bits, while exposing other gates to define the ON bits.

After the bit programming implant ($^{11}B^+$, 1×10^{13} ions/cm^2, 180 keV) was performed and the photoresist mask was stripped, a line scan using the Therma-Probe system was performed in the exact location shown by the arrow in Fig. 8(a). The printout of this scan is displayed in Fig. 8(b). The regions of this scan over which the modulated reflectance signal is greater than 3.8×10^4, and thus off scale in this figure, are 2-μm-wide polycrystalline silicon gate areas. In the first and last "valley" segments (2-6 μm and 37-40 μm on the horizontal axis) in Fig. 8(b), the lower dotted line (the OFF state) is the signal obtained from the source/drain regions that were protected from the programming implant by the photoresist mask. In the central region of this scan (between 13 and 30 μm on the axis) we see the increased signal level in the source/drain regions surrounding the two unmasked gates that were implanted to the ON state. With a suitable calibration, the measured modulated reflectance signals can be converted to

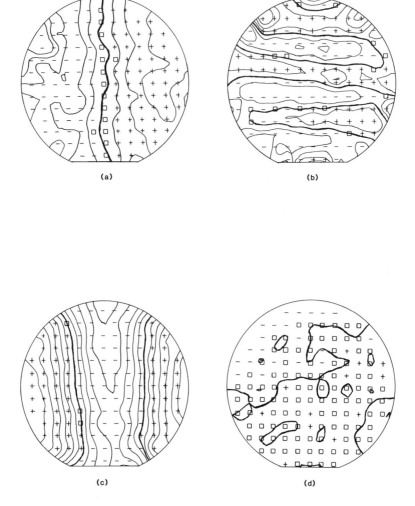

Fig. 7 (a) Contour map of an ion dose uniformity as derived from the modulated reflectance signal for a Si wafer with a target implant of $^{11}B^+$ at 8×10^{11} ions/cm^2. Contour interval is 0.5% and standard deviation is 1.17%. (b) Scan lock-up pattern displayed in a contour map for a low-dose, 6×10^{11} ions/cm^2, boron-implanted wafer. (c) Ion channeling pattern displayed in a contour map on an implanted bare Si wafer. Contour interval is 1% and standard deviation is 3.42%. (d) Contour map showing the elimination of the ion channeling seen in (c) by implanting through an oxide layer. Contour interval is 1% and standard deviation is 0.23%.

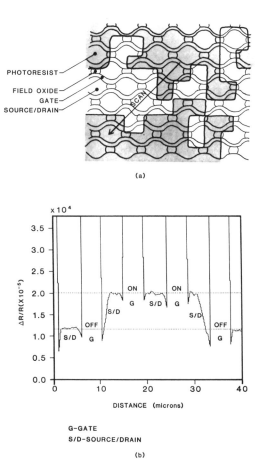

(a)

G-GATE
S/D-SOURCE/DRAIN

(b)

Fig. 8 (a) Depiction of a small region of a high-density memory cell structure of an implant-programmed ROM integrated circuit. After the implant and mask-strip steps, a modulated reflectance line scan was performed along the 40-μm distance indicated by the arrow. (b) Modulated reflectance signal along the 40-μm scan depicted in (a). The scan shows the different signals obtained from two memory cells programmed to the ON state (implanted) and from two cells that were shielded from the implant (OFF state) by the photoresist mask.

actual dose values. The micron-scale data presented in this example illustrate the ability to monitor the implant process directly on the actual patterned IC devices.

In summary the noncontact, nondestructive, and rapid nature of these modulated reflectance measurements, combined with the low-dose sensitivity and micron-scale resolution, makes this technique particularly suited to production implant monitoring of actual patterned IC wafers as well as test wafers.

VI. Dry Etch Inspection

Reactive ion etching (RIE) and plasma etching (PE) are vital processes for the attainment of densely-packed, micron-scaled structures for very large scale integrated (VLSI) circuits. However, it is widely recognized that undesirable modifications of semiconductor or insulator materials may accompany the use of these dry etch processes. For example, during the RIE process, samples are exposed to high energy ions, UV photons and x-rays, all of which can result in radiation damage (Pang, Rathman, Silversmith, Mountain and DeGraff, 1983) in the form of non-annealable structural defects in gate oxide or Si/SiO_2 interface regions, deep level traps or surface states. In terms of IC device performance, these effects can cause transistor threshold voltage shifts, poor subthreshold performance, increased junction leakage, decreased capacitor charge retention time, degradation of minority carrier lifetime, barrier shifts in Schottky diodes and reduction of the integrity of trench isolation structures (Pang, 1984; Dieleman and Sanders, 1984). Contamination from sputtering of the chamber parts or of the oxide mask may add to these problems. In addition, polymer material that may be deposited from carbon-containing gases has been found to create oxygen-induced-stacking faults (Ogden, Bradley and Watts, 1974). All of the above can lead to significant yield reduction.

A number of techniques have been developed to study the effects of RIE or PE on single crystal silicon. MOS capacitors can be fabricated and capacitance-voltage plots generated to obtain threshold shifts or interface state densities (Ephrath and Bennett, 1982). Generation lifetimes can be measured. Deep level transient spectroscopy can be used to determine trap levels in silicon, but in this technique wafers are subjected to temperature extremes (100-300°C) during the measurements (Matsumoto and Sugano, 1982). Alternatively, a decorative, wet chemical etch such as a Secco (Secco d'Aragona, 1972) or Wright (Wright Jenkins, 1976) etch can be used to highlight dislocations in silicon. These etches are, however, destructive in nature. Channeling-mode Rutherford back scattering, reflected high energy electron diffraction, electron spin resonance and measurement of the current-voltage characteristics of Schottky diode structures (Fonash, 1985) have also been employed to study radiation damage in silicon. Each of the above methods is either destructive to the examined wafers, or lacks of speed or ease of use necessary to constitute an effective, practical method for detecting RIE or PE damage during dry etch process development or during the integrated circuit manufacturing process.

Recently we have shown that measurement of laser-induced modulated optical reflectance can nondestructively detect the radiation damage in Si due to RIE or PE with good sensitivity (Smith and Rosencwaig, 1986). This method also measures etch uniformity over the wafer surface and can detect the presence of deposited polymer material on micron-scale device features on patterned wafers. We illustrate some of these capabilities in the study below where we have used the modulated reflectance method to measure RIE damage at the bottom of 6-micron deep trench structures as a function of RIE process parameters such as dc bias and pressure. Such a parametric study allows the RIE process variables to be adjusted to minimize damage to the silicon surface.

The trench samples used are p-type <100> silicon substrates with n^+ buried layer and n^- epi silicon as illustrated in Fig. 9. The masking material for the silicon trench etch is 4% P-doped CVD oxide etched in a commercial oxide etcher. Trench openings are 1.2 microns wide, with one 10-micron wide trench surrounding the die. This wider trench is used for the modulated reflectance measurements. Prior to silicon etch, samples are cleaned in RCA solution and dipped in 10:1 HF.

The main etch step is a low pressure, medium power process during which 80% of the trench is etched. As will be described below, two of the parameters varied in the experiments to determine their effect on damage are the relative proportion of the constituent gases and the total flow in this etch step. A "clean up" etch step is incorporated to remove the damaged Si produced by the "main" step while completing the etch. As experimental parameters in this "clean up" step, the dc self-bias voltage is varied between -60 and -250 volts and the pressure between 15 and 100 mTorr to determine their effect on silicon damage.

For all the measurements presented in the next section, a Therma-Probe scan of 50-micron length is made along the bottom of a wide trench (Fig. 9) on each sample wafer. The average value of the modulated reflectance, $\Delta R/R$, signal from each wafer is displayed on the following graphs. Longer (diameter) scans and whole-wafer contour maps are also available in this method.

Fig. 10 illustrates the results of varying the D.C. self-bias voltage from -60 to -250 V in the "clean up" step of the silicon etch process. The purpose of this step is to remove the first few hundred Angstroms of silicon, damaged by the previous high-bias main etching step, from the bottom of the trench while rounding out the trench bottom corners by operating at a slightly higher pressure. The results, in general agreement with previous work by Pang, Horwitz,

Fig. 9 Diagram of the structure of the trench wafer samples. Also shown is the measurement laser beam focused on the wider trench bottom.

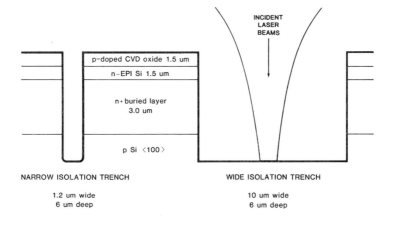

INCIDENT LASER BEAMS

p–doped CVD oxide 1.5 um

n–EPI Si 1.5 um

n+ buried layer 3.0 um

p Si <100>

NARROW ISOLATION TRENCH

1.2 um wide
6 um deep

WIDE ISOLATION TRENCH

10 um wide
6 um deep

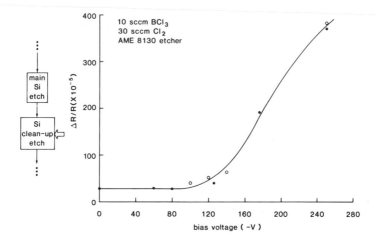

Fig. 10 Modulated reflectance signal, $\Delta R/R$, indicating the level of RIE-induced damage, versus bias voltage for the "clean up" step of a silicon etch process. The open and closed symbols denote data recorded on two different runs separated in time by several weeks.

Rathman, Cabral, Silversmith and Mountain (1983), show that the level of RIE damage as measured by the modulated reflectance technique increases markedly with increasing bias voltage above a certain threshold bias voltage, in this case about -120 volts. The data points in Fig. 10, the open and closed symbols representing measurements taken on two separate runs in the same etcher but several weeks apart, illustrate the reproducibility of the damage produced by the etch process and measured by the Therma-Probe system. Note that at low bias voltage, the measured $\Delta R/R$ signals are very close to $3\pm0.3\times10^{-4}$, the value generally obtained on non-damaged silicon in other studies (see Section IV).

The effect of varying pressure in the "clean up" silicon etch step is shown in Fig. 11. As the pressure is increased under constant flow conditions, the mean free path of the reactive ions is shortened, and the mechanisms of etch become more chemical in nature, rather than physical (sputtering). Thus damage to the crystal lattice should lessen. This is reflected in our data, where the general trend shows a decrease in crystal damage with increasing pressure. The anomalous data point at a pressure of 15 mTorr may be due to an inadequate supply of reactant gas at such low operating pressure.

The effect of the etch chemistry on crystal damage has also been investigated. Fig. 12 shows the dependence of the $\Delta R/R$ signal on the percentage of BCl_3 in a BCl_3/Cl_2 mixture at two different values of total flow rate. The substantial variation of the indicated damage in these data is surprising and needs further investigation.

The above results illustrate the capability of modulated reflectance measurements to provide a quick, nondestructive means for optimizing the silicon etc. process to minimize damage. Since device wafers can be used, another

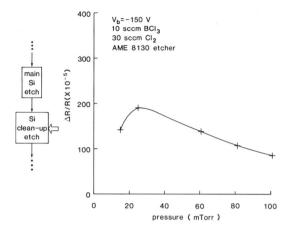

Fig. 11 Modulated reflectance signal versus pressure in the "clean up" step of a silicon etch process.

application of this technique is as a routine monitor of critical process steps that may contribute to lower yields. Fig. 13 shows the $\Delta R/R$ signals obtained at eight process steps in the trench-formation sequence: starting material, after oxide etch, after pre-silicon-etch clean, etc. Any deviation from a standard set of $\Delta R/R$ signals can be immediately observed and the process problem rectified before device yields are impacted. For example, wafers can be recleaned after oxide etch if not "in spec". Any drift in the equipment or process that affects the level of silicon damage can be identified immediately rather than later at the nonsalvageable stage of wafer probe.

This work demonstrates that etch-induced damage to silicon wafers is readily detected and quantified using the modulated reflectance technique. This method is rapid, noncontact, nondestructive, and has 1-micron spatial resolution. No special patterns or devices need be fabricated or test structures created; rather, actual device wafers can be used. The etch damage can be measured as a function of etcher bias, pressure, flow and gas composition. Such parametric studies allow for the adjustment of process etch parameters to minimize the damage. We also demonstrate the use of modulated reflectance measurements to monitor process and equipment stability at critical steps in an etch sequence.

VII. Detection of Electronic Surface States

Our experiments on incoming bare silicon wafers show that typical modulated reflectance signals, $\Delta R/R$, at 1 MHz lie in the range of $0.3-2\times10^{-3}$. Our data indicate that the higher signal levels are indicative of the small amount of residual damage present in the surface regions of the wafers that result from the chemopolishing and scrubbing processes used in the final stages of the

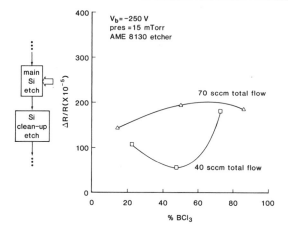

Fig. 12 Modulated reflectance signal versus percentage of BCl_3 in a BCl_3/Cl_2 mixture at two values of total flow rate, in the "main step" of a silicon etch process.

manufacture of the wafers. As stated above, theoretical calculations predict that the modulated reflectance signal should generally increase with the extent of damage, in the form of surface states or of lattice disorder in the near surface region of the silicon wafer. These theoretical predictions are in agreement with experimental results on both incoming wafers and on ion-implanted wafers. In Fig. 14 we see that the lowest signal corresponds to an incoming wafer that has been thermally annealed and has several hundred Angstroms of thermal oxide grown on it. This wafer would be expected to have the least amount of damage, both from the thermal annealing and from the consumption of any damaged silicon region by the growth of the oxide film. Our experiments on ion-implanted wafers (Smith *et al.* 1985; 1985; 1985; 1986) show that the modulated

Fig. 13 Modulated reflectance signal measured at the bottom of 10-μm wide trenches on product wafers at eight steps in the trench-formation process sequence.

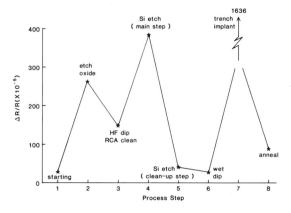

reflectance signal monotonically increases with the dose of implanted ions into the wafer. Furthermore when an implanted wafer is thoroughly annealed the $\Delta R/R$ signal decreases dramatically. All of these results indicate that the modulated reflectance signal tends to increase with the amount of damage in the near surface region of the wafer.

Returning to Fig. 14, we see that except for the wafer with the thermal-oxide film, all other incoming "bare" wafers tend to exhibit a temporal dependence of the signal, that is, the magnitude (and often the phase) of the $\Delta R/R$ signal changes if the pump and probe laser beams are allowed to continually illuminate the same spot on the silicon wafer. Of considerable importance is the fact that the time-dependent curves can be fit with high accuracy to an expression of the form,

$$f(t) = A\{1+B\exp(Ct)\mathrm{erfc}(\sqrt{Ct})\}\,, \qquad (14)$$

Fig. 14 Modulated reflectance as a function of time on a number of different silicon samples: □ - Si with 300 Å thermal-oxide film, Δ - p-Si 5 Ω-cm (#1), ♦ - p-Si 50 Ω-cm, ✕- p-Si 5 Ω-cm (#2), ✕- p-Si 5 Ω-cm (#3, 3 ✕ polishing pressure). The upper plot (a) shows the long time behavior and the lower plot (b) focuses on the behavior near t=0.

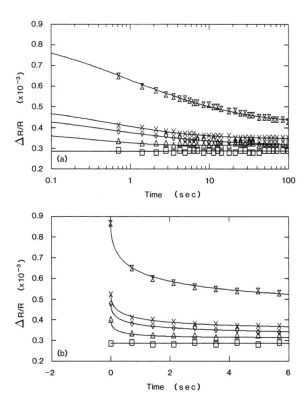

thus indicating that this laser-induced temporal effect is partly diffusive in nature. This particular solution of the diffusion equation (Carslaw and Jaeger, 1959) corresponds to having a source term which decays exponentially in time as $\exp(-Ct)$. We believe that this is the result of the gradual removal or annealing of the defect surface states from the illuminated region by the photogenerated electron-hole plasma. We will henceforth refer to this laser-induced temporal effect on the $\Delta R/R$ signal as a surface state annealing effect.

Table II

Actual $t=0$ and predicted $t=\infty$ values for the modulated reflectances shown in Fig. 14.

Sample	$\Delta R/R$ at $t = 0$ $(\times 10^{-3})$	$\Delta R/R$ at $t = \infty$ $(\times 10^{-3})$
Si with 300 Å thermal-oxide film	0.285	0.286
p-Si 5 Ω-cm (#1)	0.397	0.326
p-Si 50 Ω-cm	0.477	0.312
p-Si 5 Ω-cm (#2)	0.523	0.345
p-Si 5 Ω-cm (#3) (3 × polishing pressure)	0.867	0.404

Using Eq. (14) we are then able to predict the $t = \infty$ value of the $\Delta R/R$ signal and this value is indicated in Table II for the various wafers. We note that the $t=\infty$ signal as well as the $t=0$ signal are lowest for the silicon wafer with the least amount of damage, that is, the wafer with the thermal oxide film. On the other hand, both the $t = 0$ and $t = \infty$ signal levels for the other wafers vary considerably and these variations appear to be related to the amount of near-surface damage present and to the amount of intrinsic disorder from the as-grown dopant concentrations. Thus a heavily doped (not shown) p-Si wafer with 0.01 Ω-cm resistivity exhibits the highest $t = \infty$ signal. We attribute the increased $t = \infty$ signal for the heavily doped 0.01 Ω-cm wafer to the effects of the high doping concentration on the electron-hole diffusivity D (with the understanding of course that the recombination lifetime τ and the surface recombination velocity S may also be affected).

However if we compare the 5 Ω-cm (#1) wafer with the 50 Ω-cm wafer, we see that their $t = \infty$ signals are essentially identical, indicating that a change in doping concentration by a factor of 10 does not noticeably change the intrinsic $\Delta R/R$ signal when the doping concentration is low. Nevertheless these two curves have different $t = 0$ values, possibly indicating different amounts of near surface damage.

In fact if we compare three 5 Ω-cm wafers (#1,2 and 3) that have received different polishing treatments with wafer #1 receiving that polishing treatment that is considered to produce minimal damage while wafer #3 received the most damage, it is clear that the $t = 0$ signal is especially sensitive to the presence of this damage while the $t = \infty$ signal is less so. We are able to account for this observation by assuming that chemopolishing results in both some lattice disorder and, more significantly, in the presence of additional extrinsic, i.e., impurity or defect surface states. Since our laser beams are generating temperatures of only ~ 10°C, we would not expect any annealing effect on the lattice disorder damage. However, as we postulate in Sec. III, it may be possible that any extrinsic non-dangling bond surface states may be altered, by charge neutralization, bond reconfiguration or promotion into another state such as a bulk defect state, by the presence of the photogenerated electron-hole plasma. In particular the energy for the alteration of the extrinsic surface states may come from a local electron-hole recombination event. This may be analogous to the photogeneration of recombination defects in amorphous silicon (Staebler and Wronski, 1977; 1980; Aker and Fritzsche, 1983). At any rate, we would thus expect that while the $t = 0$ signal is indicative of the presence of both extrinsic surface states and lattice disorder, the $t = \infty$ signal would be mostly due to the amount of lattice disorder present, with most of the extrinsic surface states "annealed" out by the laser beams. This explanation is consistent with the observation that silicon wafers with a thermally grown oxide film exhibit no surface state annealing effect. It is well-known that a good Si/SiO$_2$ interface will have only dangling bond interface states and that these would not be altered by the presence of an electron-hole plasma or by local electron-hole recombination events. On the other hand extrinsic, i.e. impurity or defect-related, surface states may be altered by the electron-hole plasma and by electron-hole recombination events.

1.   FREQUENCY DEPENDENCE

In order to model the frequency dependence of the temporal behavior of the modulated reflectance, we assume the time dependent form in Eq. (14) for the number of defect states as a function of time. That is, we assume that the β in Eq. (12) has a temporal dependence given by;

$$\beta(t) = B\exp(Ct)\text{erfc}(\sqrt{Ct}) \tag{15}$$

Using this time dependent form for β in Eq. (12), we have analyzed data at frequencies of 1, 3, and 10 MHz taken on 3 different silicon wafers. The results are shown in Figs 15-17 with the single set of parameters α, B, C, τ_1, and τ_2 obtained for each wafer displayed in Table III. In Fig. 15 are the data and fitted curves for a p-type 50 Ω-cm wafer. At 1 MHz the data show a simple monotonic decay with time which are well represented by the fitted curve. However, at the higher frequencies we observe a small peak in the theoretical curves

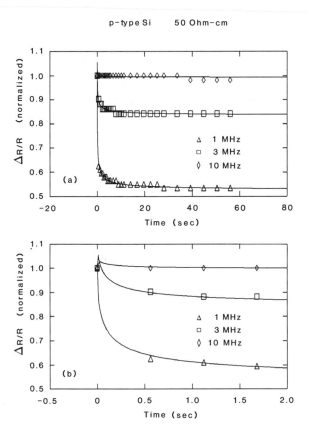

Fig. 15 Modulated reflectance as a function of time at different modulation frequencies for a p-Si 50 Ω-cm sample. The upper plot (a) shows the long time behavior and the lower plot (b) focuses on the behavior near *t* = 0.

occuring at very early times. Although we are unable to obtain data to support this behavior for this low-doped wafer, the effect is expected for the parameters obtained and furthermore is actually observed on the more heavily doped p-type sample shown in the next figure. Also, as predicted, the temporal behavior is apparently weakening with increasing modulation frequency.

In Fig. 16 we show data from a p-type 0.01 Ω-cm silicon wafer which exhibits a non-monotonic temporal dependence. Again the solid curves are a fit to the data that fall well within the experimental uncertainty which is essentially represented by the size of the plotted symbols. For these data the positive contribution to the time dependence is actually observed at 1 MHz, implied at 3 MHz, and apparently gone by 10 MHz. The larger value for B (i.e., the number

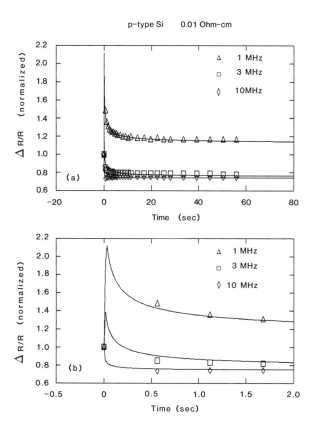

Fig. 16 Modulated reflectance as a function of time at different modulation frequencies for a p-Si 0.01 Ω-cm sample. The upper plot (a) shows the long time behavior and the lower plot (b) focuses on the behavior near $t = 0$.

Table III

Parameters of Eqs. (12) and (15) obtained by fitting to the data shown in Figs. 15, 16, and 17.

Sample	α	B	C sec^{-1}	τ_1 μsec	τ_2 μsec
p-Si 50 Ω-cm	0.571	-0.546	6.82	0.055	0.042
p-Si 0.01 Ω-cm	0.475	-1.06	25.0	0.010	0.029
n-Si 0.01 Ω-cm	0.733	0.603	0.512	0.063	0.008

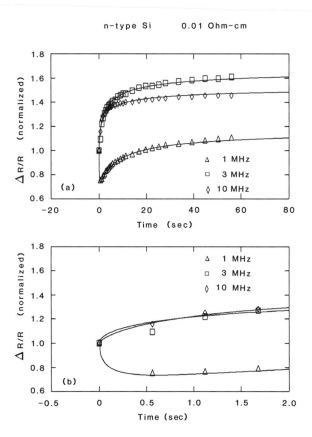

Fig. 17 Modulated reflectance as a function of time at different modulation frequencies for an n-Si 0.01 Ω-cm sample. The upper plot (a) shows the long time behavior and the lower plot (b) focuses on the behavior near $t = 0$.

of defect states) obtained for these data is consistent with our contention that the more heavily doped sample will have a higher defect state density than the lower doped 50 Ω-cm sample shown in the previous figure.

The prediction of the model that an increasing signal with time is possible is exhibited by the data in Fig. 17. This sample which is an n-type 0.01 Ω-cm wafer has a fitted value for B that is similar to the corresponding p-type sample but as expected from the model, is of opposite sign. The non-monotonic behavior is also expected for the lifetimes obtained and the weakening of the temporal effect with increasing frequency is as expected.

2. EPITAXIAL SILICON FILMS

In Fig. 18 we show data obtained on some thick (~ 10 μm) Si epitaxial films with the predicted $t = \infty$ values indicated in Table IV. Such films usually are of higher crystalline quality than as-grown bulk Si. Thus it is not surprising that the $t = \infty$ signals for both a low-doped (50 Ω-cm) epitaxial film and for a much higher doped (0.21 Ω-cm) epitaxial film are both lower than that obtained for a low-doped (50 Ω-cm) bulk Si wafer. On the other hand the $t = 0$ signal for the high-doped (0.21 Ω-cm) epitaxial film is noticeably greater than that for the lower doped (50 Ω-cm) bulk wafer. Since epitaxial films are not subjected to a polishing process, this indicates that the presence of a higher doping concentration in the film leads to a higher concentration of extrinsic surface states. The presence of extrinsic surface states in epitaxial films is also probably related to the high concentration of hydrogen and hydrogen complexes in the surface atomic layers of an epitaxial film that results from the epitaxial growth reaction.

Fig. 18 Modulated reflectance as a function of time on different samples: ♦ - 50 Ω-cm epitaxial Si, Δ - 50 Ω-cm p-Si, □ - 0.21 Ω-cm epitaxial Si. The upper plot (a) shows the long time behavior and the lower plot (b) focuses on the behavior near $t = 0$.

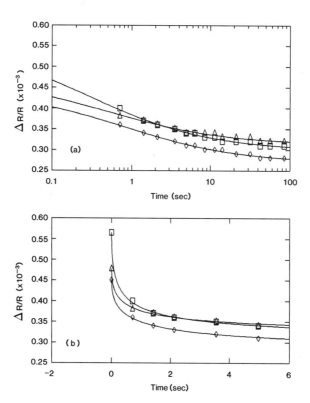

Table IV

Actual $t = 0$ and predicted $t = \infty$ values for the modulated reflectances shown in Fig. 18.

Sample	$\Delta R/R$ at $t = 0$ $(\times 10^{-3})$	$\Delta R/R$ at $t = \infty$ $(\times 10^{-3})$
p-Si 50 Ω-cm	0.477	0.312
epi-Si 50 Ω-cm	0.450	0.268
epi-Si 0.21 Ω-cm	0.565	0.297

3.   THERMAL EFFECTS

In Fig. 19 we show the temporal behavior of the $\Delta R/R$ signal for a 0.21 Ω-cm epitaxial Si film before and after a thermal anneal at a fairly low temperature of 200°C. We note that while such low temperatures have little noticeable effect on the $t = \infty$ signal, this is not true for the $t = 0$ signal. In fact after only 30 minutes at 200°C, the $t = 0$ signal decreases substantially. However, of even greater interest is the fact that if the wafer is then reexamined after 72 hours at room temperature in air, most of the original $t = 0$ signal returns. Further experiments on various bulk and epitaxial wafers have confirmed these findings. Thus thermal annealing in air at moderately low temperatures tends to have the same effect on the extrinsic surface states as does the laser-induced surface state annealing. This indicates that the activation energy needed for neutralizing or transforming these extrinsic surface states is indeed quite small, in keeping with the concept that the surface state annealing effect is a result of an electron-hole recombination phenomenon. The fact that the thermal annealing of these states is reversible is also in keeping with the diffusion aspects of the surface state annealing effect.

4.   EFFECT OF ION IMPLANTATION

In Fig. 20 we show how the $t = 0$ and $t = \infty$ $\Delta R/R$ signals vary with boron or arsenic implant into Si wafers. Of considerable importance is the observation that the $t = \infty$ signal increases rapidly with increasing implant dose, thus indicating that the major portion of the $\Delta R/R$ signal arises from lattice disorder. Although some of the implant generated signal is due to extrinsic surface states, the data in Fig. 20 indicate that the contribution to the total signal from these implant-generated extrinsic surface states does not increase as rapidly as does the contribution from the lattice disorder itself.

Epi   0.21 Ohm-cm

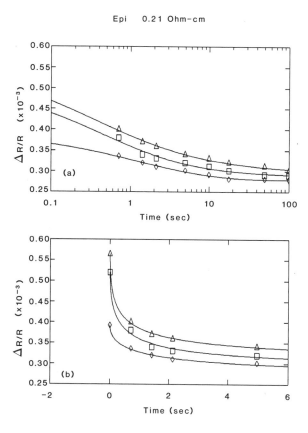

Fig. 19 Modulated reflectance as a function of time showing effects of heat treatment on a 0.21 Ω-cm epitaxial Si sample: ♦ - after 30 minutes at 200°C, □ - 72 hours after thermal treatment, △ - before heat treatment. The upper plot (a) shows the long time behavior and the lower plot (b) focuses on the behavior near $t = 0$.

As we noted before, a Si wafer with a thermal oxide film exhibits no surface state annealing behavior. However, this is not true of such a wafer which is then implanted. The implantation process disrupts the Si/SiO_2 interface and allows the creation of extrinsic interface states associated with either localized structural defects or with implanted dopant ions and other impurities at the Si/SiO_2 interface. Thus, for example, similar $t = 0$ and $t = \infty$ data are obtained for implants on oxidized wafers as on bare wafers.

However, implanted oxide coated wafers exhibit quite different behavior from implanted bare wafers after a high temperature anneal. While bare wafers will initially exhibit no temporal dependence or surface state annealing behavior

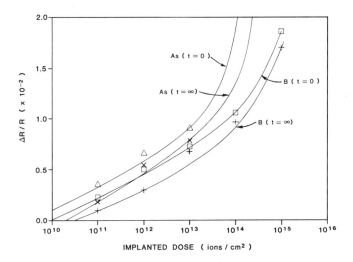

Fig. 20 Modulated reflectance as a function of implanted dose at $t = 0$, and $t = \infty$.

immediately after the high temperature anneal, this behavior will manifest itself a few days later. This result is similar to that observed for unimplanted epitaxial films or polished bulk wafers. However, wafers that have been implanted through a screen or gate oxide permanently lose all evidence of any surface state annealing behavior after a high temperature thermal anneal. This indicates that once the high temperature post-implant thermal anneal heals the Si/SiO$_2$ interface, the extrinsic interface states are either removed or acquire too high an activation energy to be altered by the laser beams. This is consistent with our prior observations of no temporal dependence in Si wafers having a good thermal oxide film.

5.   DURATION OF SURFACE STATE ANNEALING EFFECT

We have investigated the duration of the surface state annealing effect by first irradiating a fairly large area on a Si wafer until its modulated reflectance signal reaches its steady state condition (i.e. the $t = \infty$ level). We then perform some quick checks on the signal from this region as a function of time. These quick checks are performed in a manner so as to minimize as much as possible any additional surface state annealing effect from the checks themselves. Our results indicate that the signal slowly comes back to its initial $t = 0$ level in diffusive fashion. However, whereas the surface state annealing effect occurs within minutes, the return of the signal to its original $t = 0$ level usually takes several days.

This is similar to the results obtained for thermal annealing, thus indicating that the annealed state is in both cases a metastable state that will in time revert back to its original extrinsic surface state at room temperature.

6.   SPATIAL EXTENT OF ANNEALING EFFECT

Of considerable interest is the spatial extent of this surface state annealing effect. This experiment is performed by illuminating a micron sized spot on the silicon wafer for 60 seconds until a reasonable annealed state is achieved, then rapidly moving a set distance from this location and recording the $t = 0$ signal at the new location. By repeating this procedure many times we are able to map out the spatial extent of the annealing effect after a 60 second exposure at a given site. The data obtained indicate that in Si wafers with high levels of structural damage such as heavily implanted wafers, the annealing effect appears to be confined to the irradiated site. However for wafers that have little or no structural damage such as starting Si wafers or epitaxial films, the surface state annealing effect appears to extend 50-100 μm beyond the initial illuminated point after 60 seconds. This is most surprising since three-dimensional calculations readily show that neither the D.C. nor the A.C. thermal waves and electron-hole plasma waves extend much beyond 5-10 μm from the illuminated spot.

These data indicate that the laser annealed surface states either diffuse away from the illuminated area under their own concentration gradient, or that a neutralizing factor diffuses out from the illuminated area. This neutralizing factor may be a trapped charge from the photogenerated carriers that can hop or diffuse a considerable distance away from the irradiated area provided the Si crystal has little or no structural damage.

7.   SUMMARY ON SURFACE STATE DETECTION

Our laser-induced modulated reflectance data on crystalline silicon, epitaxial silicon, and ion implanted silicon show that the modulated reflectance signal is, in general, dependent on the time of exposure to the pump and probe beam illumination. This effect, which is reversible, appears to be a new phenomenon, certainly for crystalline and epitaxial silicon, and related to the presence and temporal evolution of electronic surface states. Using a phenomenological trapping and diffusion model, we are able to explain the temporal behavior observed at different modulation frequencies with the effect achieving its maximum at low modulation frequencies and vanishing in the high frequency limit. To support our conjecture that electronic surface states are involved, we have examined samples which have undergone different surface treatments as well as a number that have been ion implanted. This temporal behavior, which we have termed a surface state annealing effect, is present to varying degrees in all samples and appears to depend on the amount of surface damage. In addition, the effect completely disappears for samples with a thermally grown oxide as well as on samples that have been implanted through a screen or gate oxide and then subjected to a high temperature anneal. Finally, our data on bare silicon wafers having low levels of lattice damage show that the surface state annealing effect can extend well beyond the area of illumination indicating that the diffusion process may be along the surface.

In conclusion, we want to emphasize that this surface state annealing effect is reproducible and appears to be well correlated with surface conditions. Thus we believe that, in addition to its potential for providing more information about electronic surface states in silicon (and perhaps in other semiconductors as well), the temporal behavior of the laser-induced modulated reflectance can be used in practice as a nondestructive method for semiconductor surface characterization.

VIII. Conclusions

Thermal waves can be detected by several different methods. Many of these methods have limited applicability for characterizing and inspecting semiconductor materials and devices because they either require some physical contact with the sample, which can be contaminating or damaging, or they are unable to provide the needed spatial resolution or sensitivity. The laser-induced modulated reflectance technique appears to address these difficulties. In non-semiconducting materials, or when the photon energy is greater than the bandgap energy, the modulated reflectance signal is primarily a result of thermoreflectance effects and thus is directly affected by the propagation of thermal waves. Under these purely thermal wave experimental conditions, the modulated reflectance signal will provide measureable information only when there are significant variations in the sample's thermal parameters, such as the thermal conductivity or volume specific heat. In semiconductors, however, there are many manufacturing processes that can result in significant changes in important electrical parameters of the sample while producing only insignificant changes in the thermal parameters. Thus a purely thermal wave method will not be sensitive to these changes. Unlike other thermal wave detection methods, the modulated reflectance technique provides an opportunity to detect variations in local electrical parameters as well. This is accomplished by performing the modulated reflectance measurement with an intensity-modulated pump laser having a photon energy greater than the bandgap energy. Under these conditions, electron-hole plasma waves as well as thermal waves are generated. The plasma waves are sensitive to local variations in such electrical parameters as carrier mobility, carrier concentration, recombination lifetime and so on. By operating with a sufficiently intense pump laser, the optical reflectivity of the sample can be directly affected by these plasma waves through the Drude effect.

We have demonstrated that when a major component of the modulated reflectance signal is due to these plasma waves, excellent sensitivity is obtained for detecting and quantifying the extent of lattice damage or disorder produced in silicon wafers from the chemopolishing, ion implantation and dry etching processes. This damage or disorder, while sufficient to affect the electrical properties of the silicon, is usually not sufficient to measurably affect the thermal properties. Under the appropriate experimental conditions, the modulated reflectance technique is able to detect this damage and thus be used as a highly sensitive, noncontact method for inspecting important, but previously difficult to monitor, processes in the manufacture of integrated circuits. Furthermore, by using highly focussed pump and probe laser beams, these inspections can be performed with micron-scale resolution directly on patterned product wafers.

Finally we have recently discovered that the temporal behavior of the modulated reflectance signal appears to be very useful for detecting and studying the electronic surface states in silicon that are defect or damage related. In addition to providing more information about these electronic surface states, the temporal behavior of the laser-induced modulated reflectance can be used in practice as a nondestructive method for semiconductor surface characterization.

We believe that as this thermal wave technique is further developed, its value as a tool for the characterization and inspection of semiconductor materials and devices will become appreciable.

IX. References

Aker B. and Fritzsche H. (1983), J. Appl. Phys. **54**, 6628.

Albers W.A. (1969), Phys. Rev. Lett. **23**, 410.

Aspnes D.E. (1970), Solid State Commun. **8**, 267.

Aspnes D.E. (1980), in *Handbook on Semiconductors*, Vol. **2**, (Moss, ed.), North-Holland, New York, p. 109.

Aspnes D.E. and Frova A. (1969), Solid State Commun. **7**, 155.

Batz B. (1966), Solid State Commun. **4**, 241.

Batz B. (1972), in *Semiconductors and Semimetals*, (Willardson and Beer, eds.) Academic Press, New York, p. 315.

Berglund C.N. (1966), J. Appl. Phys. **37**, 3019.

Bonch-Bruevich A.M., Kovalev V.P., Romanov G.S., Imas Ya.A., and Libenson M.N. (1968), Sov. Phys. Tech. Phys. **13**, 507.

Carslaw H.S. and Jaeger J.C. (1959), *Conduction of Heat in Solids*, Clarendon, Oxford, p. 70.

Cerdeira F. and Cardona M. (1969) **7**, 879.

Dieleman J. and Sanders F.H.M. (1984), Solid State Tech. **27**(4), 191.

Ephrath L.M. and Bennett R.S. (1982), J. Electrochem. Soc. **129**, 1822.

Fisher W.A. and Amick J.A. (1966), J. Electrochem. Soc. **113**, 1054.

Fonash S.J. (1985), Solid State Tech. **28**(4), 201.

Gay J.G. and Klauder L.T., Jr. (1968), Phys. Rev. **172**, 811.

Geraghty P. and Smith W.L. (1986), Mat. Res. Soc. Symp. Proc. **68**, 387.

Golin J.R., Schell N.W. and Glaze J.A. (1985), Solid State Tech. **28**(6), 155.

Guidotti D. and van Driel H.M. (1985), Appl. Phys. Lett. **47**, 1336.

Kashiwagura N., Harada J. and Ogino, J. (1983), J. Appl. Phys. **54**, 2706.

Keenan W.A., Johnson W.H. and Smith A.K. (1985), Solid State Tech. **28**(6), 143.

Kotake H. and Takasu S. (1981), J. Mater. Sci. **16**, 767.

Kugimiya K.V. and Matsumoto M. (1981), in *Proc. 21st Symp. on Semicond. and IC Tech.*, Electrochem. Soc. Japan, Tokyo, p. 12.

Kundsen J.F. (1963), in *Advances in X-ray Analysis*, Vol. **7** (Mueller, Mallet and Fay, eds.) Plenum, New York, p. 159.

Lompre L.A., Liu J.M., Kurz H., and Bloembergen N. (1984), Appl. Phys. Lett. **44**, 3.

Masur R.G. and Gruber G.A. (1981), Solid State Tech. **24**(11), 64.

Matsumoto H. and Sugano T. (1982), J. Electrochem. Soc. **129**, 2823.

Nahory R.E. and Shay J.L. (1968), Phys. Rev. Lett. **21**, 1569.

Nicollian E.H. and Brews J.R. (1982), *MOS Physics and Technology*, Wiley, New York.

Nilsson N.G. (1969), Solid State Commun. **7**, 479.

Ogden R., Bradley R.R. and Watts B.E. (1974), Phys. Stat. Sol. **26**, 135.

Opsal J., Rosencwaig A., and Willenborg D.L. (1983), Appl. Opt. **22**, 3169.

Opsal J. and Rosencwaig A. (1985), Appl. Phys. Lett. **47**, 498.

Opsal J., Taylor M.W., Smith W.L. and Rosencwaig A. (1987), J. Appl. Phys., Phys. **61**, 240.

Pang S.W. (1984), Solid State Tech. **27**(4), 294.

Pang S.W., Horwitz C.M., Rathman D.D., Cabral S.M., Silversmith D.J. and Mountain R.W. (1983), Proc. Electrochem. Soc. **83-10**, 84.

Pang S.W., Rathman D.D., Silversmith D.J., Mountain R.W. and DeGraff P.D. (1983), J. Appl. Phys. **54**, 3272.

Ravi K.V. (1972), J. Appl. Phys. **43**, 1785.

Rosencwaig A. (1985), in *VLSI Electronics: Microstructure Science*, Vol. **9** (Einspruch, ed) Academic Press, Orlando, p. 227.

Rosencwaig A., Opsal J. and Willenborg D.L. (1983), Appl. Phys. Lett., **43**, 166.

Rosencwaig A., Opsal J., Smith W.L., and Willenborg D.L. (1985), Appl. Phys. Lett. **46**, 1013.

Rosencwaig A. and Opsal J. (1986), IEEE Trans. UFFC, **UFFC-33**, 516.

Secco F. d'Argona (1972), J. Electrochem Soc. **119**, 948.

Seraphin B.O. (1972), in *Semiconductors and Semimetals*, (Willardson and Beer, eds.) Academic Press, New York, p. 1.

Smith W.L., Powell R.A., and Woodhouse J.D. (1985), Proc. Soc. Photo-Opt. Instrum. Eng. **530**, 188.

Smith W.L., Rosencwaig A., and Willenborg D.L. (1985), Appl. Phys. Lett. **47**, 584.

Smith W.L., Taylor M.W., and Schuur J. (1985), Proc. Soc. Photo-Opt. Instrum. Eng. **530**, 201.

Smith W.L., Hahn S., and Arst M. (1986), in *Proc. Fifth Intl. Symp. of Silicon Mat. Sci. and Tech. Semiconductor Silicon 1986* (Huff, Abe and Kolbesen, eds.) Electrochem. Soc. Pennington, NJ, p. 206.

Smith W.L. and Rosencwaig A. (1986), Bull. Amer. Phys. Soc. **31**, 273. Also, Application Note 200.02, Therma-Wave, Inc. (March 1986).

Smith W.L., Rosencwaig A., Willenborg D.L., Opsal J. and Taylor M.W. (1986), Solid State Tech. **29(1)**, 85.

Staebler D.L. and Wronski C.R. (1977), Appl. Phys. Lett. **31**, 292.

Staebler D.L. and Wronksi C.R. (1980), J. Appl. Phys. **51**, 3262.

Takasu S. and Matsushita Y. (1975), in *Proc. 6th Conf. on Solid State Devices, Tokyo, 1974*; Jap. J. Appl. Phys. Supplement **44**, 259.

Turner N.L., Current M., Smith T.C. and Crane D. (1985), Solid State Tech. **28(3)**, 163.

Wang E.Y., Albers W.A., and Bleil C.E. (1967), in *II-VI Semi-Conducting Compounds*, (Thomas, ed) Benjamin, New York, p. 136.

Wright M. Jenkins (1976), in *Proc. of ECS Meeting, Washington DC*, Electrochem. Soc. Pennignton, NJ, p. 63.

Yasuami S. (1980), J. Electrochem. Soc. **127**, 1404.

Yasuami S., Harada J. and Wakamatsu K. (1979), J. Appl. Phys. **50**, 6860.

Yasuami S. and Harada J. (1981), J. Appl. Phys. **52**, 3989.

MEASUREMENTS OF THERMAL PROPERTIES OF SEMICONDUCTORS USING PHOTOTHERMAL WAVE DETECTION

Helion Vargas

Instituto de Fisica
Universidade Estadual de Campinas
13100 Campinas, SP, BRAZIL

and

Luiz C.M. Miranda

Laboratorio Associado de Sensores e Materiais
Instituto de Pesquisas Espaciais
12200 S.J. Campos, SP, BRAZIL

I. Introduction

As we have seen elsewhere in this book Photoacoustic (PA) spectroscopy looks directly at the heat generated in a sample due to nonradiative de-excitation processes following the absorption of light. In the conventional PA experimental arrangement a sample enclosed in an air-tight cell is exposed to a chopped light beam. As a result of the periodic heating of the sample, the pressure in the chamber oscillates at the chopping frequency and can be detected by a sensitive microphone coupled to the cell. The resulting PA signal depends not only on the amount of heat generated in the sample (i.e., on the optical absorption coefficient and the sample's light into heat conversion efficiency) but also on how this heat diffuses through the sample. The quantity that measures the rate of heat diffusion in the sample is called the thermal diffusivity α_s. Apart from its own intrinsic importance, its determination gives the value of the thermal conductivity K_s, if the density ρ and the specific heat at constant pressure C_s are known, since

$$\alpha_s = \frac{K_s}{\rho_s C_s} \tag{1}$$

The importance of α as a physical quantity to be monitored is due to the fact that, like the optical absorption coefficient, it is unique for each material. This can be appreciated by the tabulated values of α, presented by Touloukian *et al.* (1973), for a wide range of materials, such as metals, minerals, foodstuffs, biological specimens and polymers.

The experimental methods for determining the thermal diffusivity are of two types, depending on whether the measured heat flow is transient (Parker *et al.*, 1961) or periodic (Danielson and Sidles, 1969). The latter method was introduced more than a century ago by Angström (1963), and consists basically of a rod that is periodically heated at one point and whose temperature oscillation is measured at another point. The phase lag between the thermal oscillation at any two points gives a precise determination of α, as was demonstrated by Hirschman *et al.* (1961). They measured the phase of the non-illuminated rear-end surface oscillation in relation to the phase of the square-wave radiation on the front-end surface of a thin disc. The PA effect is related to the second method and has been proved to be a suitable and precise technique for measuring the thermal diffusivity, because it is a simple, direct and highly sensitive technique for probing the sample surface temperature.

The importance of simple thermal diffusivity measurements can be seen in their application to semiconductors. Power dissipation in microelectronic and optoelectronic devices is an important mechanism in limiting device performance. For example, in semiconductor diode lasers, heat generated in the active layer (e.g. GaAs) must pass through a substantial thickness of a confining layer of higher band gap and lower refractive index (e.g., $Ga_{1-x}Al_xAs$) to reach the heat sink. The room-temperature value of the thermal diffusivity of the materials used in these devices is therefore an important parameter in their optimization (Afromowitz, 1973).

II. Diffusivity Measurements by the Photoacoustic Effect

Various methods have been developed to measure the thermal diffusivity by means of the PA effect. Adams and Kirkbright (1977) have used the PA method to obtain α values for copper and glass by plotting the phase angle ϕ of the PA signal as a function of the square root of angular chopping frequency ω. Their assumption that $\phi = la_s$, where $a_s = (\omega/2\alpha)^{1/2}$ and l is the sample's thickness, is an approximation for the case of a thermally thick sample ($la_s \gg 1$). Similar studies using the PA signal phase data have been reported in the literature. For example, the thermal properties of polymer foils have been measured using this phase method by Korpiun *et al.* (1983) and Lachaine and Poulet (1984), whereas Kordecki *et al.* (1985) and Swimm (1983) have applied it to the thermal investigation of metglass ribbons and multilayer thin-film coatings, respectively.

An equivalent procedure (Charpentier *et al.*, 1982) is to measure, instead of the phase, the attenuation of the magnitude of the thermal oscillation as a function of $\sqrt{\omega}$, for the case of rear-surface illumination. The angular frequency ω, for which the slope of the attenuation curve diverges from the value -3, is typical of thermally thick samples and indicates the $la_s = 1$ point at which extraction of the thermal diffusivity may be made.

Another method alternative to the Angström technique is to vary the distance x_o between the point where the heat is generated and one point at which the thermal oscillation is measured. Cesar *et al.* (1983) adapted this method to photoacoustics, by varying the position of a laterally incident laser beam on a transparent sample. The plot of both the photoacoustic phase and magnitude versus the distance x_o provided the thermal diffusivity by a data fit to expressions with, respectively, a linear and exponential dependence on $(-a_s x_o)$.

For samples not necessarily thermally thick, but having a large optical absorption coefficient β (which means that the conditions $\beta l \gg 1$ and $\beta \gg a_s$ are valid), an attractive and simple method has been introduced by Yasa and Amer (1979). This method consists of measuring the attenuation of the rear-surface illumination signal magnitude (S_R) relative to the front-illumination signal magnitude (S_F). Using the appropriate formula they calculated the thermal diffusivity from the slope of the value of the curve S_F/S_R for several modulation frequencies. In all these methods, the thermal diffusivity is obtained by recording the PA signal (amplitude and phase) as a function of the modulation frequency. Pessoa *et al.* (1986) have recently demonstrated the usefulness of a single frequency modulation method for measuring the thermal diffusivity of semiconducting samples. The method is an extension of the Yasa and Amer (1979) work and consists of measuring the relative phase-lag $\Delta\phi = \phi_F - \phi_R$, at a single modulation frequency, between the rear-surface (R) and front surface (F) illuminations respectively, as follows:

Using the thermal diffusion model of Rosencwaig and Gersho (1976) for the production of the PA signal, the ratio S_F/S_R of the signal amplitude and phase-lag $\Delta\phi$ for front- and rear-surface illuminations (c.f. Fig. 1) are given by (Pessoa *et al.*, 1986)

$$S_F/S_R = (I_F/I_R)\,|\cosh^2(la_s) - \sin^2(la_s)|^{\frac{1}{2}} \tag{2}$$

and

$$\tan(\Delta\phi) = \tanh(la_s)\tan(la_s) \tag{3}$$

where I_F (I_R) is the absorbed light intensity for the front (rear) illumination. In arriving at Eqs. (2) and (3) we have assumed that the sample is optically opaque to the incident light (i.e., all the incident light is absorbed at the surface) and that the heat flux into the surrounding air is negligible. These approximations correspond to the actual experimental situation and represent no limitation of the method. In principle, either Eq. (2) or Eq. (3) would give us the value of α_s from a single modulation frequency measurement. However, since Eq. (2) depends explicitly on the ratio I_F/I_R (i.e., one needs precise power monitoring and identical surface conditions on both sides of the sample), the value of the thermal diffusivity in the signal amplitude ratio measurement is obtained from the slope of the curve S_F/S_R as a function of the modulation frequency as suggested by Yasa and Amer (1979). In contrast, Eq. (3) exhibits no explicit dependence on the absorbed power and surface conditions so that a single modulation frequency measurement is sufficient to derive the thermal diffusivity.

To demonstrate the usefulness of this two-beam phase-lag method, Pessoa *et al.* (1986) have performed the thermal diffusivity measurement of several semiconducting samples. Their experimental arrangement is schematically

Fig. 1 Schematic arrangement for the two-beam photoacoustic measurements of the thermal diffusivity.

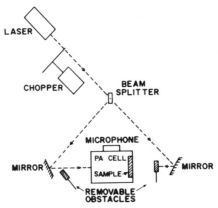

shown in Fig. 1. A chopped Argon ion laser beam is divided by a beamsplitter, and each beam is directed to opposite sides of the photoacoustic cell. The plane-shaped sample is fixed with silicon gel to a brass ring that has a 4-mm-diam hole through which the rear-side laser beam is incident. This corresponds to an air backing. Some of the heat generated in the sample, resulting from the light absorption, is lost to the brass ring, but this is of little effect considering the short thermal-diffusion length of the sample. In order to check the reproducibility of their results, the phase measurements were carried out at several modulation frequencies in the interval between 25 to 160 Hz. The semiconducting samples used were: (1) Ge(100) n-type, 784 μm thick; (2) Si(111), p-type, 345 μm thick; (3) GaAs(100), n-type, 568 μm thick; (4) GaSb, 420 μm thick; (5) InP(100), n-type 442 μm thick; (6) InAs, 550 μm thick; (7) PbTe, 320 μm thick. For these semiconductors the large optical-absorption coefficient (i.e., $\beta > 10^4$ cm^{-1}) for the Argon laser light ensures the condition of optical opaqueness implicit in Eqs. (2) and (3).

Table I

Comparison of the thermal diffusivity values obtained from the two-beam phase-lag method with the values quoted in the literature.

Material	Measured (cm^2/s)	Standard Deviation	Literature Values of α	Reference
Ge	0.35	0.01	0.346	Touloukian *et al.* (1973)
Si	0.94	0.03	0.880	Touloukian *et al.* (1973)
GaAs	0.24	0.01	0.21-0.26	Touloukian *et al.* (1973)
GaSb	0.24	0.01	0.24	Stegmeier (1961)
InP	0.45	0.02	0.46	Stegmeier (1961)
InAs	0.22	0.01	0.19	Stegmeier (1961)
PbTe	0.012	0.02	0.012	McNeill (1962)

In Table I, the results for the thermal diffusivity are summarized. In Fig. 2 the dependence of the phase-lag $\Delta\phi$ as a function of $la_s = l(\pi f/\alpha_s)^{1/2}$ is shown as given by Eq. (3). Also shown in Fig. 2 are the experimental values for the Ge and Si samples, using the observed balues given in Table I. It follows from this plot that the phase-lag method is valid for all values of lag, i.e., for both thermally thick ($la_s \gg 1$) and thermally thin ($la_s \leq 1$) samples. In contrast, the Adams and Kirkbright (1977) method is valid only for the case of thermally thick samples ($la_s > 1.4$), when the slope of $\Delta\phi$ with respect to $a_s l$ is one (i.e., dashed line of Fig. 2). The good agreement between the observed values of the thermal diffusivity and the literature values shows that the proposed two-beam photoacoustic phase measurement is a simple and accurate method for obtaining the thermal diffusivity. Its advantages over the conventional photoacoustic techniques is that the thermal diffusivity is obtained from a single-chopping frequency measurement, it is not limited to thermally thick samples, and does not depend on the beam power calibrations and the sample surface conditions.

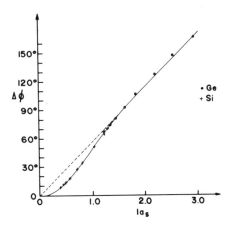

Fig. 2 Plot of $\Delta\phi$ versus la_s, Eq. (3), showing the experimental values obtained for Ge and Si using the observed values of α.

III. Other Photothermal Techniques

In addition to the gas-microphone PA techniques mentioned above, several other photothermal techniques have been developed for the thermal characterization of semiconductors. These include photothermal beam deflection (Kuo *et al.*, 1985; Yacoubi *et al.*, 1985) and photopyroelectric detection (Mandelis, 1984; Ghizoni and Miranda, 1985). This latter technique consists of using a thin pyroelectric (e.g., polyvinylidene difluoride, PVF_2) film in intimate contact with a solid sample on which a monochromatic light beam whose intensity is sinusoidally modulated at an angular frequency ω is incident. Following the absorption of the incident light, the nonradiative de-excitation processes within the solid cause the sample temperature to fluctuate and, through heat diffusion to the surrounding pyroelectric film, the sample-pyroelectric film-interface temperature fluctuates. As a result of this temperature fluctuation a pyroelectric voltage is produced in the film which is proportional to the rate of change of the temperature rise in the pyroelectric thin film (Mandelis, 1984). The photopyroelectric approach for measuring the thermal diffusivity is schematically shown in Fig. 3. The light from a chopped laser beam is focused by means of a cylindrical lens along one face of the sample normal to the y axis, cf. Fig. 3, thereby generating at a lateral distance x_o from the pyroelectric film-sample border ($x=0$) a localized periodic heat source. As a result of the localized heating at $x = x_o$ a thermal wave is set in the back face (at the $y=0$ plane) of the sample (Mandelis and Zver, 1985; John *et al.*, 1986). This thermal wave diffuses along the x direction and eventually reaches the PVF_2 detector at $x=0$, attenuated exponentially as $\exp(-a_s x_o)$, where $a_s = (\pi f/\alpha_s)^{1/2}$ is the thermal diffusion coefficient. Thus, by measuring the pyroelectric signal, at a fixed modulation frequency, as a function of the heating beam offset x_o, the thermal diffusivity α_s is readily obtained from the coefficient of x_o in the

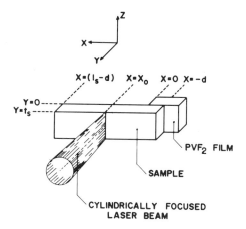

Fig. 3 Schematic configuration for the photopyroelectric lateral heating method for measuring the thermal diffusivity.

exponential. To ensure that along the $x = x_o$ plane the sample is being uniformly heated one may consider two possibilities. One of them is to use a heating laser wavelength such that the sample is optically transparent. In this case, only a fraction βt_s (β is the bulk optical absorption coefficient at the incident laser wavelength and t_s is the sample thickness) of the incident light intensity is converted into heat, and the heat deposition in the $x = x_o$ plane may be assumed uniform. If, however, the sample is optically opaque at the heating laser wavelength, which is the more likely situation for the cases of metallic and semiconducting samples and the common heating lasers available, the uniform heating may be achieved by working with thermally thin samples such that the thermal diffusion length $(\alpha_s/\pi f)^{1/2}$ is much longer than the sample thickness t_s. For a thermally thin sample it can be shown that the periodic temperature fluctuation T_s in the $x = x_o$ plane is given by

$$T_s(x_o,t) = \frac{\beta' I_o}{t_s k_s \sigma_s^2} e^{j\omega t},$$ (4)

where β' is the surface absorption coefficient of the sample, I_o is the heating laser intensity, k_s is the sample thermal conductivity, and $\sigma_s = (1+j)a_s$. With use of Eq. (4) as a boundary condition in the heat-flow equation along the x direction and neglecting the heat diffusion into the surrounding air one gets

$$T_s(x,t) = \frac{\beta' I_o}{t_s k_s \sigma_s^2} \left[\frac{e^{\sigma_s(d+x)} + e^{-\sigma_s(d+x)}}{e^{\sigma_s(d+x_o)} + e^{-\sigma_s(d+x_o)}} \right] e^{j\omega t},$$ (5)

where d is the length of the thermal-contact region between the PVF$_2$ film and the sample, as shown in Fig. 3. The temperature rise ΔT in the PVF$_2$ film is obtained by taking the average temperature fluctuation in the detector-sample region ($-d < x < 0$), namely,

$$\Delta T = \frac{\beta' I_o}{t_s k_s \sigma_s^3 d} \left[\frac{e^{\sigma_s d} - e^{-\sigma_s d}}{e^{\sigma_s (d+x_o)}} \right] e^{-\sigma_s(d+x_o)} e^{j\omega t} \, ,$$

which for $d\sigma_s \gg 1$ (large detector-sample thermal-contact region) reduces to

$$\Delta T \approx \frac{\beta' I_o}{t_s k_s \sigma_s^2 d} e^{-x_o \sigma_s} e^{j\omega t} \, , \tag{6}$$

Eq. (6) together with Eq. (4) indicates that the pyroelectric signal should decay exponentially as a function of the beam offset x_o, with the coefficient of x_o in the exponential being the thermal diffusion coefficient a_s.

The above method for determining α_s was tested by measuring the photopyroelectric signal of a silicon wafer (6 mm long, 4 mm wide, and 185 μm thick) as a function of x_o. The heating laser beam was provided by a 4 mW He-Ne laser. The laser beam modulated by a variable frequency chopper was focused on the 4-mm-wide face of the silicon wafer by means of a cylindrical lens which provided a focused strip 50 μm wide. The sample was in thermal contact with a 28 μm-thick Pennwalt Kynar PVF$_2$ film (5.5 mm long and 4 mm wide) supported by a printed circuit plate. The length d of the sample-PVF$_2$ thermal-contact region was 3.2 mm. To optimize the sample-PVF$_2$ thermal contact, Ghizoni and Miranda (1985) used a few microns thin layer of thermally conducting grease. The output voltage from the PVF$_2$ detector was connected to a preamplifier and later fed into a lock-in amplifier. The whole setup was mounted on a micrometer positioner so that it could be moved back and forth along the x direction. In this way the heated position x_o could be accurately varied. Fig. 4 shows the amplitude of the pyroelectric signal as a function of x_o at a modulation frequency of 29 Hz. Here we note that, because of the finite waist of the focused laser beam the illumination position is defined relative to the beam waist. The thermal diffusivity value obtained from the data fitting was $\alpha_s = 0.83$ cm^2/s which compares well with the literature value of 0.88 cm^2/s.

IV. Conclusions

In this chapter we have reviewed the photothermal wave detection techniques applied to the thermal characterization of solid samples. Special emphasis was given to the photoacoustic and photopyroelectric measurements of the thermal diffusivity of semiconductors. The simplicity and accuracy of the photothermal detection may render these techniques standard methods for the thermal characterization of semiconductors and semiconductor devices.

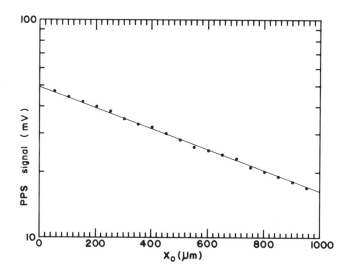

Fig. 4 Semilogarithmic plot of the photopyroelectric signal (PPS) of a p-type silicon wafer, 185 μm thick, as a function of the illumination position x_o, with use of a He-Ne laser modulated at 29 Hz. The dots refer to the experimental data, whereas the solid curve is the least-square fitting of the data to an expression of the form $S = S_o \exp(-a_s x_o)$.

V. References

Adams M.J. and Kirkbright G.F. (1977) Analyst. **102**, 281.

Afromowitz M.A. (1973) J. Appl. Phys. **44**, 1292.

Angström, M.A.J. (1863) Philos. Mag. **25**, 181.

Cesar C.L., Vargas H., Filho J.M. and Miranda L.C.M. (1983) Appl. Phys. Lett. **43**, 555.

Charpentier P., Lepoutre F., and Bertrand L. (1982) J. Appl. Phys. **53**, 608.

Danielson G.C. and Sidles P.H. (1969) In *Thermal Conductivity*, (R.T.Tye, ed.,), p. 151, Academic Press, New York.

Ghizoni C.C. and Miranda L.C.M. (1985) Phys. Rev. **B32**, 8392.

Hirschman A., Dennis J., Derken, W. and Monahan T. (1961) In *International Developments in Heat Transfer*, Part IV, p. 863. ASME, New York.

John P.K., Miranda L.C.M. and Rastogi A.C. (1986), Phys. Rev. **B34**, 4342.

Kordecki R., Bein B. and Pelzi J. (1985). *Proc. 4th International Meeting on Photoacoustic, Thermal and Related Sciences*, Montreal, paper WA 3.1.

Korpiun P., Merte B., Fritsch G., Tilgner R. and Lüscher E. (1983) Colloid & Polymer Sci., **261**, 312.

Kuo P.K., Lin M.J., Reyes C.B., Favro L.D., Thomas R.L., Zhang S.Y., Inglehart L.J., Fournier D., Boccara A.C., and Yacoubi N. (1985). *Proc. 4th International Meeting on Photoacoustic, Thermal and Related Sciences*, Montreal, paper WB 3.1.

Lachaine A. and Poulet P. (1984) Appl. Phys. A45, 953.

Mandelis A. (1984). Chem. Phys. Lett. 108, 388.

Mandelis A. and Zver M.M. (1985). J. Appl. Phys. 57, 4421.

McNeill D.J. (1962) J. Appl. Phys. 33, 597.

Parker W.J., Jenkins R.J., Butler C.P. and Abbot G.L. (1961) J. Appl. Phys. 32, 1679.

Pessoa O. Jr., Cesar C.L., Patel N.A., Vargas H., Guizoni C.C. and Miranda L.C.M. (1986) J. Appl. Phys. 59, 1316.

Rosencwaig A. and Gersho A. (1976) J. Appl. Phys. 47, 64.

Stegmeier E.F. (1961) *Thermal Conductivity*, in *International Developments in Heat Transfer*, ASME, New York.

Swimm R.T. (1983) Appl. Phys. Lett. 42, 955.

Touloukian Y.S., Powell R.W., Ho C.Y. and Nicolasu M.C. (1973) *Thermal Diffusivity*, IFI/Plenum, New York.

Yacoubi N., Girault B., Alibert C., Erman M., Jarry P., Theeten J.B. (1985) *Proc. 4th. International Meeting on Photoacoustic, Thermal and Related Sciences*, Montreal, paper WD 9.1.

Yasa Z. and Amer N. (1979) *Topical Meeting on Photoacoustic Spectroscopy*, Ames, Iowa, paper WA 5-1.

—————————————— III ——————————————

Novel Photothermal Wave Techniques,

Instrumentation and Semiconductor

Applications

7

PHOTOPYROELECTRIC SPECTROSCOPY OF SEMICONDUCTORS

Hans Coufal

IBM Almaden Research Center
San Jose, California 95120-6099
U.S.A.

and

Andreas Mandelis

Photoacoustic and Photothermal Sciences Laboratory
Department of Mechanical Engineering
University of Toronto
Toronto, M5S 1A4, CANADA

I. Introduction

Whenever a sample is excited by a source of energy, part of the incident energy is absorbed and causes, after radiationless de-excitation, local heating of the sample. Laser processing of semiconductors and optical recording of data rely on laser induced temperature transients. These industrial applications stimulated the interest in techniques that allow time resolved measurements of the sample temperature during laser exposure. Such temperatures have been determined, for example, from Raman scattering experiments, from the velocity of electrons or atoms emitted from hot surfaces, or derived from the temperature dependence of optical or lattice constants. Recently electron-diffraction techniques were employed in studies of the crystalline-liquid phase transition at the surface and thin film thermocouples were utilized to determine the surface temperature directly. Among all these methods only the thermocouple with its excellent signal/noise ratio allows the measurement of the transient sample temperature during one single laser pulse. All other methods rely on repetitive probing and assume that excitation at a number of locations by a series of laser pulses results in an identical signal. This assumption is not necessarily valid. Optical images of samples frequently show substantial contrast indicating that light absorption might vary substantially across the sample. The same holds true for other sample parameters. In similar fashion fluctuations in laser power or pulse shape affect repetitive probing. Of the above detection schemes thermocouples are the only one with single shot capability. The response of thermocouples is, however, slow compared to that of the other methods. To study processes such as laser annealing, marking or optical recording a method for surface temperature measurement is desirable with a time resolution sufficient to resolve the shape of a laser pulse combined with a sensitivity enabling single shot experiments. A large variety of detection schemes has evolved in the area of photo- or optoacoustics (Rosencwaig, 1980; Tam, 1983). A characteristic of these techniques is the use of time dependent excitation, *i.e*, the source of energy or the absorption of the sample is amplitude or frequency modulated. Therefore a time dependent temperature profile develops in the sample. Due to thermal expansion, sound waves are generated. If the sample is in contact with another medium (gases or liquids are commonly used) sound is coupled into that medium. In that case heat diffusion extends also into a thin layer of the said medium providing an additional mechanism for sound generation. Modulated or pulsed light sources, ion or electron beams are commonly utilized as excitation sources. The wavelength or energy of the incident radiation can be varied to perform spectroscopy. The reflectivity of the sample surface and the absorptivity of the material can be imaged and the thermal and acoustic properties of the object can be mapped by scanning a focussed excitation across the sample under study. The response of the sample to a well defined transient source of heat and sound can be analyzed to derive thermal and acoustic material parameters of the sample. Quite a large number of techniques have been developed to detect these transient thermal processes or the associated acoustic waves. Terms like photo- or optoacoustic are used when light serves as stimulus and acoustic detectors such as microphones or piezoelectric transducers are employed. Other detection schemes, frequently referred to as photothermal techniques, use the deflection of a probe laser by thermal or acoustic gradients or displacements. The increase in black body radiation associated with the increase in sample temperature can be detected by a suitable infrared detector. In all of the above detection schemes, only a small fraction of the energy deposited in the sample is actually detected. Despite detectors that are approaching the theoretical limits, the ultimate sensitivity of the complete photothermal system consisting of excitation source, sample and detector is not necessarily reached. The time resolution that can be achieved in a photothermal or acoustic set up is limited by the transit time of the signal generation processes across excitation and probe volume. In an idealized laser beam deflection experiment with ultra fast excitation source and photo detector, the time resolution would still be limited by the diffusion of the thermal energy across the probe laser beam. With a beam waist of the order of 1 μm and for a high refractive index medium such as CCl_4 this transit time is on the order of μs.

An alternative detection scheme for thermal wave phenomena is the use of pyroelectric transducers. A suitable pyroelectric detector is a true thermal receiver limited in sensitivity only by temperature noise (Cooper, 1962). If a thin film sample is deposited directly onto a pyroelectric substrate, essentially all of the absorbed energy contributes towards the signal thus enabling a superior signal/noise ratio. The time resolution with an idealized pyroelectric detector is limited only by the thermal diffusion from the sample into the detector. With a 10 nm thick metal film directly deposited onto the pyroelectric, the thermal transit time would be of the order of 10 ps. For a thin film sample, pyroelectric detection has therefore the unique potential of combining the highest possible sensitivity with a time resolution better than any other photothermal detection scheme. A pyroelectric sensor has to be attached directly to the sample under study, in a fashion similar to a piezoelectric sensor. Due to the diffusive character of thermal waves, the detector has to be, however, in close proximity to the excited area. Therefore, delay lines, which can be conveniently used in the case of sound waves to suppress electromagnetic interference propagating at the speed of light, are not feasible. For this reason, pyroelectric detection, requires improved shielding and is restricted to samples which are thermally transparent or where the sensor can be close to the excitation area.

In Section II of this Chapter the signal generation in a simplified one-dimensional pyroelectric detector will be discussed and the use of such detectors as thermometers or calorimeters in the time (pulsed laser-induced excitation) and frequency (broadband light source or CW laser excitation) domain will be presented. In Section III emerging applications to semiconductors with respect to thermal and spectroscopic property measurements will be reviewed, as well as characteristic novel applications to non-semiconducting solids with a view to highlight the strong thermal-wave detection potential of the photopyroelectric effect.

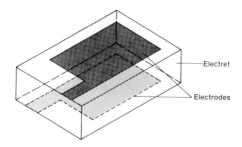

Fig. 1 Schematic of pyroelectric sensor element.

II. Signal Generation

A pyroelectric material can be characterized by its pyroelectric coefficient $p(\mathbf{r},t)$, where $\mathbf{r}$ and t describe the spatial and temporal dependence, respectively. A change $\Theta(\mathbf{r},t)$ of the temperature distribution in the pyroelectric, relative to a reference temperature distribution, $T_0(\mathbf{r},t_0)$,

$$\Theta(\mathbf{r},t) = T(\mathbf{r},t) - T_0(\mathbf{r},t_0), \tag{1}$$

causes a change in polarization. In a pyroelectric film of thickness l, metallized on both faces, as shown in Fig. 1, such a temperature change induces an electric charge $q(t)$

$$q(t) = \frac{1}{l}\int p(\mathbf{r},t)\,\Theta(\mathbf{r},t)\,d\mathbf{r} \tag{2}$$

on the electrodes (Lang, 1974). Assuming that the pyroelectric coefficient p is constant, the observed charge

$$q(t) = \frac{p}{l}\int \Theta(\mathbf{r},t)\,d\mathbf{r} \tag{3}$$

is then proportional to the average temperature change in the film. For a thin pyroelectric sensor, such as the one in Fig. 1, once attached to a large sample, the average temperature change in the pyroelectric element is approximately equal to the temperature change at the interface between transducer and sample. The transducer in that case produces a charge proportional to this temperature change and can therefore be used as a *thermometer*. The time resolution of such a thermometer is limited by the time required to reach thermal equilibrium within the pyroelectric. Once equilibrium is reached all of the volume of the pyroelectric element contributes to the signal. A signal maximum is then observed. Even at this signal maximum only a small fraction of the energy absorbed by the sample actually reaches the detector. Extreme time resolution or utmost sensitivity will not be obtained with such a pyroelectric thermometer. A pyroelectric calorimeter, however, combines all of the above features in one element.

1.   TIME-DOMAIN SIGNAL GENERATION

The change $\Theta(\mathbf{r},t)$ of the temperature distribution in Eq. (1) is associated with a change ΔU in the heat content of the pyroelectric film

$$\Delta U(t) = \int c\rho \Theta(\mathbf{r},t)\, d\mathbf{r}, \tag{4}$$

where c is specific heat and ρ is the density of the pyroelectric material. Assuming that the material parameters are temperature independent and that the pyroelectric material is homogeneous the induced charge, Eq. (3), is then proportional to the change of heat content of the pyroelectric film, *i.e.*,

$$q(t) = \frac{p}{c\rho l}\, \Delta U(t). \tag{5}$$

If heat loss from the pyroelectric element into the surroundings can be neglected, the element is in fact a pyroelectric *calorimeter*. This condition can be met for example by placing a thin sample on a large, self supporting piece of pyroelectric foil (Coufal and Peterson, 1985). The time resolution of the detector itself is only limited by the time required to convert an input of thermal energy into a change of polarization, a process that takes place in picoseconds. The time resolution of the complete sample-detector assembly is then only limited by the thermal propagation time τ_t from the sample surface to the interface with the detector. The temperature at a given distance l from a pulsed heat source reaches its maximum at a time τ_t after the excitation. For one dimensional heat flow the thermal transit time τ_t is determined by (Carslaw and Jaeger, 1959)

$$\tau_t = \frac{l^2}{2\alpha}, \tag{6}$$

where α is the thermal diffusivity of the sample. With a suitable thin film sample directly deposited onto one of the electrodes of the detector, a time resolution of several hundred picoseconds can be achieved. With such a sample, essentially all the energy absorbed by the sample will diffuse into the detector, resulting in the largest possible signal and the best time resolution at the same time. Fig 2 shows the temporal shape of a pulse from a typical excimer laser

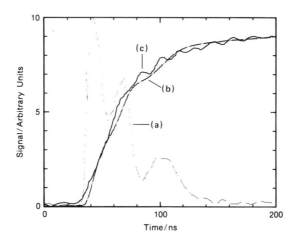

Fig. 2 Rise time of a β-PVDF calorimeter excited by a single 0.5 mJ pulse from an excimer laser. Shown are the laser pulse (a) as recorded with a fast vacuum photodiode, the energy (b) calculated by numerical integration of Eq. (a) and the response of the calorimeter (c) with superimposed ultrasonic ringing [From Coufal, Grygier, Horne and Fromm (1987)]

(Pulse energy 0.5 mJ), recorded with a fast photodiode, and the response of a pyroelectric calorimeter (coated with a 120 nm thick silver film) to that laser pulse. The time response of the calorimeter is almost identical to the numerically integrated light flux. The time resolution is evidently of the order of nanoseconds. Considering the fact that due to the high reflectivity of silver only a total of 0.005 mJ of laser energy is absorbed, the noise floor corresponds to nanojoules of absorbed energy. This unique combination of sensitivity with time resolution makes the calorimetric mode, Eq. (5), preferable to the thermometric one, Eq. (3), in particular when considering that with suitable calibration a calorimeter can always be used as thermometer!

One of the key assumptions in the one dimensional model of the signal generation process was that the pyroelectric material is homogeneous and that all relevant material parameters (p, c and ρ) are temperature independent. Ferroelectrics are, however, characterized by the fact that they depolarize substantially when heated above their Curie temperature, T_c. Even below T_c a slow decay of "permanent polarization" is observed. Besides this aging effect a reversible temperature dependence is observed. For large temperature transients this effect has obviously to be taken into account. For small temperature amplitudes at a base temperature well below T_c, the assumption of a temperature independent pyroelectric coefficient is normally justified. The same holds true for the temperature dependence of the other material parameters. This fact can be verified easily for the conditions of a particular experiment by measuring the power dependence of the observed pyroelectric signal.

The spatial dependence p(r) of the pyroelectric coefficient in the detector material can be determined, for example, by thermal (Sessler, West and Gerhard, 1982) or acoustic (De Reggi, Guttman, Mopsik, Davis and Broadhurst, 1978) pulse techniques. Frequently a thermal diffusion length l_d, which is defined as the square root of the mean square distance of thermal energy from the location of a transient point source of heat, at a time t after the heat pulse, is used to characterize thermal diffusion. For a one-dimensional thermal diffusion process the thermal diffusion length, l_d, is given by (Carslaw and Jaeger, 1959; Mandelis and Royce, 1979)

$$l_d = \sqrt{2\alpha t}. \tag{7}$$

Whenever critical dimensions of the detector are small compared to the thermal diffusion length of interest or the corresponding time resolution t, p(r) and other material parameters can be assumed to be constant. At the surface of a pyroelectric material a thin layer is depolarized. This depolarization can be neglected, for example, if the thermal diffusion length, that corresponds to the time resolution of the detection electronics, is much longer than the thickness of that layer. For a thin sheet of a pyroelectric plastic material such as β-Polyvinylidene difluoride (β-PVDF or PVF_2) with a thickness of less than 20 μm this condition will be normally met for modulation frequencies below one MHz or a corresponding time resolution of less than 500 ns.

Due to the polar nature of the pyroelectric effect, all pyroelectric materials are at the same time piezoelectric. Spurious signals due to radiation-induced sound waves can sometimes be observed as fast oscillations superimposed on the slowly varying thermal signal (Fig. 2). Both signal contributions can be separated by taking advantage of their quite different characteristics. Ultrasonic waves are essentially unattenuated, propagating waves with wavelengths of kilometer to millimeter for frequencies between one hertz and one megahertz. The transit time of a sound pulse through a distance is proportional to that distance. The velocity of sound determines, *via* the acoustic impedances, the coupling of sound waves from one medium into another. Thermal waves, however, are critically damped waves, attenuated by more than two orders of magnitude within one wavelength, which is typically between millimeter and micrometer for the frequencies considered here. The transit time of a thermal pulse across a certain distance is proportional to the square of that distance according to Eq. (6). The diffusion of thermal waves through an interface is governed by the thermal diffusivities of the materials involved. These fundamental differences between acoustic and thermal waves can be used to distinguish between thermal and acoustic signal contributions. Frequently, the coupling of acoustic waves from the sample under study into the pyroelectric detector can be avoided altogether, by a sample-coupler-detector arrangement that takes advantage of these differences. Where this is not feasible other techniques have been devised to subtract acoustic signal contributions (Coufal, 1986).

Depending on the electronic circuitry used for detection of the signal (Coufal, Grygier, Horne, and Fromm, 1987), the induced charge can be monitored directly, *via* the voltage generated by the element, or by short-circuiting the element and measuring the induced current. In the latter case the first derivative with respect to time of Eqs. (3) and (5) would apply. For non-ideal, *i.e.*, real detection electronics, the observed signal is, however, affected by the impedance of sensor and electronics. The pyroelectric element can be described as an ideal current source with a parallel capacitance, C. The signal from this element is detected by a voltage detection circuit with a load resistor R. The observed voltage is

$$V(t) = \frac{p}{c\rho lC}\left[\Delta U(t) - \frac{1}{\tau}\exp\left(-\frac{t}{\tau}\right)\int_0^t \exp\left(\frac{t'}{\tau}\right)\Delta U(t')dt'\right] \qquad (8)$$

with the electronic time constant τ, that is

$$\tau = RC, \qquad (9)$$

of the system consisting of sensor element and electronics. The observed pulse shape has two components: To the true calorimetric response, $\Delta U(t)$, a second term is superimposed that represents the effects of the time constant of the electronics. For a very large time constant τ, the second term in expression (8) becomes negligible and an indeal calorimetric response identical to Eq. (5) is observed.

Fig. 3 Observed (a) and calculated (b) time dependence of the calorimeter signal for a calorimeter with a capacitance of 600 pF recorded by a transient digitizer with an input impedance of 3 MΩ. For the numerical simulation these values and a reflectivity R=99%, a thermal diffusivity $\alpha = 5.2\times10^{-4}$ cm^2/s, and a pyroelectric coefficient $p = 2.5$nC/cm^2 were used [From Coufal *et al.* (1987)].

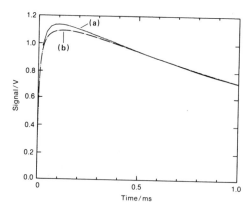

Fig. 3 shows the signal from a typical calorimeter with a capacitance of 600 pF recorded with a transient digitizer with an input impedance of 3MΩ. The time constant $\tau = 1.8$ ms, according to Eq. (8), agrees well with the signal of Fig. 3. The second curve in Fig. 3 is the result of a numerical simulation. The heat diffusion problem is solved using a finite difference technique to calculate the time dependent temperature distribution $\Theta(\mathbf{r},t)$. From these data the energy content of the calorimeter ΔU is obtained, Eq. (4), and the signal derived as a function of time, Eq. (8). Experimental and simulation data agree very well demonstrating that the assumptions during the discussion of the signal generation and detection process were justified.

2. FREQUENCY-DOMAIN SIGNAL GENERATION

For clarity all of the above arguments were presented only for the simplest of cases: a single, ultra-short laser pulse exciting an ideal calorimeter with negligible thermal diffusion time from the sample into the calorimeter, but infinite diffusion time from the calorimeter to the surroundings. The signal generation process was treated in the time domain assuming that all material parameters and electronic components are independent of time. The same analysis can be carried out in principle for the other cases, such as, periodic or random excitation, frequency dependent electronic parameters or material parameters that are a function of temperature or location in the detector material. With frequency dependent studies being extremely popular for applications, such as depth profiling or thermal diffusivity measurements, a detailed general one-dimensional photopyroelectric theory has been developed to account in a quantitative fashion for experimental observations of optical and thermal effects in the frequency-domain (Mandelis and Zver, 1985). Simplified specialized models have also been developed to account for thermal property measurements using the photopyroelectric effect (Coufal, 1984b; Ghizoni and Miranda, 1985; John, Miranda and Rastogi, 1986; see also Chapter 6). These specialized models are, in principle, special cases of the more general theory by Mandelis and Zver (1985), whose main features are discussed below.

Fig. 4 One-dimensional geometry of a frequency-domain photopyroelectric system. See text for details.

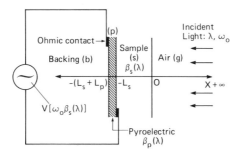

A one-dimensional geometry of a photopyroelectric system is shown in Fig. 4. A solid sample, of thickness L_s, is irradiated by monochromatic light of wavelength λ, whose intensity is modulated at angular frequency ω_o by a chopper. The sample has optical absorption coefficient $\beta_s(\lambda)$ and is in intimate contact with a pyroelectric transducer thin film of thickness l. The optical absorption coefficient and pyroelectric coefficient of the detector are $\beta_p(\lambda)$ and p, respectively. The sample/transducer system is supported by a transparent backing material of thickness large, compared to L_s or l. The photopyroelectric cell unlike the photoacoustic cell, does not require hermetic closure (open geometry), and the incident light is assumed to illuminate the sample surface uniformly. Light absorption by the sample/transducer system and nonradiative energy conversion to heat increases the temperature of the pyroelectric thin film. This temperature increase results in a potential difference between the upper and lower surfaces of the transducer due to the pyroelectric effect. This voltage $V[\omega_o, \beta_s(\lambda)]$ amounts to an electrical signal, which is measured in the external circuit through the connection of Ohmic leads to the pyroelectric as shown in Fig. 4.

The charge accumulated in the pyroelectric, due to a change $\Theta(t)$ in temperature is given by (Holeman, 1972)

$$q(t) = p\Theta(t) \tag{10}$$

For a thin pyroelectric film of thickness l, exposed to a sinusoidally varying temperature field, the average charge induced due to the pyroelectric effect is

$$= (p/l)\text{Re}\left[\left[\int_{Thickness, \, l} \Theta(x)dx\right] e^{i\omega_o t}\right] \tag{11}$$

The average pyroelectric voltage under no load conditions is given by

$$V = <q>/C \tag{12}$$

Where C is the capacitance per unit area of the thin film. For two parallel charged plates of thickness l and dielectric constant K, Eq. (12) becomes

$$V(\omega_o) = \left[\frac{P \, l\theta_p(\omega_o)}{K\varepsilon_o}\right] \exp(i\omega_o t) \tag{13}$$

where

$$\theta_p(\omega_o) \equiv \frac{1}{l}\int_{Thickness, \, l} \Theta_p(\omega_o, x)dx \tag{14}$$

and ε_o is the permittivity constant of vacuum (8.85418×10^{12} C/V-m). $\Theta(\omega_o, x)$ is the temperature field in the bulk of the pyroelectric, a result of heat conduction processes through the radiation absorbing solid. For the geometry of Fig. 4 the field $\Theta(\omega_o, x)$ can be found from the solution of coupled, one-dimensional thermal transport equations. Allowing for the finite optical absorption coefficients β_s and β_p, and assuming negligible optical reflection and radiative heat transfer coefficients on the sample surface and pyroelectric - sample interface (Belvin and Geist, 1973), the appropriate heat diffusion equations have the form

$$\frac{d^2}{dx^2}\,\theta_g(\omega_o, x) - \left[\frac{i\omega_o}{\alpha_g}\right]\theta_g(\omega_o, x) = 0 \; ; \; x \geq 0 \tag{15a}$$

$$\frac{d^2}{dx^2}\theta_s(\omega_o, x) - \left[\frac{i\omega_o}{\alpha_s}\right]\theta_s(\omega_o, x) = -(I_o\beta_s\eta_s/2k_s)\exp(\beta_s x) \; ;$$

$$-L_s \leq x \leq 0 \tag{15b}$$

$$\frac{d^2}{dx^2}\theta_p(\omega_o, x) - \left[\frac{i\omega_o}{\alpha_p}\right]\theta_p(\omega_o, x) = -(I_o\beta_p\eta_p e^{-\beta_s L_s}/2k_p)$$

$$\times \exp[\beta_p(x+L_s)] \; ; \; -(l+L_s) \leq x \leq -L_s \tag{15c}$$

and

$$\frac{d^2}{dx^2}\theta_b(\omega_o, x) - \left[\frac{i\omega_o}{\alpha_b}\right]\theta_b(\omega_o, x) = 0 \; ; \; x \leq -(l+L_s) \tag{15d}$$

In Eqs. (15a-d), a harmonic dependence of all temperatures on time was assumed:

$$\theta_j(\omega_o, x; t) = \theta_j(\omega_o, x)e^{i\omega_o t} \; ; \quad j = g, s, p, b \tag{16}$$

The following parameters were also defined: α_j, the thermal diffusivity of $j (= g, s, p, b)$; k_j, the thermal conductivity of j; η_s and η_p, the nonradiative conversion efficiencies for the absorbing solid and pyroelectric, respectively. I_o is the light source irradiance incident at the solid sample surface. Eqs. (15a-d) are coupled via boundary conditions of temperature and heat flux continuity at all interfaces. As such, they can be solved (Mandelis and Zver, 1985) and yield a few important special cases according to the opacity or transparency of the sample. Assuming the experimentally realistic case of an optically opaque pyroelectric detector (due to the metallic nickel coating of both Kynar^{TM} PVDF surfaces), a classification scheme similar to the photoacoustic case adopted by

Rosencwaig and Gersho (1976) can be used. Specifically, all cases evaluated below have been classified according to the relative magnitudes of three characteristic lengths in the solid and the pyroelectric, namely, i) thickness L_s or l; ii) optical absorption depth μ_{β_s} or μ_{β_p}, defined as

$$\mu_{\beta_j} \equiv \beta_j^{-1} \quad ; \tag{17}$$

and iii) thermal diffusion length μ_s or μ_p, defined from Eq. (18):

$$\mu_j \equiv (2\alpha_j/\omega_o)^{\frac{1}{2}} \tag{18}$$

Optically Opaque and Thermally Thick Pyroelectric

This case is likely to occur experimentally at high chopping frequencies ω_o and/or for thick transducers. In this limit $\mu_{\beta_p} \ll l$, $\mu_p < l$, and $\mu_{\beta_p} < \mu_p$.

Case 1: Optically Opaque Sample ($\mu_{\beta_s} \ll L_s$)

1.A. Thermally thin sample ($\mu_s \gg L_s$; $\mu_s \gg \mu_{\beta_s}$)

The photopyroelectric signal is

$$V(\omega_o) = A \left[\frac{\eta_s \alpha_p}{k_p(1+b_{gp})\omega_o} \right] \exp(-i\pi/2) \tag{19}$$

where

$$A \equiv pI_o/2k\varepsilon_o, \quad b_{gp} \equiv k_g\sqrt{\alpha_p}/k_p\sqrt{\alpha_g} \tag{20}$$

Eq. (19) shows that the photopyroelectric voltage is independent of β_s. This behavior can be termed *photopyroelectric saturation*. The signal depends on thermal properties of both the gas and the pyroelectric, it varies with chopping frequency as ω_o^{-1}, and its phase lags by 90° that of the light intensity modulating device (the chopper).

1.B. Thermally thick sample ($\mu_s < L_s$; $\mu_s > \mu_{\beta_s}$).

Assuming that g=air, the signal is

$$V(\omega_o) = A \left[\frac{\eta_s \alpha_p}{k_p(1+b_{sp})\omega_o} \right] \exp \left[-\left(\frac{\omega_o}{2\alpha_s} \right)^{\frac{1}{2}} L_s \right]$$

$$\times \exp \left\{ -i \left[\frac{\pi}{2} + \left(\frac{\omega_o}{2\alpha_s} \right)^{\frac{1}{2}} L_s \right] \right\} \tag{21}$$

In this limit, the photopyroelectric signal is saturated with respect to β_s as in the previous case. The voltage amplitude $|V(\omega_o)|$, however, is extremely small and decreases more rapidly than ω_o^{-1}, while the voltage phase lag increases with the square root of ω_o. The thermal properties of the contact gas have now been replaced with those of the solid. Eq. (21) indicates that in this limit the photopyroelectric signal can be used, in principle, to determine the thickness L_s of the sample, if its thermal diffusivity α_s is known, or *vice versa*. However, practical difficulties may arise due to the very small magnitude of the signal.

1.C. Thermally thick sample ($\mu_s \ll L_s$; $\mu_s < \mu_{\beta_s}$).

Here, the experimentally useful limit can be obtained

$$V(\omega_o, \beta_s) = A\beta_s \left[\frac{\eta_s b_{sg} \alpha_s \sqrt{\alpha_p}}{k_p(1+b_{sp})\omega_o^{3/2}} \right] \exp\left[-\left[\frac{\omega_o}{2\alpha_s} \right]^{1/2} L_s \right]$$

$$\times \exp\left\{ -i\left[\left[\frac{\omega_o}{2\alpha_s} \right]^{1/2} L_s - \frac{\pi}{4} \right] \right\} \qquad (22)$$

where

$$b_{sp} \equiv k_s \sqrt{\alpha_p} / k_p \sqrt{\alpha_s} \qquad (23)$$

This case is of spectroscopic interest, as the photopyroelectric voltage is out of the saturation and proportional to β_s. Therefore, the technique can be used in this limit as a spectroscopy, yielding signal information similar to absorption spectra (Mandelis, 1984; Coufal, 1984c).

Case 2: Optically Transparent Sample ($\mu_{\beta_s} > L_s$)

2.A. Thermally thin sample ($\mu_s \gg L_s$; $\mu_s > \mu_{\beta_s}$).

This approximation results in the following expression:

$$V(\omega_o, \beta_s) = A\left[\frac{\eta_p + (\eta_s - \eta_p)\beta_s L_s}{k_p(1+b_{gp})\omega_o} \right] \exp(-i\pi/2) \qquad (24)$$

The photopyroelectric voltage is proportional to $\beta_s L_s$, provided that $\eta_s = \eta_p$. In the experimentally common case where $\eta_s \sim \eta_p \sim 1$ and $b_{gp} < 1$, the signal carries neither optical nor thermal information about the sample, and is entirely generated by direct light absorption in the pyroelectric.

2.B. Thermally thin sample ($\mu_s > L_s$; $\mu_s < \mu_{\beta_s}$).

Further, if we assume $\mu_s^2 > \mu_{\beta_s} L_s$, the general expression for the pyroelectric voltage reduces to:

$$V(\omega_o, \beta_s) = A\alpha_p \left[\frac{\eta_p + (\eta_s - \eta_p)\beta_s L_s}{k_p \omega_o} \right] \exp(-i\pi/2) \qquad (25)$$

This case is similar to 1.A. in that, for $\eta_p \sim \eta_s \sim 1$, no optical or thermal information about the sample is obtained. If, however, we assume $\mu_s^2 \ll \mu_{\beta_s} L_s$, then

$$V(\omega_o, \beta_s) = A\alpha_p \left[\frac{(\eta_p + \eta_s \beta_s L_s)(1-\beta_s L_s)}{k_p \omega_o} \right] \exp(-i\pi/2) \qquad (26)$$

Here, if $\eta_p \sim \eta_s \sim 1$, the photopyroelectric voltage will be proportional to $1-(\beta_s L_s)^2$, with a ω_o^{-1} frequency dependence. This limit is non-linear in β_s and the spectral information from the system will be similar to distorted transmission spectra.

$\qquad$ *2.C. Thermally thick sample* $(\mu_s < L_s; \mu_s \ll \mu_{\beta_s})$.

The photopyroelectric voltage then is

$$V(\omega_o, \beta_s) = A(1-\beta_s L_s)\left[\frac{\eta_p}{k_p(1+b_{sp})\omega_o}\right]\; \exp(-i\pi/2) \qquad (27)$$

$\qquad$ This limit is of great experimental interest, because the photopyroelectric voltage amplitude is proportional to $1-\beta_s L_s$. Therefore, the technique can be used as a spectroscopy, with a signal information similar to a transmission spectrum. It must be emphasized that cases 1.C., and 2.C. are the only spectroscopically important cases, which give *direct* and *undistorted* information about the optical absorption coefficient of the sample material, as the result of thermal and optical transmission, respectively (Coufal, 1984b; Mandelis, 1984).

Optically Opaque and Thermally Thin Pyroelectric

$\qquad$ There are six special cases involving relationships between μ_s, μ_{β_s}, and L_s identical to those examined in cases 1.A through 2.C, above. For each new case with a thermally thin pyroelectric transducer, the simplified expression for $V(\omega_o, \beta_s)$ is structurally similar to the respective expression in the case of a thermally thick detector examined previously, with the following substitutions:

Thermally thick pyroelectric	$\rightarrow$	*Thermally thin pyroelectric*
(a) $1/\sigma_p$	$\rightarrow$	l,
(b) $k_p \sigma_p$	$\rightarrow$	$k_b \sigma_b$

$\qquad$ The full expression for the film-thickness-averaged photopyroelectric voltage has been evaluated numerically for the amplitude $|V(\omega_o)|$ and phase $\Phi(\omega_o)$ of the signal. The following numerical values were used for these calculations: $I_o = 1$ W/m^2, p $= 3 \times 10^{-5}$ C/m^2–$^\circ$K, $K = 12, \varepsilon_o = 8.854 \times 10^{-12}$ C/V–m, $l = 28$ μm, $\alpha_p = 5.4 \times 10^{-8}$ m^2/sec, $k_p = 0.13$ W/m–$^\circ$K, $k_g = 2.38 \times 10^{-2}$ W/m– $^\circ$K, $\alpha_g = 1.9 \times 10^{-5}$ m^2/sec (g=air), and L_s=50 μm. All other parameters were allowed to vary.

$\qquad$ The values for p,K,l,α_p and k_p were obtained from the Kynar^{TM} Piezo Film Technical Manual, Pennwalt Corp. King of Prussia, Penn. (1983).

$\qquad$ Fig. 5A shows the theoretical spectral response to a Gaussian absorption band centered at 500 nm, with the chopping frequency as a parameter. A spectral inversion of the absorption amplitude peak at higher frequencies is observed, due to the interplay in thermal energy contribution between the sample and the pyroelectric. No inversion occurs in the phase data, which correlates

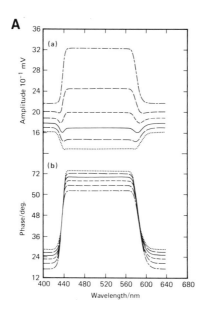

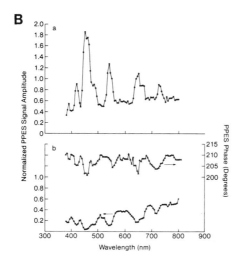

Fig. 5A Photopyroelectric spectroscopic response to a Gaussian absorption band with $0 \leq \beta_s \leq 10^5 cm^{-1}$ as a function of the light modulation frequency: A) Amplitude, and B) Phase; o f = 10 Hz, Δ f = 15 Hz, + f = 20 Hz, x f = 25 Hz, $\blacklozenge$ f = 30 Hz, $\uparrow$ f = 35 Hz [From Mandelis and Zver (1985)]; **B)** Photopyroelectric spectrum of Ho_2O_3 powder mixed with water. Chopping frequency 11 Hz (a); and 50 Hz (b). Notice spectral inversions in (b). [From Mandelis (1984)].

with the amplitude for f < 20 Hz, and anticorrelates with it for higher frequencies. These calculations show that the photopyroelectric voltage is governed by the interplay between optical absorption in the sample and thermal response of the pyroelectric transducer. Fig. 5B shows experimental results demonstrating the spectroscopic nature of the photopyroelectric detection in the thermally thin limit and verification of the spectral inversion features predicted by the Frequency-Domain theory (Mandelis, 1984; Coufal, 1984b).

Time- and frequency domain description of the signal generation process are basically equivalent. The fast rise time of the sensor translates into large bandwidth, good fidelity of the time domain signal into flat frequency response of the detector. Depending on the emphasis of an experiment, the time domain studies will be advantageous whenever a pulsed excitation source is employed. This is normally the case when high transient temperatures or time resolution are of interest. Otherwise excitation with amplitude modulation of a CW radiation source offers the advantage of simple instrumentation and low thermal gradients in the sample under study. The following Section highlights the salient features of the pyroelectric calorimeter with a number of typical applications.

III. Applications to Semiconductors and Other Condensed Matter

Pyroelectric detectors have been used for quite some time for the detection of light (Sil'vestrova, 1960; Perls, Diesel and Dobrov, 1958). Thermal waves generated *via* the absorption of modulated visible light were utilized to determine the pyroelectric characteristics of materials (Chynoweth, 1956). It thus became evident that a pyroelectric element could be used as a detector for all radiation-induced heating processes throughout the entire electromagnetic spectrum (Steiner and Yamashita, 1963; White and Wieder, 1963; White, 1964). The main application at that time was in wide band radiometers and detectors for pulsed lasers in the infrared. Considerable effort went into developing suitable absorbers, methods for the characterization of the detector properties and models describing the overall response of the detector (Peterson, Day, Gruzensky and Phelan, 1974). Based on the excellent understanding of pyroelectric detectors a sensitivity limited only by thermal noise (Cooper, 1962) was achieved and a time resolution of 500 ps reported (Roundy and Byer, 1972). PVDF based pyroelectric sensors were "recognized fairly early as attractive choices for applications where a high detectivity coupled with a broad spectral response and a large area are required" (Day, Hamilton, Gruzensky and Phelan, 1976). Despite the fact that this development is closely related to photoacoustics, it went largely unnoticed by the photoacoustics community. Only recently the first applications of pyroelectric detection for genuine thermal wave phenomena start to emerge.

1. THERMAL PROPERTIES

a. Time-Domain Measurements

The most straightforward application of a pyroelectric detector is of course, the measurement of the temperature of a thin film directly deposited onto the element. Due to the high time resolution, laser-induced transient heating effects can be observed directly in real time. The first application of the pulsed laser photopyroelectric effect to the study of thermal and thermodynamic phenomena (melting, boiling and recrystallization) in thin semiconducting Te films has been reported by Coufal and Lee (1984). These authors observed the first order phase transition during optical recording. Data in Figs. 6 and 7 show the characteristics of a first order phase transition. Due to the latent heat of the phase transition an increase in laser power does not result in a proportional temperature increase. In addition melting, boiling and crystallization of the film show up directly as intervals with constant temperature. Phase transitions of a variety of materials were also studied in the frequency domain (Mandelis, Care, Chan and Miranda, 1985). With first order phase transitions the direct time domain observation seems to be preferable. For second order phase transitions frequency domain experiments, however, have shown unique potential.

Tellurium being a semiconductor has, of course two different relaxation mechanisms, a fast vibrational relaxation with a ps lifetime and a subsequent electronic relaxation with a lifetime of the order of tens of nanoseconds. Fig. 7d then would be consistent with instantaneous melting during the laser pulse by vibrational relaxation and subsequent heating of the molten Te to the boiling point by electronic de-excitation. To a lesser degree these two relaxation mechanisms seem to be visible also at other fluences. To test this interpretation of the data an experiment at a wavelength of 2.2 μm was conducted and the data showed no wavelength dependence of the observed melting process. The time

Fig. 6 Pyroelectric signal during annealing of 30 nm thick Te-films with a XeC*l* excimer laser for three different laser fluences: (a) 0.7 mJ/cm^2, (b) 7 mJ/cm^2 and (c) 70 mJ/cm^2 [From Coufal and Lee (1984)].

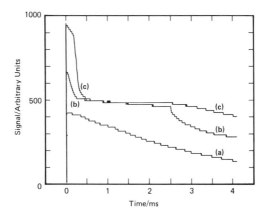

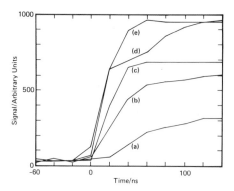

Fig. 7 Initial temperature during annealing of 30 nm thick Te-films with a XeCl excimer laser for 5 different fluences: (a) 0.7 mJ/cm^2, (b) 2 mJ/cm^2, (c) 7 mJ/cm^2, (d) 24 mJ/cm^2 and (e) 70 mJ/cm^2. The time resolved pulse shape of the laser is shown in Fig. 2. [From Coufal and Lee (1984)].

dependence was shown to be entirely due to a low intensity tail of the laser pulse. Despite its low intensity, it contributes substantially towards the total energy deposited in the sample due to its long duration (Coufal and Lee, 1987).

In a similar fashion to the pulsed experiments using Te films, the temperature of a metal Cu film on a pyroelectric transducer film was studied during pulsed laser induced desorption of Xe atoms from that surface (Hussla, Coufal, Träger and Chuang, 1986a) under UHV conditions. The momentum of the desorbed atoms was determined with a time of flight measurement (Fig. 8). A comparison of the maximum surface temperature, as determined with the pyroelectric transducer, and the translational temperature, as derived from Maxwell-Boltzmann distributions fitted to the observed time of flight data, show excellent agreement between both methods and shed light into desorption kinetics of adsorbates (Hussla *et al.*, 1986b).

With the excellent time resolution of pyroelectric detectors the feasibility of pulsed measurements of the thermal diffusivity became possible. Starting from pyroelectric materials (Negran, 1981) the method was soon applied to non-pyroelectric materials on a pyroelectric detector (Yeack, Melcher and Jha, 1982) and was recently refined to allow measurements of the thermal diffusivity of polymer films with a thickness of less than 1 μm (Coufal and Hefferle, 1986). In these experiments the propagation of a thermal pulse across the sample is monitored (Fig. 9). Due to the sensitivity of pyroelectric transducers thermal pulses with low amplitudes can be detected readily. A minimum of sample heating is therefore required and thermal gradients in the sample are small. This is of particular concern with poor thermal conductors or temperature sensitive samples. With pulsed excitation the thermal diffusivity of a sample can be determined within several milliseconds. Time resolved studies of the thermal diffusivity of thin films are thus possible. Where time resolution is of no importance and for

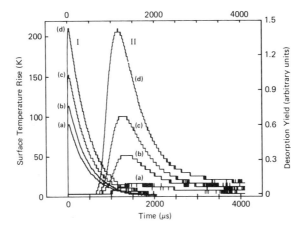

Fig. 8 Time resolved signals of (I) surface temperature rise ΔT (left axis) as determined with a pyroelectric calorimeter and (II) desorption yield of xenon measured with a mass spectrometer after a flight distance of 24 cm (right ordinate). Single laser induced heating and desorption events are shown for four laser fluences: (a) 1.05 J/cm², (b) 1.35 J/cm², (c) 1.75 J/cm² and (d) 2.35 J/cm². For details see Hussla, Coufal, Träger and Chuang (1986b).

thicker samples frequency domain methods, measuring the attenuation or the phase shift or the time delay of a thermal wave, have been demonstrated successfully (Coufal and Hefferle, 1985; Ghizoni and Miranda, 1985; John, Miranda and Rastogi, 1986).

Fig. 9 Pyroelectric signal for a calorimeter excited by a 10μJ pulse from an XeCl excimer laser: (a) bare calorimeter, (b) calorimeter coated with a 0.65 μm and (c) 0.80 μm thick Novolac film [From Coufal and Hefferle (1985)].

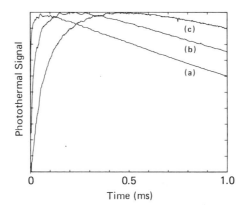

b. Frequency-Domain Measurements

Photopyroelectric applications to semiconductors include the measurement of thermal properties as well as spectroscopic investigations. Details of frequency-domain measurements of thermal diffusivity are also discussed in Chapter 6 of this volume (Ghizoni and Miranda, 1985). In those measurements the thermal diffusivity of a Si wafer was determined using a modified theoretical model of the sample /PVF_2 system in the limit of a thermally thin and optically opaque sample (Section II.2, Case 1.A.). The value of 0.83 cm^2/s for α_{Si} found from those measurements is in very good agreement with the literature value of 0.88 cm^2/s (Touloukian, Powell, Ho, and Nicolaou, 1973). John *et al.* (1986) performed similar photopyroelectric measurements on an silicon wafer, a cover plate and a Cu plate. These authors found a modulation frequency dependence of the signal from optically opaque and thermally thick samples in agreement with predictions from the general theory presented in Section II.2 (Case 1.B.). Their results on the thermal diffusivity of Cu foils were also in very good agreement with earlier photopyroelectric measurements by Mandelis (1984). For spin coated resist films, however, a thermal diffusivity quite different from bulk data was found (Coufal and Hefferle, 1985).

A unique feature of thermal wave diffusion is the potential to obtain a nondestructive depth profile of the thermal properties of a sample. The attenuation and/or phase shift of thermal waves at different frequencies allows, at least in principle, the reconstruction of this profile. A wide dynamic range and a large bandwidth of the detector are required for that type of application. A feasibility study demonstrated that pyroelectric detection meets these criteria (Coufal, 1984a) and how this technique can be applied to compare two light absorbing layers in a sample in order to obtain absolute quantum yields (Coufal, 1984b). Thermal wave imaging using photopyroelectric detection, or photopyroelectric scanning microscopy, of semiconductors has been demonstrated by Faria, Ghizoni, and Miranda (1985). These workers used the photopyroelectric lateral heating method (Chapter 6) for imaging a surface and a sub-surface defect consisting of a 250 μm-wide slot on a 600 μm-thick silicon wafer. They exploited the dependence of the pyroelectric signal from an optically opaque, thermally thin sample on the material thermal diffusivity to map out defect regions where this quantity was different from that of pure silicon. As a matter of historical interest, thermal wave imaging was the technique used to first introduce pyroelectric transducers to the field of photoacoustics (Luukala, 1980; Petts and Wickramasinghe, 1981; Baumann, Dacol and Melcher, 1983).

The conversion of solar energy into electrical energy in semiconductor devices is of considerable commercial and ecological interest. Calorimeters have been employed to complement electrical measurements of the relevant properties of photovoltaic cells. Faria, Ghizoni, Miranda and Vargas (1986) have used a pyroelectric sensor to measure the conversion efficiency of a p-n junction semiconductor Si solar cell and have compared their results with conventional gas-microphone photoacoustic data from the same device, Fig. 10. The photopyroelectric technique was found to be sensitive to the dissipation of load current through the internal resistance R_s of the solar cell. The pyroelectric signal was affected by the Joule heating reaching the back device contact, whereas the photoacoustic signal contained contributions from the front surface only. These mechanisms were found to be consistent with experimental differences between the two conversion efficiencies as shown in Fig. 10 and they establish photopyroelectric detection as an important complementary scheme to conventional photoacoustics in the study of electronic phenomena in semiconductor devices.

2. SPECTROSCOPY

With a sample directly deposited onto a pyroelectric element the observed signal is of course, directly proportional to the absorbed energy or I(1-R), I being the incident intensity and R the reflectivity of the sample (Burns, Dacol and Melcher, 1983). The change in the reflectivity of an oxidized Cu surface during reduction can thus be used to monitor the course of this chemical reaction (Mandelis, 1984). With rare earth oxide doped films the sensitivity of this technique and its susceptibility to spectral artifacts was demonstrated (Coufal, 1984b; Mandelis, 1984), and explained theoretically (Mandelis and Zver, 1985). With thick samples the influence of the thermal wave phase shift on spectral features was studied (Mandelis, 1984; Mandelis and Zver, 1985) and a technique to derive quantum yields established (Coufal, 1984b).

Fig. 10 Photopyroelectric γ_{pp} (solid) and photoacoustic γ_{PA} (dashed) conversion efficiencies as a function of the load resistance R_L, for a 500 μm-thick p-n junction Si solar cell, at 18 Hz under ~5.2 mW He-Ne laser illumination. At 350 Ω the PA-determined conversion efficiency is ~ 12%, whereas the PP-determined optimal conversion efficiency is ~ 5% at 400 Ω [From Faria, Ghizoni, Miranda and Vargas (1986)].

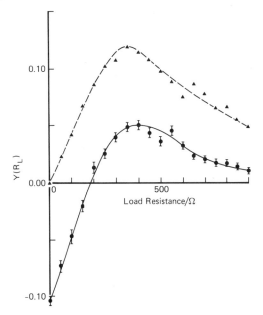

The work by Coufal (1984c) established the excellent potential of photo-pyroelectric spectroscopy to measure small absorption coefficients in solids. Taking advantage of this feature, recently Tanaka, Sindoh and Odajima (1987) utilized transparent pyroelectric PVDF films sandwiched between transparent CuI electrodes as photopyroelectric sensors of the absorption profile of evaporated As_2S_3 semiconducting thin films. Fig. 11 shows the measured photopyroelectric absorption spectrum of As_2S_3, as well as results obtained by the use of a thermocouple and conventional transmission measurements. The photopyroelectric method was capable of measuring the absorption coefficient of the thin film down to 10 cm^{-1}. The sensitivity was comparable to those obtained by photoacoustic and photothermal deflection spectroscopies, and was found to be much superior to that obtained by the thermocouple and the transmission methods.

For highly reflecting films R can be fairly small and very difficult to determine by conventional reflection spectroscopy. As is typical for photothermal

Fig. 11 Spectral dependence of the absorption coefficient in evaporated As_2S_3 films, measured by photopyroelectric spectroscopy (solid line). For comparison, results obtained by thermocouple photothermal spectroscopy (dashed line) and by conventional transmission measurements (dotted line) are also shown [From Tanaka, Sindoh and Odajima (1987)].

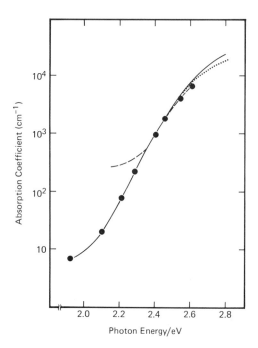

spectroscopy, an increase of the incident intensity might increase the total signal due to absorption to the point where detection by photothermal techniques can be readily achieved. The same is also true for studies of adsorbates on a reflecting metal surface (Coufal, Chuang and Träger, 1983) making this detection scheme ideal for spectroscopic studies of adsorbates. Using polarization modulation to compensate for the residual absorption of a silver surface, as little as a few thousandths of adsorbate are detectable (Coufal, Träger, Chuang and Tam, 1984).

Recently, metal surfaces with well defined periodicity were produced on calorimeters using a holographic technique (Grygier, Knoll and Coufal, 1986). Pyroelectric foils, coated with a photo resist, were exposed to an interferogram generated by a laser. After developing the resist the surface had a periodic profile. Overcoating the resist with a silver film resulted in a corrugated silver surface with the same periodicity. Such metal surface allows to couple light into the plasmon surface polariton and to study the plasmon dispersion and the influence of dielectric adsorbates and thin films conveniently.

A review of pyroelectric detection of photothermal wave processes would be incomplete without mentioning work in other spectral ranges such as the detection of magnetic resonance phenomena (Melcher and Arback, 1982) or other radiation induced thermal phenomena as for example the detection of ultraheavy nuclei (Simpson and Tuzzolino, 1984).

IV. Conclusions

Pyroelectric detectors have been demonstrated to be extremely powerful detectors for thermal wave phenomena in semiconductors and other condensed matter. With pulsed excitation thermal transients can be studied directly in the time domain with utmost sensitivity and time resolution. Using conventional single frequency or noise amplitude modulation, frequency domain experiments can be performed over an extremely wide dynamic range and bandwidth. Photopyroelectric spectroscopy appears, at this time, to be very promising in the measurement of low optical absorption coefficients in semiconducting thin films, such as amorphous chalcogenides, in spectral regions unaccessible by conventional transmission spectroscopy. The unique potential of pyroelectric detection for thermal transients and spectroscopy of surfaces and thin films has also been exemplified. Sensors have been shown to be compatible with UHV conditions as well as most gases and liquids. They are usable over a wide temperature range. The low cost of pyroelectric plastic materials, such as PVDF, combined with good electrical properties and the fact that the alignment of the detector is trivial make pyroelectric sensors an excellent choice for most photothermal applications with thin film samples.

V. References

Baumann, T., Dacol, F. and Melcher, R.L., (1983) *Proc. IEEE Ultrasonics Symposium*, pp. 832-836.

Belvin, W.R. and Geist, J. (1973). Appl. Opt. **13**, 1171.

Burns, G., Dacol, F. and Melcher, R.L., (1983) J. Appl. Phys., **54**, 4228.

Carslaw, H.S. and Jaeger, J.C., (1959) *Conduction of Heat in Solids*, Oxford University Press, Oxford.

Chynoweth, A.G., (1956) J. Appl. Phys., **27**, 78.

Cooper, J., (1962) Rev. Sci. Instr., **33**, 92.

Coufal, H., (1984a) J. Photoacoust. **1**, 413.

Coufal, H., (1984b) Appl. Phys. Lett., **45**, 516.

Coufal, H., (1984c) Appl. Phys. Lett., **44**, 59.

Coufal, H., (1986) IEEE Trans., UFFC, **UFFC-33**, 507.

Coufal, H., Chuang, T. and Träger, F. (1983) J. Phys. (Paris) C-**6**, 297.

Coufal, H. and Lee, Y., (1984) "*Time Resolved Calorimetry of 30 nm Te-Films During Laser Annealing*", in *Laser Processing and Diagnostics* (D. Bäuerle, ed.), Springer Series in Chemical Physics, Vol. **39**, pp. 25-28, Springer, Heidelberg.

Coufal, H. and Lee, Y., (1987) Appl. Phys. A (in the press).

Coufal, H., Träger, F., Chuang, T.J. and Tam, A.C., (1984) Surf. Sci., **145**, L504.

Coufal, H. and Hefferle, P., (1985) Appl. Phys. A **38**, 213.

Coufal, H. and Peterson, S.R., (1985) IBM Tech. Disclosure Bull., **27**, 5538.

Coufal, H. and Hefferle, P., (1986) Can. J. Phys., **64**, 1200.

Coufal, H., Grygier, R., Horne, D. and Fromm, J., (1987) J. Vac. Sci. Technol. (in the press).

Day, G.W., Hamilton, C.A., Gruzensky, P.M. and Phelan, R.J., (1976) Ferroelectrics, **10**, 99.

DeReggi, A.S., Guttman, C.M., Mopsik, F., Davis, G.T. and Broadhurst, M.G., (1978) Phys. Rev. Lett., **40**, 413.

Faria, I., Ghizoni, C. and Miranda, L., (1985) Appl. Phys. Lett., **47**, 1154.

Faria, I., Ghizoni, C., Miranda, L. and Vargas, H., (1986) J. Appl. Phys., **59**, 3294.

Ghizoni, C. and Miranda, L., (1985) Phys. Rev. B **32**, 8392.

Grygier, R., Knoll, W. and Coufal, H., (1986) Can. J. Phys., **64**, 1067.

Holeman, B.R., (1982). Infrared Phys. **12**, 125.

Hussla, I., Coufal, H., Träger, F. and Chuang, T., (1986a) Can. J. Phys., **64**, 1070.

Hussla, I., Coufal, H., Träger, F. and Chuang, T., (1986b) Ber. Bunsenges. Phys. Chem., **90**, 241.

John, P.K., Miranda, L.C.M. and Rastogi, A.C. (1986) Phys. Rev. **B 34**, 4342.

Lang, S.B. (1974) *Sourcebook of Pyroelectricity*, Gordon and Breach, London.

Luukala, M., (1980) *"Photoacoustic Microscopy at Low Modulation Frequencies"* in *Scanned Image Microscopy* (E. Ash, ed.), Academic Press, New York, pp. 273-289.

Mandelis, A., (1984) Chem. Phys. Lett., **108**, 388.

Mandelis, A. and Royce, B.S.H., (1979). J. Appl. Phys. **50**, 4330.

Mandelis, A. and Zver, M.M., (1985) J. Appl. Phys., **57**, 4421.

Mandelis, A., Care, F., Chan, K.K. and Miranda, L.C.M., (1985) Appl. Phys. A **38**, 117.

Melcher, R.L. and Arbach, G.V., (1982) Appl. Phys. Lett., **40**, 910.

Negran, T.J., (1981) Ferroelectrics, **34**, 31.

Perls, T.A., Diesel, T.J. and Dobrov, W.I., (1958) J. Appl. Phys. **29**, 1297.

Peterson, R.L., Day, G.W., Gruzensky, P.M. and Phelan, R.J., (1974) J. Appl. Phys., **45**, 3296.

Petts, C.R. and Wickramasinghe, H.K., (1981) *Proc. IEEE Ultrasonics Symposium*, pp. 832-836.

Rosencwaig, A., (1980) *Photoacoustics and Photoacoustic Spectroscopy*, in *Chemical Analysis*, (P.J. Elving and J.D. Winefordner, eds.), vol. **57**, J. Wiley, New York.

Rosencwaig, A. and Gersho, A. (1976). J. Appl. Phys. **47**, 64.

Roundy, C.B. and Byer, R.L., (1972) Appl. Phys. Lett., **21**, 512.

Sessler, G.M., West, J.E. and Gerhard, G., (1982) Phys. Rev. Lett., **48**, 563.

Sil'vestrova, I.M., (1960) Izv. Akad. Nauk UZ.SSR, Ser. Fiz. Mat. Nauk., **24**, 1213.

Simpson, J.A. and Tuzzolino, A.J., (1984) Phys. Rev. Lett., **52**, 601.

Steiner, W.H. and Yamashita, E., (1963) Proc. IEEE, **51**, 1144.

Tam, A.C., (1983) in *Ultrasensitive Spectroscopic Techniques*, (D. Kliger, ed.), Academic Press, New York, pp. 1-107.

Tanaka, K., Sindoh, K., and Odajima, A., Rept. Progr. Polym. Phys. Jpn. (in the press).

Touloukian, L.R., Powell, R.W., Ho, C.Y. and Nicolaou, M.C., (1973). *Thermal Diffusivity* Plenum, IFI, New York.

White, D.J., (1964) J. Appl. Phys., **35**, 3536.

White, D.J. and Wieder, H.H., (1963) J. Appl. Phys., **34**, 2487.

Yeack, C.E., Melcher, R.L. and Jha, S.S., (1982) J. Appl. Phys. **53**, 3947.

8

PULSED LASER PHOTOACOUSTIC AND PHOTOTHERMAL DETECTION

Andrew C. Tam

IBM Almaden Research Center
650 Harry Road
San Jose, California 95120-6099, USA

I. Introduction

Spectroscopic measurements utilizing photoacoustic (PA) or other kinds of photothermal (PT) detection have continued to gain importance in the last ten years. The reasons include the following. First, PA or PT measurements are simple and can be performed on "difficult" samples like highly opaque or light-scattering materials where conventional transmission measurements either fail or require extensive sample preparation procedures. Second, high sensitivites for detecting optical absorption of highly transparent samples have been demonstrated. Third, depth-profiling measurements are possible, providing a nondestructive evaluation of layered structures. Fourth, extension to remote sensing applications is possible for certain PT measurements.

This chapter reviews, in some detail, two specific instrumentation and measurement methods useful for semiconducting systems, namely, high-sensitivity pulsed PA techniques for spectroscopic and other applications (Section II) and pulsed photothermal radiometry for remote spectroscopic and other measurements (Section III). For pulsed PA spectroscopy, simplified signal generation theories are presented, and examples of applications in trace detection, energy transfer and materials testing are discussed. Pulsed photothermal radiometry can provide the absolute optical absorption coefficient as well as the thickness of a sample film in a "single-ended" measurement arrangement, and thus, extension to remote sensing is possible.

II. Photoacoustic Generation and Applications

1. OVERVIEW

Photoacoustic generation and detection have opened up new directions in spectroscopy, trace detection, excitation transfer and ultrasonic testing applications. Numerous reviews have been published (e.g., Rosencwaig, 1980; Patel and Tam, 1981; Tam, 1983; Tam, 1986). This section reviews some basic theoretical and experimental studies of the photoacoustic generation mechanisms, as well as provides examples of spectroscopic and other applications.

The optical generation of sound and its detection can be classified into two usual types: a continuous wave (CW) modulated technique whereby the excitation beam is modulated near 50% duty cycle, and the acoustic signal generated inside the sample or in an adjacent fluid is detected by a lock-in amplifier or a correlator; and a pulsed technique whereby the excitation beam is modulated at very low duty cycle (e.g., $\leq 10^{-5}$) but with high peak intensity of short duration, and the acoustic signal is detected usually by a boxcar averager or a transient digitizer. In the CW case the signal is typically analyzed in the frequency domain: Amplitude and phase of one or several Fourier components are measured and filters can be used to suppress noise. In the pulsed technique, however, the signal is acquired and analyzed in the time domain so that time gating techniques for noise suppression can be used. Pulsed excitation can be

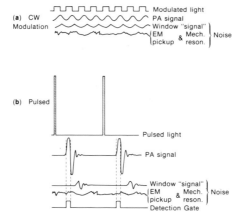

Fig. 1 Schematic comparison of (a) the CW modulated photoacoustic method and (b) the pulsed photoacoustic method.

regarded as frequency-multiplexed probing with many Fourier components simultaneously. The features of the two PA techniques are indicated in Fig. 1. If the same average optical power is used on a weakly absorbing sample, the pulsed technique typically gives higher detection sensitivity compared to the CW modulated technique. Other types of high-sensitivity detection also using pulsed laser excitation and gated detection have already been successful in other fields, for example, single-atom detection (Hurst, Payne, Karmer and Young, 1979).

This chapter is mainly concerned with the pulsed PA technique and those applications where the pulsed technique is advantageous compared to the CW modulated technique. However, we must point out that there are situations where the CW modulated technique may be more advantageous, e.g., cases of strongly absorbing materials or materials that can be damaged by intense pulsed excitation, and cases where well-defined modulation frequencies are needed to have a fixed thermal diffusion length (e.g., for depth-profiling applications).

2. THEORY

PA generation can be due to diversified processes. Some of the possible PA generation mechanisms are shown in Fig. 2, where the PA generation efficiency η (i.e., acoustic energy generated/light energy absorbed) generally increases downward for the mechanisms listed. For electrostriction or thermal expansion mechanisms, η is small, typically on the order of 10^{-12} to 10^{-8}, while for breakdown mechanisms, η can be as large as 30% (Teslenko, 1977). We will mainly consider the case of thermal expansion generation where η is small.

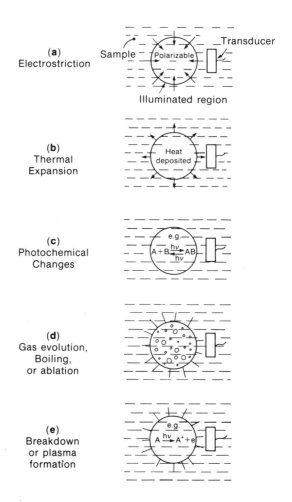

Fig. 2 Some common mechanisms for photoacoustic generation. Efficiency usually increases downwards.

We will discuss a simplified case of pulsed PA generation in a fluid with pulses short enough so that thermal diffusion can be neglected (this usually means light pulses much shorter than one millisecond). Consider a long cylindrical source with small radius R_s (see Fig. 3a), i.e., $R_s < c\tau_L$ where c is the sound velocity in the medium and τ_L is the laser pulse width. The initial expansion of the "source radius" R immediately after the laser pulse is denoted as ΔR. The source radius R is larger than R_s due to the acoustic propagation during τ_L, and so at the end of the excitation pulse, $R = c\tau_L$. The expansion Δ is given by:

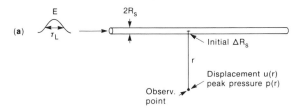

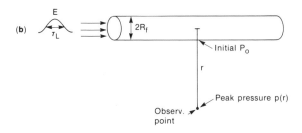

Fig. 3 Pulsed PA generation for weak absorption in an infinite medium; (a) laser beam radius R_s smaller than $c\tau_L$; (b) laser beam radius R_f larger than $c\tau_L$.

$$\pi(R + \Delta R)^2 l - \pi R^2 l = \beta V \Delta T \tag{1}$$

where the initial temperature rise is

$$\Delta T = \frac{E\alpha l}{\rho V C_p}. \tag{2}$$

Here l is the length of the PA source (assumed long), β is the thermal expansion coefficient, $V = \pi R^2 l$ is the source volume, E is the laser pulse energy, α is the absorption length (with $\alpha L \ll 1$), ρ is the density and C_p is the specific heat at constant pressure. Combining Eqs. (1) and (2) and assuming $\Delta R \ll R$ (true in all cases we are considering), we have

$$\Delta R = \frac{\beta E \alpha}{2\pi R \rho C_p} = \frac{\beta E \alpha}{2\pi c \tau_L \rho C_p}. \tag{3}$$

The peak displacement $U_s(r)$ at the observation point at distance r from the PA source (for $r_s \ll l$) varies as $r^{-\frac{1}{2}}$ because of conservation of acoustic energy, as described by Landau and Lifshitz (1959) for a cylindrical acoustic wave:

$$U_s(r) = \Delta R (R/r)^{1/2} = \frac{\beta E \alpha}{2\pi\rho C_p (c\tau_L r)^{1/2}} \ . \tag{4}$$

The peak acoustic pressure $P_s(r)$ at position r is related to the acoustic displacement $U_s(r)$ by

$$P_s(r) \approx c\rho U_s(r)/\tau_L \ . \tag{5}$$

Substituting Eq. (4) into (5), we obtain the peak PA pressure observed at r for small source radius,

$$P_s(r) \approx \frac{\beta E \alpha c^2}{2\pi C_p (c\tau_L)^{3/2} r^{1/2}} \ . \tag{6}$$

The opposite case of a large laser radius, i.e., $R_f > c\tau_L$ (see Fig. 3b) is also simple. Here, a large radius means that the heated volume does not have time to expand isobarically immediately after the laser pulse; instead a pressure increase P_0 is produced at the cylinder surface immediately after the laser pulse is absorbed:

$$P_0 = \rho c^2 \beta \Delta T = \frac{\rho c^2 \beta E \alpha}{\pi R_f^2 \rho C_p} \tag{7}$$

where ρc^2 is the bulk modulus of the medium and Eq. (7) are obtained from the consideration that the stress P_0 and the strain $\beta \Delta T$ are related by the bulk modulus. Again, the peak acoustic pressure $P_f(r)$ for the cylindrical wave scales as $r^{1/2}$, so that

$$P_f(r) = P_0 (R_f/r)^{1/2} = \frac{\beta c^2 E \alpha}{\pi R_f^{3/2} C_p r^{1/2}} \ . \tag{8}$$

Comparing Eqs. (6) and (8), we see that

$$\frac{P_f(r)}{P_s(r)} \approx \left[\frac{c\tau_L}{R_f} \right]^{3/2} < 1 \tag{9}$$

which shows that a large source radius produces a weaker PA pulse compared to a small source radius, with all other conditions being identical. This is intuitively appealing, since for the case of large radius ($R_f > c\tau_L$), the contributions from different positions in the source do not add up coherently.

Both Eqs. (6) and (8) imply that the peak acoustic pressure P is linearly dependent on the laser pulse energy E, which means that the acoustic energy E_{ac} varies as E^2. Hence, the PA generation efficiency η is

$$\eta = \frac{E_{ac}}{E} \propto E .$$

(10)

Thus, higher PA efficiency occurs for higher laser energy, and this is true for all cases of PA generation by a thermal expansion mechanism.

Rigorous theories of PA generation by thermal expansion mechanisms have been given by White, (1963), Hu (1979), Liu (1982) and others. Rigorous theories of PA generation by thermal expansion and by electrostriction have been given by Lai and Young (1982), Heritier (1983) and Brueck, Kidal and Belanger (1980). Here, we briefly outline Lai and Young's theory for the weak absorption case (pulsed PA generation in a fluid neglecting thermal diffusion). The basic equations of the PA generation are the equation of motion:

$$\rho \ddot{u} = - \nabla p$$

(11)

and the equation of expansion:

$$\nabla \cdot u = -\frac{p}{\rho c^2} + \beta T - \frac{\gamma I}{2 n c_L \rho c^2}$$

(12)

where u(r,t) is the acoustic displacement at distance r from the axis of the PA cylindrical source, p(r,t) is the acoustic pressure, T is the temperature rise due to the laser pulse of an intensity I(r,t), γ is the electrostrictive coefficient, n is the refractive index of light and c_L is the velocity of light in vacuum. We use the notation that one or two dots above a quantity indicate a first or second time derivative, respectively. Taking the second time-derivative of Eq. (12), we get

$$\frac{1}{\rho c^2} \frac{\partial^2 p}{\partial t^2} + \nabla \cdot \ddot{u} = \beta \ddot{T} - \frac{\gamma}{2 n c_L \rho c^2} \frac{\partial^2 I}{\partial t^2} .$$

(13)

Substituting Eq. (11) into Eq. (13) and also using

$$\ddot{T} = \frac{\alpha \dot{I}}{\rho C_p}$$

(14)

we get the following inhomogeneous wave equation for the acoustic pressure:

$$\left[\frac{1}{c^2} \frac{\partial^2}{\partial t^2} - \nabla^2 \right] p = \left[\frac{\alpha\beta}{C_p} \frac{\partial}{\partial t} - \frac{\rho\gamma}{2nc_L\rho c^2} \frac{\partial^2}{\partial t^2} \right] I \ . \tag{15}$$

The solution of Eq. (15) may be simplified by introducing a potential function $\phi(r,t)$ which satisfies the following reduced wave equation (Lai and Young, 1982):

$$\left[\frac{1}{c^2} \frac{\partial^2}{\partial t^2} - \nabla^2 \right] \phi = I(r,t) \ . \tag{16}$$

Eqs. (15) and (16) imply the following: the acoustic pressure p can be written as the sum of a thermal term p_{th} and an electrostriction term p_{el}, given by

$$P_{th} = \frac{\alpha\beta}{C_p} \frac{\partial\phi}{\partial t} \ , \tag{17}$$

and

$$P_{el} = - \frac{\rho\gamma}{2nc_L\rho c^2} \frac{\partial^2\phi}{\partial t^2} \ , \tag{18}$$

with

$$p = p_{th} + p_{el} \ . \tag{19}$$

Eqs. (17) and (18) have the following two important implications. First, p_{el} is proportional to the time-derivative of p_{th}, i.e.,

$$p_{el} \propto dp_{th}/dt \ . \tag{20}$$

Hence, p_{el} will pass through zero when p_{th} is at a peak pressure. Thus, the effect of p_{el} can be minimized by using a boxcar integrator to detect the PA signal with the boxcar gate set at a suitable time and gate-width. Second, the peak magnitudes $| p_{el} |$ and $| p_{th} |$ are related by:

$$\frac{| p_{el} |}{| p_{th} |} \approx \frac{\rho\gamma C_p}{2nc_L\rho c^2\alpha\beta} \frac{1}{\tau_{PA}} \tag{21}$$

where τ_{AP} is the width of the PA pulse. If we put $\tau_{PA}=1$ μs, and substitute values for the other parameters in Eq. (21) for typical liquids like water of ethanol, we conclude that

$$\frac{|P_{el}|}{|P_{th}|} \leq \frac{10^{-5}}{\alpha} . \tag{22}$$

where α is expressed in cm^{-1}. This means that the electrostrictive pressure is small compared to the thermal expansion pressure, unless α is smaller than $\approx 10^5$ cm^{-1}; however, even in this low absorption case, the electrostrictive pressure effect can be strongly suppressed (Lai and Young, 1982) by suitably setting the boxcar gate for detection, as indicated in Eq. (20).

Quantitative solutions of Eqs. (16)-(18) have been obtained and some results are indicated in Fig. 4

3. PA SIGNAL DETECTION

Many types of piezoelectric ceramics or crystals are commercially available, e.g., lead zirconate titanate (PZT), lead metaniobate, lithium niobate, crystalline quartz, etc., and reviews on these transducers exist in the literature (*Physical Acoustics,* 1979). For PA detection, the piezoelectric element usually must have two metalized surfaces as electrodes, and should be mounted in a suitable manner. Various ways of transducer mounting for PA detection have been published (Tam and Patel, 1979a; Farrow, Burnham, Auzanneau, Olsen, Purdie and Eyring, 1978; Wickramasinghe, Bray, Jipson, Quate and Salcedo, 1978; Kohanzadeh, Whinnery and Carroll, 1975; Lahman, Ludewig and Welling, 1977; Oda, Sawada and Kamada, 1978; Sigrist and Kneubuhl, 1978). The sensitivity of a PZT transducer is typically ≈ 3 V/atm. This is much smaller than that of a sensitive microphone (e.g., B&K model 4166) with a sensitivity $\approx 5 \times 10^3$ V/atm. However, PZT transducers are useful for pulsed PA studies in condensed matter and particularly in semiconductors, because of the much faster risetimes and much better acoustic impedance matching compared to microphones.

Some highly insulating polymer films can be poled in a strong electric field at elevated temperatures or can be subjected to charged beam bombardment so that they become polarized and exhibit piezoelectric characteristics. Such films include polyvinylidene difluoride (PVF_2), teflon, mylar, etc., with PVF_2 being the most commonly used. There is strong interest in the use of PVF_2 film as transducers (Bui, Shaw and Ziteli, 1976; Tam and Coufal, 1983) for acoustic imaging because of the acoustic nonringing characteristics of the film (its mechanical Q factor is much lower than that of PZT), fast risetime, structural flexibility, and good acoustic impedance matching to liquids like water. A disadvantage of PVF_2 is that its sensitivity is typically one order of magnitude lower than that of conventional piezoelectric transducers such as PZT or lithium niobate. Details of the use of PVF_2 transducers in photothermal/photopyroelectric measurements are discussed in Chapter 7.

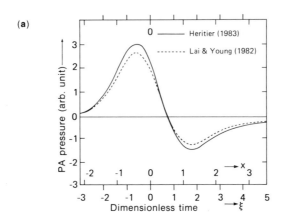

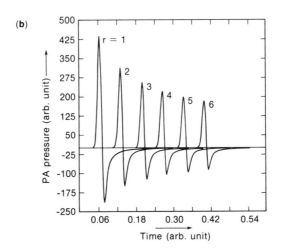

Fig. 4 Pulsed PA signal shapes according to various theoretical calculations. (a) Theoretical PA pressure profiles according to Lai and Young (1982) and to Heritier (1983). The amplitudes of the two profiles are made slightly different to show clearly the shape of each. The retarded dimensionless time used by Lai and Young is ξ, and by Heritier x. (b) Theoretical PA profiles according to the calculation of Patel and Tam (1981) for several propagation distances r (in arbitrary units) from a line PA source.

Some examples of microphone or piezoelectric transducer-coupled PA cells suitable for semiconductor PA measurements in samples are illustrated in Fig. 5. Other types of transducers for detecting pulsed acoustic signals have been described in the literature. For example, Aindow *et al.* (1981) and Hutchins and Tam (1986) have used capacitance transducers for detection of PA pulses generated in metallic samples by pulsed Nd:YAG lasers. Olmstead *et al.* (1983)

have used a continuous probe laser beam directed at the sample surface to detect the surface distortions due to the thermal or acoustic effects excited by a pulsed laser. A continuous probe laser beam has also been used to detect the transient PA profile in condensed phase samples such as semiconductor silicon wafers (Sontag and Tam, 1985a): such an acoustic surface profile causes the probe beam to undergo a transient deflection proportional to the spatial derivative of the acoustic profile. This method has been used successfully to measure the echo time τ_l corresponding to reflections of the observed longitudinal pulse, by Fourier transform analysis for several silicon wafers with different orientations, doping levels and thicknesses l. Longitudinal velocities $v_L = 2l/\tau_l$ for these samples at 23°C have been measured and shown in Table I.

Table I
Longitudinal velocities v_L determined for different silicon wafers.

Sample	Resistivity (Ω cm)	Orientation	Thickness (μm)	v_L (m/s)
1 (*n* type)	10.2	111	264	8830±25
2 (*n* type)	1.25	111	293	8851±25
3 (*p* type)	0.10	111	267	8836±25
4 (*n* type)	38.1	100	340	8139±25
5 (*p* type)	0.01	100	330	8154±25
6 (*n* type)	12.0	100	612	8159±25

Fig. 5 Examples of various PA cells suitable for semiconducting samples: (a) simple gas-microphone cell; (b) separated-chamber cell (with Helmholtz resonance possibility); (c) simple direct piezoelectric detection cell.

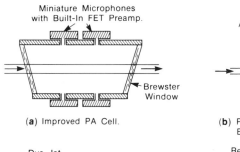

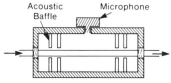

(a) Improved PA Cell.

(b) PA Cell with Window Absorption Effect Minimised.

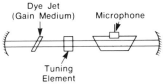

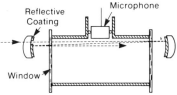

(c) Intra-Cavity Laser PA Cell.

(d) Multi-Pass, Resonant PA Cell.

The above are examples of detection of the acoustic profile generated by the pulsed heating by the excitation beam in condensed matter. Of course, the pulsed heating also generates a temperature gradient close to the excitation beam, which dissipates slowly by thermal diffusion. Many methods of detecting the ensuing thermal refractive index gradient are possible. Several examples can be mentioned here, such as thermal lensing (Swofford, Long and Allrecht, 1976); probe-beam deflection (Boccara, Fournier, Jackson and Amer, 1980; Loulerque and Tam, 1985; Sontag and Tam, 1985b; Tam, Sontag and Hess, 1985); and "mirage" detection with gas coupling (Murphy and Aamodt, 1981; Royce and Benziger, 1986; Low and Morterra, 1986).

4.   PA APPLICATIONS

Suitable laser sources can be advantageously used to obtain detectable PA signals in cases of weak absorption. Kreuzer (1971) first reported such an application by using a 15 mw HeNe laser at 3.39 μm and showed that 10 parts-per-billion of CH_4 in air are detectable. Linear absorption spectroscopy of small absorption features in liquids (Patel and Tam, 1981) and in solids (Hordvik and Schlossberg, 1977) have been detected by using pulsed laser sources. Furthermore, pulsed laser sources with high peak power for excitation provide a means through which nonlinear optical absorptions like multiphoton absorption (Cox, 1978; Tam and Patel, 1979b) or stimulated Raman scattering (Siebert, West and Barrett, 1980; Patel and Tam, 1979) in gases as well as in condensed matter can be observed. These higher order optical absorption processes are readily observable with the pulsed PA detection scheme in semiconductors, but not with CW modulated PA methods (see also Chapter 15). Furthermore, kinetic studies with a time resolution on the order of nanoseconds can be conducted readily with pulsed techniques only. A review of pulsed PA techniques for weak-absorption spectroscopy has been given by Patel and Tam (1981).

PA monitoring of de-excitation or energy transfer has also been used extensively. Following optical excitation, four de-excitation branches (with various de-excitation times) are possible: luminescence, photochemistry, photoelectricity, and heat (which may be generated directly or through energy transfer collisions). Under sufficiently simple conditions, the PA signal, whose magnitude depends on the branching ratio for heat production, can provide information on another de-excitation channel. For example, if luminescence and heat are the only two competing channels, the PA signal magnitude is then complementary to the efficiency of luminescence decay. If the latter efficiency can be changed (e.g., by varying the quenching condition), PA monitoring at different quenching conditions can yield an absolute value for the luminescence efficiency without the requirement of any absolute measurements (Adams, Highfield and Kirkbright, 1977; Quimby and Yen, 1978). Similarly, if photo-carrier generation and heat are the only two competing channels as frequently is the case in semiconductors, the photo-carrier generation efficiency may be obtained by PA measurements at different generation conditions, demonstrated by Cahen (1978) who studied a solar cell under different external loading conditions, and by Tam (1980) who studied an organic photoconductive film under different applied electric fields. PA monitoring of photochemical activities is best illustrated by the work of Cahen *et al.*, (1978), who showed that the PA signal from "poisoned" chloroplasts is larger than that from "active" chloroplasts, because there is no photochemical energy conversion in the former. Deexcitation times can often be measured by monitoring the phase (for modulated excitation at close to 50% duty cycle) or the time development (for pulsed excitation) of the PA signal. Hunter *et al.*, (1974) have studied the effect on the phase of the PA signal due to "fast" heat release and "slow" heat release when optical excitations produce both singlet and triplet states.

Many PA material testing applications are possible because the photothermal heating of a sample and the generation and propagation of the PA waves all depend on the thermoelastic and physical properties of the sample. Hence, by monitoring the PA signal, we may be able to probe or measure various physical properties: e.g., acoustic velocities, (Tam, Zapka, Chiang and Imaino, 1982; Zapka and Tam, 1982a), thickness (Tam and Coufal, 1983), crystallinity and phase transitions (McClelland and Kniseley, 1979), nonuniformity of parts used in industry (Coufal, Lam and Tam, 1983), porosity in loose powders (Imaino and Tam, 1983), temperatures and flows in fluids or flames (Zapka and Tam, 1981; Zapka, Pokrowsky and Tam, 1981), and materials anisotropy (Tam and Leung, 1984a). The shape of the pulsed PA profile can also be quantitatively monitored (Sullivan and Tam, 1984). Extremely narrow acoustic pulses can be produced optically in condensed matter as well as in gases. The propagation of such ultra-narrow acoustic pulses can be monitored by contact methods using transducers, (Tam, 1984) or by noncontact methods using optical detection techniques (Sontag and Tam, 1986). Acoustic velocities, dimensions, surface defects and the ultrasonic absorption spectrum can thus be detected (Tam and Ayers, 1986). By focusing the light beam, some of these physical properties may be measured locally, and by raster scanning the beam over the sample, new types of "PA imaging" can be performed, as first demonstrated by Wong *et al.* (1978). The use of a light beam for thermal/acoustic excitation can be replaced by any other type of energetic beams (e.g., electrons, ions, etc.). This was demonstrated by Cargill, (1980) for example, who constructed an electron-beam-excited acoustic microscope to detect invisible subsurface features in solids.

An important extension of PA techniques into basic ultrasonic absorption spectroscopy measurements in fluids has recently been demonstrated by Tam and Leung (1984b). These workers used pulsed PA generation and quantitative probe-beam-deflection for acoustic profile measurements. This optical ultrasonic measurement technique is useful, because the absorption spectrum of high frequency sound (frequency f much larger than 1 MHz) in gases has traditionally been very difficult to measure, due to the lack of a controllable high-frequency source with good acoustic coupling to the gas, and of a sensitive detection method with fast response. Tam and Leung (1984b) have given a first demonstration of an all optical Fourier-multiplexed method to measure ultrasonic absorption spectra of gases and mixtures near normal temperature and pressure at frequencies up to tens of MHz. This technique is new in two respects: (a) It is all optical and is noncontact, and hence applicable to hostile environments; and (b) It is a frequency-multiplexed technique relying on pulse shape and magnitude measurement and Fast Fourier Transform (FFT) analysis, and hence is applicable to "real-time" monitoring under rapidly changing conditions.

5.   SUMMARY OF PA MEASUREMENTS

This section presented some illustrative considerations of the theory and applications of pulsed photoacoustics, which is a type of photothermal effect, and has generated much renewed interest in the past decade for the following reasons. Intense light sources and sophisticated acoustic detection have become available. The PA technique has been shown to be a sensitive and easy-to-use spectroscopic technique for gases as well as for condensed samples. PA methods have many unique applications, e.g., spectroscopic studies of opaque or powdered materials, studies of energy conversion processes, imaging of invisible or subsurface features in solids and industrial materials testing and quality control. Furthermore, future "exotic" applications like air-to-ocean communications or cosmic ray detection appear possible, while some of these techniques await valuable applications to semiconducting materials characterization.

The diverse applications of photacoustics are summarized in Table II. The generalized concept of photoacoustics (where acoustic generation is due to any type of energetic beam) and its applications are indicated in Fig. 6. The rapid growth of PA activities has led to many reviews such as Rosencwaig (1980); Patel and Tam (1981); Tam (1983) *and* (1986); West, Barrett, Siebert and Reddy (1983); Zharov and Letokhov (1986); and the present volume on semiconductor applications.

III. Pulsed Photothermal Radiometry (PTR)

1.   OVERVIEW

Photothermal Radiometry (PTR) is a sensitive technique for noncontact spectroscopy and inspection; a modulated beam of photons (or other particles) is used to produce temperature transients, and the corresponding transients in the infrared thermal radiation emitted from the sample are analyzed. There are four common variations of PTR techniques that have been reported in the literature as indicated in Table III. These variations can be classified according to the excitation mode (continuously modulated or pulsed) and to the detection mode (transmission or back-scattered). The excitation beam (which can be photons, electrons, microwaves, etc.) is usually either continuously modulated with about 50% duty cycle, or pulsed modulated with very low duty cycle, but high peak power. The detection can, in principle, be in any direction but is simplest for observation backwards from the excitation spot (called back-scattering PTR here), or "end-on" through the sample thickness with respect to the excitation spot (called transmission PTR here). Most workers have used either the back-

Table II

Summary of Various Applications of Photoacoustic (PA) Techniques

Type	Principle of Operation	Remarks
1. PA spectroscopy	Absorption spectrum is obtained by varying the excitation wavelength to produce a corresponding variation in acoustic response. Constant quantum efficiency for thermal de-excitation is usually assumed.	Contact acoustic detection includes the use of gas coupled microphones and piezoelectric transducers. Noncontact acoustic detection includes a variety of probe beam deflection techniques.
2. PA monitoring of de-excitation processes	The quantum efficiency for thermal de-excitation is varied (e.g., by changing concentrations, temperature, electric field, etc.) to make inference on the quantum efficiency of a complementary channel.	Contact acoustic detection has been generally used; complementary channels to thermal de-excitation include luminescence, photochemistry, photoelectricity, and energy transfer.
3. PA sensing of physical properties	The thermal or acoustic waves generated in the sample are used to sense subsurface flaws, layered structure, material composition or crystallinity, sound velocities, flow rates, temperatures, etc.	Contact acoustic detection has been generally used to sense subsurface features. Noncontact acoustic detection for various velocities and temperature sensing measurements involve the use of probe beams.
4. PA generation of mechanical motions	PA pulses or shock waves (e.g., due to boiling or breakdown) may generate mechanical motions efficiently.	Liquid droplet formation and ejection from nozzle and laser-induced structural vibrations in metal beams and plates have been demonstrated.

Table III

Examples of Various Modes of Photothermal Radiometry

Source	Detection	
	Transmission	**Back-scattering**
Continuous modulated	Cowan (1961) Busse (1980) Busse and Eyerer (1983) Busse and Renk (1983)	Hendler and Hardy (1961) Nordal and Kanstad (1979,80,81) Bults *et al.*, (1982) Kanstad *et al.*, (1983) Luukkala (1980) Vanzetti and Traub (1983)
Pulsed	Parker *et al.*, (1961) Deem and Wood (1962) Taylor (1972)	Tam and Sullivan (1983) Leung and Tam (1984a,b) *and* (1985) Cielo (1984) Imhof *et al.*, (1984)

Fig. 6 Schematic of the generalized PA generation and detection technique.

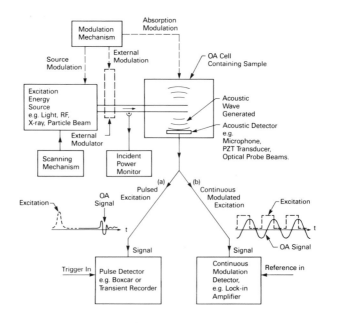

scattering geometry or the transmission geometry, although Luukkala *et al.* (1982) have demonstrated PTR with the observation spot being laterally displaced from the excitation spot. Only back-scattering PTR (*not* transmission PTR) can be used for thick or bulky materials, for samples with inaccessible back surface, or for single-ended remote sensing. This section is mainly concerned with the recent experimental and theoretical developments of pulsed PTR (PPTR) with back-scattering detection. Applications of modulated PTR to semiconductors are discussed in detail in Chapter 3. PTR is useful for many types of noncontact measurements, including absolute absorption spectroscopy, excitation spectroscopy, and the monitoring of thickness, layered structures, thermal diffusivity, thermal contacts, and associated properties. Furthermore, PTR being a noncontact technique, it can be applied to "difficult" samples like opaque materials, powders, aerosols, gels, skins *in-vivo,* and so on. PPTR performed in the back-scattering mode with pulsed laser excitation appears to be useful for single-ended remote-sensing of samples that may be distances on the order of ≈km away, i.e., back-scattering PTR may provide a new PT LIDAR (*LI*ght *D*etection *A*nd *R*anging).

2. PULSED PTR IN BACKSCATTERING

In this section, we shall mainly discuss the relatively new technique of pulsed PTR (PPTR) in the back-scattering configuration, as indicated in Fig. 7. Various pulsed excitation sources can be used, for example a short-duration (FWHM = 8 ns) N_2 laser beam with less than 1 mJ energy at 337 nm. The back-scattered PPTR measurement is shown for beam path (a) in Fig. 7, which also shows transmission PPTR as indicated in beam path (b). The thermal IR radiation from the sample is refocused onto a HgCdTe detector with a rise time of 0.5 μs, and the PPTR signal in back-scattering ($S_B(t)$) or in transmission ($S_T(t)$) is stored in a transient recorder (Tektronix 7854 scope with 7D20 plug-in). Maximum sample heating by the laser beam of a spot size of several mm is generally limited to ≈10°C to avoid any sample damage or any nonlinear radiometric effects, since the radiometric signal magnitude is proportional to the surface temperature rise ΔT only for small ΔT.

Tam and Sullivan (1983) have applied this PPTR in back-scattering for novel absolute absorption spectroscopy of opaque solids and liquids. This is possible because the magnitude of the absorption coefficient α at the excitation beam wavelength is related to the "steepness" of the initial temperature profile produced in the sample. Large α corresponds to a steeper temperature profile near the sample surface, causing faster cooling of the surface and hence faster decay of the back-scattering PPTR signal. By analyzing the shape of the PPTR signal from bulk samples, Tam and Sullivan (1983) showed that α can be obtained under some conditions if the thermal diffusivity D is known.

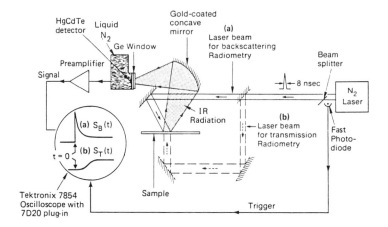

Fig. 7 Experimental setup for both single-ended back-scattering (laser beam represented by solid lines) and double-ended transmission (laser beam represented by dashed lines) pulsed photothermal radiometry measurements.

Another very important application (Tam and Sullivan, 1983) of potential value to the characterization of semiconductor structures is the remote sensing of the layered samples by PPTR. For demonstration purposes, these authors coated a black rubber substrate with a 45 μm thick polyester film that was transparent in the wavelength of the excitation beam. For this layered structure, the pulsed PTR exhibited two peaks: a prompt peak that decays with the same rate as the bare substrate but with decreased magnitude, and a delayed peak occurring at a time t_d after the firing of the laser. It was found that t_d is related to the thickness l and the thermal diffusivity D_l of the coating by $t_d = l^2/4D_l$, indicating that the prompt peak is due to IR thermal radiation from the irradiated substrate and attenuated by the coating, and the delayed peak is due to IR thermal radiation from the top coating surface due to heat diffusion.

Tam and Sullivan (1983) also demonstrated that PPTR is useful for sensing powder aggregation which affects the "effective" thermal conductivity. The pulsed laser was used to irradiate black carbon-loaded epoxy powders of different degree of compactness. It was found that the PPTR signal for the loose powder stops decaying after some time indicating that inter-particle heat transport is slow, but the PPTR signal for a compact powder exhibits a continuous decay as in a neat solid. Similar results were observed experimentally and explained theoretically using modulated microphonic photoacoustic detection by Mandelis and Lymer (1985).

3.   THEORY OF PULSED PTR

In this section we present a theoretical investigation of the PPTR signal profile (Leung and Tam, 1984; 1985). For a sample of thickness L, thermal diffusivity D, absorption coefficient α at the excitation wavelength and α' at the detection wavelength, the PPTR signal shape excited by a pulsed laser of duration τ_0 is dependent on different parameters at different delay time periods (see Fig. 8). For simplicity in the present discussion, we assume that τ_0 is short, and the IR detector rise-time is fast. The theoretical PPTR signals $S_B(t)$ and $S_T(t)$ are given by

$$\begin{bmatrix} S_B(t) \\ S_T(t) \end{bmatrix} = \frac{AK}{L} \left\{ (1 - e^{-\alpha L})(1 - e^{-\alpha' L}) \right.$$

$$\left. \pm 2 \begin{bmatrix} 1 \\ e^{-\alpha' L} \end{bmatrix} \sum_{n=1}^{\infty} \left[\frac{1 - (-1)^n e^{-\alpha L}}{1 + \frac{n^2 \pi^2}{\alpha^2 L^2}} \right] \left[\frac{1 - (-1)^n e^{+\alpha' L}}{1 + \frac{n^2 \pi^2}{\alpha'^2 L^2}} \right] e^{-n^2 t / \tau_L} \right\} \quad (23)$$

where the equation for the upper (lower) quantities in the two square brackets corresponds to the upper (lower) $\pm$ signs. Here, A is a constant depending on the flash pulse energy and the thermal properties, and K is a constant depending on the emissivity and the ambient temperature.

Some results based on Eq. (23) for $S_B(t)$ are shown in Figs. 9 and 10. Fig. 9 shows the theoretical shapes of $S_B(t)$ for two samples of identical material (stainless steel) but different thicknesses: L = 0.0005 cm and L = ∞. We see that the two signals totally overlap at early itmes; however, after a "thickness diffusion time" $\tau_L = 63.3$ µsec for the L = 0.005 cm sample, the two signals start

Fig. 8 Thermal diffusion in an opaque sample of thickness L after a short pulsed heating at the surface. The local temperature increase above ambient (θ) is plotted at several times (t), and indicated by the density of the black dots. Early decay is related to spectroscopic features of the sample. Late decay is related to geometric features.

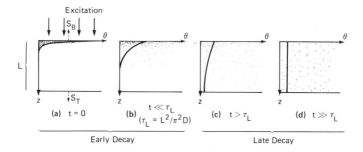

(a) t = 0   (b) $t \ll \tau_L$ (c) $t > \tau_L$ (d) $t \gg \tau_L$
                   ($\tau_L = L^2/\pi^2 D$)

Early Decay   Late Decay

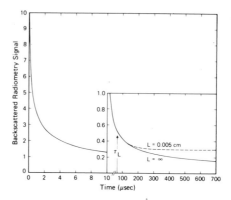

Fig. 9 Theoretical profiles of the single ended back-scattering flash radiometric signal showing the effect of sample thickness L. The dashed line is for L = 0.005 cm and the solid line is for L = ∞. In both curves α, α′ and D are taken as 2 x 10⁴ cm⁻¹, 1 x 10⁴ cm⁻¹ and 0.04 cm²/s, respectively, which are typical values for stainless steel. Note that the scale changes for the late decay in the inset.

to deviate: the thin-film signal stops decaying while the thick-film signal continues to decay. These effects have indeed been verified experimentally (Leung and Tam, 1984; 1985). Fig. 10 shows how the theoretical shapes of $S_B(t)$ depend on the absorption coefficients α and α′, with fixed thickness and thermal diffusivity; at early times (i.e., $t \ll \tau_L$), the stronger absorption coefficient causes a steeper decay of the signal. However, at late times ($t \gg \tau_L$) the decay curves for different values of α and α′ are overlapping, showing that the late decay rate is only dependent on L and D, in agreement with the analysis by Parker *et al.* (1961). In conclusion, the back-scattering PPTR signal $S_B(t)$ is sensitively dependent on α, α′ and D for $t \ll \tau_L$, and it sensitively depends on L and D for $t \gg \tau_L$. Theoretical fitting (Leung and Tam, 1985) of the observed $S_B(t)$ for a sample can provide values of α, α′, D and L. Such measurements have the distinct advantages of being single-ended, nondestructive and can be accomplished through remote sensing. In Eq. (23), we have neglected the effects due to the finite excitation pulse width and the finite IR detector rise time. Such effects have also been taken into account elsewhere (Leung and Tam, 1985).

Eq. (23) approaches simple limits (Leung and Tam, 1984; 1985) when L is large. In this case, $S_T(t) \rightarrow 0$, and

$$S_B(t) \underset{L \rightarrow \infty}{\rightarrow} \frac{AK\alpha\alpha'}{\alpha'^2 - \alpha^2} \{\alpha' e^{t/4\tau_\alpha}(1 - \mathrm{erf}\sqrt{t/4\tau_\alpha})$$

$$-\alpha e^{t/4\tau_{\alpha'}}(1 - \mathrm{erf}\sqrt{t/4\tau_{\alpha'}})\}$$

$$(24)$$

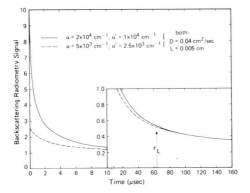

Fig. 10 Theoretical profiles of the back-scattering radiometric signal showing the effect of absorption coefficents α and α'. The dashed line is for $\alpha = 5 \times 10^3$ cm^{-1} and $\alpha' = 2.5 \times 10^3$ cm^{-1}; the solid line is for $\alpha = 2 \times 10^4$ cm^{-1} and $\alpha' = 1 \times 10^4$ cm^{-1}. In both curves, L and D are taken as 0.005 cm and 0.04 cm^2/s, respectively. Note that the scale changes for the late decay in the inset.

where $\tau_\alpha = (4\alpha^2 D)^{-1}$, $\tau_{\alpha'} = (4\alpha'^2 D)^{-1}$ and "erf" is the error function (Carslaw and Jaeger, 1959). Eq. (24) is symmetrical with respect to the absorption coefficients α and α'. It can be further simplified if one of the absorption coeffients is much larger than the other, e.g., $\alpha' >> \alpha$. In this case:

$$S_B(t) \underset{\substack{L \to \infty \\ \alpha' >> \alpha}}{-\!\!\!\longrightarrow} AK\alpha e^{t/4\tau_\alpha}(1 - \mathrm{erf}\sqrt{t/4\tau_\alpha}) , \qquad (25)$$

which is precisely the form used in earlier PPTR work (Tam and Sullivan, 1983) for obtaining absolute values of α at the excitation wavelength for "semi-inifinite" samples.

4. EXCITATION SPECTRUM

An excitation spectrum of a distant sample can be obtained with back-scattered PPTR by scanning the wavelength of the excitation laser and monitoring the corresponding peak magnitude of the back-scattered PPTR signal. An example of such an excitation spectrum is shown in Fig. 11. Such an excitation spectrum truly represents an absorption spectrum of the sample only under certain conditions, for example, when sample thickness is large and $\alpha' >> \alpha$; in this case Eq. (25) indicates that the peak value of $S_B(t)$ occuring at t = 0 is proportional to $\alpha(\lambda)$ at the excitation wavelength λ.

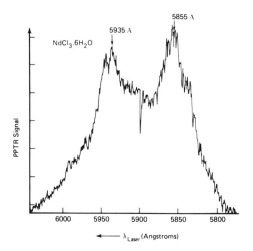

Fig. 11 PPTR excitation spectrum with back-scattering detection for $NdCl_3*6H_2O$: This is obtained by scanning the excitation pulsed dye laser over the Rhodamine 6G tuning range (5770Å to 6030Å).

5.   SUMMARY ON PTR MEASUREMENTS

PTR techniques rely on the detection of modulated infrared thermal radiation from a sample that is excited by a modulated beam of energy. The recently developed technique of pulsed PTR in back-scattering detection is suitable for remote sensing of semiconductors and other condensed phase materials. Its theoretical basis has been discussed, and its new applications have been indicated. They include absolute absorption spectroscopy, excitation spectroscopy, thickness or thermal diffusivity measurements, monitoring of layered structures, and sensing of degree of contact between materials. These applications should be valuable in semiconductor substrate material and layered device characterization such as junctions, diode lasers and quantum well structures.

IV Conclusions

"Photothermal effects" are caused by the heating of a sample after the absorption of optical energy (or, in general, the absorption of any energetic beam). After optical absorption, heating and/or other de-excitation channels (e.g., luminescence, photochemistry, photoelectricity and energy transfer) may occur. These other "de-excitation branches" complement the heating branch in the sense that all branching ratios must add up to one. The PT de-excitation is typically the dominant branch for semiconducting materials at normal temperature and pressure conditions. Photothermal heating of a sample is frequently produced with the use of laser beams, Xe arcs or other intense light sources.

Generally, these PT sources are not simply used as "very expensive Bunsen burners," but are needed for some of the following reasons: 1. PT heating can provide convenient and sensitive methods for detecting optical absorption in condensed matter; 2. information concerning de-excitation mechanisms can be obtained; and 3. very localized or very rapid photothermal heating can be achieved to provide novel measurements or produce new effects.

This chapter has reviewed two commonly used PT monitoring instrumentation techniques with high application potential to semiconductors (see Chapter 3), namely photoacoustic sensing and photothermal radiometry monitoring. The PT effect can also be detected by other means, e.g., calorimetry, (Brilmyer *et al.*, 1977; Bass *et al.*, 1979; Baumann *et al.*, 1983; Coufal, 1984) monitoring refractive index variations causing defocusing (Swofford *et al.*, 1976; Cremers and Keller, 1981; Bialkowski, 1985) or deflection (Fournier *et al.*, 1980; Fournier *et al.*, 1982; Low *et al.*, 1982; Murphy and Aamodt, 1983; Wetsel and Stotts, 1983) of probe beams, surface distortions, (Olmstead *et al.*, 1983; Sontag and Tam, 1985a; 1986) and reflectivity variations (Paddock and Eesley, 1986). The PT heating also affects the Boltzmann distribution of the molecular states in a gas, detectable by absorption spectroscopy; this can be called a Boltzmann spectroscopy method for detecting the photo-thermal effect (Zapka and Tam, 1982b).

V. REFERENCES

Adams M.J., Highfield J.G. and Kirkbright G.F., (1977) *Anal. Chem.* **49**, 1850.

Aindow A.M., Dewhurst R.J., Hutchins D.A. and Palmer S.B., (1981) *J. Acoust. Soc. Am.* **69**, 449.

Bass M.E., Van Stryland E.W. and Steward A.F., (1979) *Appl. Phys. Lett.* **34**, 142.

Baumann T., Dacol F. and Melcher R.L., (1983) *Appl. Phys. Lett.* **43**, 71.

Bailkowski S.E., (1985) *Appl. Opt.* **24**, 2792.

Boccara A.C., Fournier D., Jackson W. and Amer N.M., (1980) *Opt. Lett.* **5**, 377.

Brilmyer G.H., Fujishima A., Santhanam K.S.V. and Bard A.J., (1977) *Anal. Chem.* **49**, 2057.

Brueck S.R.J., Kidal H. and Belanger L.J., (1980) *Opt. Commun.* **34**, 199.

Bui L., Shaw H.J. and Ziteli L.T., (1976) *Electron. Lett.* **12**, 393.

Bults G., Nordal P-E. and Kanstad S.O., (1982) *Biochim. et Biophys. Acta* **682**, 234.

Busse G., (1980) *Infrared Phys.* **20**, 419.

Busse G. and Eyerer P., (1983) *Appl. Phys. Lett.* **43**, 355.

Busse G. and Renk K.F., (1983) *Appl. Phys. Lett.* **42**, 366.

Cahen D., (1978) *Appl. Phys. Lett.* **33**, 810.

Cahen D., Malkin S. and Lerner E.J., (1978) *FEBS Lett.* **91**, 339.

Cargill C.S., (1980) *Nature* **286**, 691.

Carslaw H.S. and Jaeger J.C., (1959) *Conduction of Heat in Solids,* (second edition) Clarendon, Oxford.

Cielo P., (1984) *J. Appl. Phys.* **56**, 230.

Coufal H., (1984) *Appl. Phys. Lett.* **44**, 59.

Coufal H., Lam S.T. and Tam A.C., (1983) *IBM Tech. Disclosure Bull.* **25**, 4996; and **25**, 5446.

Cowan R.D., (1961) *J. Appl. Phys.* **32**, 1363.

Cox D.M., (1978) *Opt. Commun.* **24**, 336.

Cremers D.A. and Keller R.A., (1981) *Appl. Opt.* **20**, 3838.

Deem H.W. and Wood W.D., (1962) *Rev. Sci. Instr.* **33**, 1107.

Farrow M.M., Burnham R.K., Auzanneau M., Olsen S.L., Purdie N. and Eyring E.M., (1978) *Appl. Opt.* **17**, 1093.

Fournier D., Boccara A.C., Amer N.M. and Gerlach R., (1980) *Appl. Phys. Lett.* **37**, 519.

Fournier D., Boccara A.C. and Badoz J., (1982) *Appl. Opt.* **21**, 74.

Hendler E. and Hardy J.D., (March 1961) *"Skin Heating and Temperature Sensation Produced by Infrared and Microwave Irradiation" Proc. Symp. on Temperature, Its Measurement and Control in Science and Industry* (sponsored by AIP, Instrum. Soc. of Am., and NBS), Columbus, Ohio, p. 51.

Heritier J.M. (1983). Opt. Commun. **44**, 267.

Hordvik A. and Schlossberg H., (1977) *Appl. Opt.* **16**, 101.

Hu C.L., (1979) *J. Acoust. Soc. Am.*, **46**, 728.

Hunter T.F., Rumbles D. and Stock M.G., (1974) *J. Chem. Soc. Faraday Trans. II* **70**, 1010.

Hurst G.S., Payne M.G., Karmer S.D. and Young J.P., (1979) *Rev. Mod. Phys.* **51**, 767.

Hutchins D.A. and Tam A.C., (1986) *IEEE Trans.* UFFC. **UFFC-33**, 429.

Imaino W. and Tam A.C., (1981) *Appl. Opt.* **22**, 1875.

Imhof R.E., Birch D.J.S., Thornley F.R., Gilchrist J.R., and Strivens T.A., (1984) *J. Phys. E: Sci. Instrum.* **17**, 521.

Kanstad S.O. and Nordal P-E., (1980) *Appl. Surf. Sci.* **6**, 372.

Kanstad S.O., Cahen D. and Malkin S., (1983) *Biochim. et Biophys. Acta* **722**, 182.

Kohanzadeh Y., Whinnery J.R. and Carroll M.M., (1975) *J. Acoust. Soc. Am.*, **57**, 67.

Kreuzer L.B., (1971) *J. Appl. Phys.* **42**, 2934.

Lahman W., Ludewig H.J. and Welling H., (1977) *Anal. Chem.* **49**, 549.

Lai H.M. and Young K., (1982) *J. Acoust. Soc. Am.* **72**, 2000.

Landau L.D. and Lifshitz E.M., (1959) *"Fluid Mechanics"* (J.B. Sykes and W.H. Reid, Transl.) Pergamon Press, New York.

Leung W.P. and Tam A.C., (1984) *Opt. Lett.* **9**, 93; (1984) *J. Appl. Phys.* **56**, 153; *and* (1985) **58**, 1087.

Liu G., (1982) *Appl. Opt.* **21**, 955.

Loulerque J.C. and Tam A.C., (1985) *Appl. Phys. Lett.* **46**, 457.

Low M.J.D., Morterra C., Severdia A.G. and Lacroix M., (1982) *Appl. Surf. Sci.* **13**, 429.

Low M.J.D. and Morterra C., (1986) *IEEE Trans. UFFC.* **UFFC-33**, 585.

Luukkala M., (1980) in *Scanned Image Microscopy*, (E. Ash, ed.) Academic Press, London, p. 273.

Luukkala M., Lehto A., Jaarinen J., and Jokinen M., (1982) *IEEE Ultrasonic Symposium Proc.* 591.

Mandelis A. and Lymer J.D. (1985) *Appl. Spectr.* **39**, 473.

McClelland J.F. and Kniseley R.N., (1979) *Appl. Phys. Lett.* **35**, 121.

Murphy J.C. and Aamodt L.C., (1981) *Appl. Phys. Lett.* **38**, 196.

Murphy J.C. and Aamodt L.C., (1983) *J. de Physique (Paris) Colloque* **C6**, 513.

Nordal P-E. and Kanstad S.O., (1979) *Physica Scripta* **20**, 659.

Nordal P-E. and Kanstad S.O., (1981) *Appl. Phys. Lett.* **38**, 486.

Oda S., Sawada T. and Kamada H., (1978) *Anal. Chem.* **50**, 865.

Olmstead M., Amer N.M., Fourier D. and Boccara A.C., (1983) *Appl. Phys.* **A32** 141.

Paddock C.A. and Eesley G.L., (1986) *Opt. Lett.* **11**, 273.

Parker W.J., Jenkins R.J., Butler C.P. and Abbott G.L., (1961) *J. Appl. Phys.* **32**, 1679.

Patel C.K.N. and Tam A.C., (1979) *Appl. Phys. Lett.* **34**, 760.

Patel C.K.N. and Tam A.C., (1981) *Rev. Mod. Phys.* **53**, 517.

Physical Acoustics, Vol. **XIV** (1979), (W.P. Mason and R.N. Thurston, eds.), Academic Press, New York.

Quimby R.S. and Yen W.M., (1978) *Opt. Lett.* **3**, 181.

Rosencwaig A. (1980) *"Photoacoustics and Photoacoustic Spectroscopy"* Chem. Anal. Vol. **57**, New York.

Royce R.S.H. and Benziger J.B., (1986) *IEEE Trans. UFFC.* **UFFC-33**, 561.

Siebert D.R., West G.A. and Barrett J.J. (1980b) *Appl. Opt.* **19**, 53.

Sigrist M.K. and Kneubuhl F.K., (1978) *J. Acoust. Soc. Am.* **64**, 1652.

Sontag H. and Tam A.C., (1985a) *Appl. Phys. Lett.* **46**, 725.

Sontag H. and Tam A.C., (1985b) *Opt. Lett.* **10,** 436.

Sontag H. and Tam A.C., (1986) *IEEE Trans. UFFC.* **UFFC-33,** 500.

Sullivan B. and Tam A.C., (1984) *J. Acoust. Soc. Am.* **75,** 437.

Swofford R.L., Long M.E. and Allrecht A.C., (1976) *J. Chem. Phys.* **65,** 179.

Tam A.C., (1980) *Appl. Phys. Lett.* **37,** 978.

Tam A.C. (1983) in *Ultrasensitive Laser Spectroscopy* (D. Kliger, ed.), Academic Press, New York.

Tam A.C. (1984) *Appl. Phys. Lett.* **45,** 510.

Tam A.C. (1986) *Rev. Mod. Phys.* **58,** 381.

Tam A.C. and Patel C.K.N., (1979a) *Appl. Opt.* **18,** 3348.

Tam A.C. and Patel C.K.N., (1979b) *Nature (London)* **280,** 304.

Tam A.C., Zapka W., Chiang K. and Imaino W., (1982) *Appl. Opt.* **21,** 69.

Tam A.C. and Coufal H., (1983) *Appl. Phys. Lett.* **42,** 33.

Tam A.C. and Sullivan B. (1983) *Appl. Phys. Lett.* **43,** 333.

Tam A.C. and Leung W.P., (1984a) *Appl. Phys. Lett.* **45,** 1040.

Tam A.C. and Leung W.P., (1984b) *Phys. Rev. Lett.* **53,** 560.

Tam A.C., Sontag H. and Hess P., (1985) *Chem. Phys. Lett.* **120,** 280.

Tam A.C. and Ayers G., (1986) *Appl. Phys. Lett.* **49,** 1420.

Taylor R., (1972) *High Temp., High Press.* **4,** 649.

Teslenko V.S., *Sov. J. Quant. Electron.* (Engl. Transl.) **7,** 981.

Vanzetti R. and Traub A.C., (February 1983) *"Automatic Solder Joint Inspection in Depth"*, in *Proc. 7th Annual Seminar in Soldering Technology and Product Assurance, China Lake,* (Naval Weapons Center Publ. China Lake, California). A photothermal solder-joint inspection system, called Laser Inspect, is available from Vanzetti Systems, Inc., Stoughton, Massachusetts, U.S.A.

West G.A., Barrett J.J., Siebert D.R. and Reddy K.V., (1983) *Rev. Sci. Instr.* **54,** 797.

Wetsel G.C. and Stotts S.A., (1983) *J. de Physique (Paris) Colloque* **C6,** 215.

White R.M. (1963) *J. Appl. Phys.* **34,** 3559.

Wickramasinghe H.K., Bray R.C., Jipson V., Quate C.F. and Salcedo J.R., (1978) *Appl. Phys. Lett.* **33,** 923.

Wong Y.H., Thomas R.L. and Hawkins G.F., (1978) *Appl. Phys. Lett.* **32,** 538.

Zapka W. and Tam A.C., (1981) *Appl. Phys. Lett.* **40,** 1015.

Zapka W., Pokrowsky P. and Tam A.C., (1981) *Opt. Lett.* **7,** 477.

Zapka W. and Tam A.C., (1982a) *Appl. Phys. Lett.* **40,** 310.

Zapka W. and Tam A.C., (1982b) *Opt. Lett.* **7,** 86.

Zharov V.P. and Letokhov V.S., (1986) *"Laser Optoacoustic Spectroscopy"* Springer Verlag, Berlin.

9

PHOTOACOUSTIC TECHNIQUES FOR THE MEASUREMENT OF ELECTRONIC PROPERTIES OF SEMICONDUCTORS

Richard G. Stearns and Gordon S. Kino

Ginzton Laboratory
Stanford University
Stanford, CA 94305, USA

I. Introduction

We will discuss three new photoacoustic techniques which show promise for the measurement of certain electronic properties of semiconductors. All of the techniques are based on the photoinjection of a temporally-modulated free carrier population within the semiconductor. While most, if not all, of the techniques should be applicable to a large variety of semiconductors, the experiments discussed here deal specifically with silicon.

Typically, photoacoustic measurements are indirect measurements of periodic heating. In the most common embodiment of photoacoustic detection, the absorption of intensity-modulated radiation within a sample causes the generation of thermal waves which are photoacoustically converted into acoustic waves. These waves are subsequently detected with an acoustic detector (Rosencwaig, 1977; White, 1963). The photoacoustic conversion mechanism is due to the elastic strain in the material that results from the periodic heating associated with the thermal wave. In such a direct photoacoustic measurement, the acoustic waves detected may be generated in the material being examined or in the air (or fluid) above the surface of the sample.

In an alternative type of photoacoustic measurement, an externally-generated acoustic wave is made to propagate through or near the region of the sample that is irradiated. The subsequent periodic phase perturbation of the acoustic wave is then detected (Stearns, Khuri-Yakub and Kino, 1983). In this case, it is primarily the change in the elastic constants of the material, due to the periodic heating, that produces the acoustic phase perturbation. Here again, the perturbed acoustic wave may be made either to propagate in the sample or through the air (or fluid) directly above the sample surface.

For metals and insulators, any of these photoacoustic techniques may yield thermal, mechanical and optical information concerning the sample under study. When such techniques are applied to semiconductors, further information may also be obtained concerning the modulated free carrier population that is photoinjected into the semiconductor. This follows from the fact that the introduction of a periodic free carrier population into a semiconductor will produce directly a periodic elastic strain in the sample, which in turn generates an acoustic wave. In addition, the injected free carrier population will perturb the elastic constants of the semiconductor, perturbing the phase of an externally generated acoustic wave that propagates through the region of carrier injection. Therefore, by performing these photoacoustic generation and perturbation measurements in semiconductors, information directly pertaining to the injected carrier population may be obtained.

The first two experiments described in this chapter demonstrate these electronic acoustic wave perturbation and generation concepts, which are presented in Section II. In the first experiment, (Section III) we study the phase perturbation of a Rayleigh wave propagation on a thermally oxidized silicon wafer, due to the absorption of a modulated argon-ion laser beam near the sample surface. It will be shown that through the electronic effects on the elastic constants, the photoinjected carriers contribute measurably to the acoustic phase perturbation in the silicon. Electronic properties, such as the photocarrier life-time, may be determined from the acoustic phase perturbation measurement.

Photoacoustic generation in silicon is studied in the second experiment (Section IV). Here again an Argon-ion laser beam is used to illuminate the sample. It is shown that the periodic photoinjected carrier density in a lightly doped n-type silicon sample is the predominant source of photoacoustic generation at 1 MHz. This is so because the mechanical strain is linearly dependent on the photocarrier density in the semiconductor. The dilation constant in silicon is large enough so that the electronically-generated strain dominates the periodic thermal strain at 1 MHz in the illuminated silicon.

The final experiment described in this chapter (Section V) involves the detection of the photoacoustic generation in the air above the illuminated silicon. In this experiment, unlike the other two, it is only the periodic heating at the sample surface that influences the photoacoustic signal. The most novel feature of this technique is that it is performed at 2 MHz, an acoustic frequency considerably higher than those used in conventional photoacoustic measurements in air. In order to work at this frequency in air, a novel transducer was designed, which has good sensitivity and compares favorably with infrared detection at the 2 MHz thermal modulation frequency employed. It has the additional advantage over infrared detectors, in certain applications, that it is sensitive to the temperature on the top surface of a material. In contrast, an infrared detector determines the temperature of the top surface of the first infrared absorbing layer, typically the substrate material.

Because a portion of the heating at the semiconductor surface results from the recombination of photoinjected carriers, the 2 MHz photoacoustic technique in air may be used to image variations in certain electronic properties of the semiconductor. An experiment is described in which variations in the surface recombination velocity of a silicon sample are imaged photoacoustically, with good (~7 μm) spatial resolution.

In each of the experiments, the results are compared to, and are shown to be in good agreement with, theoretical prediction.

II. The Periodic Thermal and Photocarrier Fields
Due To Laser Absorption in a Semiconductor

Before discussing the three measurement techniques separately, it is useful to describe the general nature of the temporally-modulated thermal and photocarrier fields that result when a focused laser beam is absorbed at the surface of a semiconductor such as silicon. We will assume that a laser beam intensity-modulated at an angular frequency ω, is incident on the surface of a semiconductor sample, whose thickness is d (Fig. 1). At the sample surface, the intensity distribution of the laser beam is taken to be Gaussian, with a modulated component of the form $I(\rho) = \exp(-\rho^2/R^2) I_O \exp(-i\omega t)$ (here a cylindrical coordinate system ρ, z is assumed, with radial coordinate ρ). The energy of the laser beam, $h\nu$, is assumed to be greater than the band gap energy of the

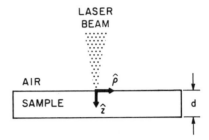

Fig. 1 Geometry of the incident laser beam and the sample. Absorption of the laser beam produces both a periodic thermal as well as a periodic photocarrier field within the semiconductor.

semiconductor, E_g. In describing the generation and diffusion of the photocarriers in the illuminated semiconductor, several approximations will be made. First, it will be assumed that at each point in the sample the excess free electron density $\Delta n(\mathbf{r})$ is equal to the excess free hole density, $\Delta p(\mathbf{r})$. It will furthermore be assumed that the carrier lifetime in the semiconductor is independent of the free carrier concentration[*]. Therefore, Auger recombination will be neglected. This is a valid assumption in silicon for free carrier densities less than $\sim 10^{18}$ cm^{-3} (Passari and Sosi, 1983; Nilsson and Svantesson, 1972). Finally, the effects of temperature gradients in the semiconductor on the carrier diffusion rate will be ignored.

Under these assumptions, the equations governing the diffusion of the photocarriers in the semiconductor are linear in both $\Delta n(\mathbf{r})$ and $\Delta p(\mathbf{r})$, and hence the components of Δn and Δp at a given temporal modulation frequency ω may be solved for independently, using as the driving source of the photocarrier generation the component of the incident radiation at the modulation frequency ω. In our experiments, the irradiated semiconductor, silicon, is typically lightly doped ($N_a = 10^{13} - 10^{15}$ cm^{-3}). For the incident laser intensities used, the resulting photocarrier population is sufficiently large that ambipolar diffusion may be assumed. In this case, the equation governing the photocarrier generation and diffusion at the modulation frequency ω in the semiconductor may be written in the form:

$$\nabla^2(\Delta n) + \left[\frac{2i}{\mu_{el}^2} \right] \Delta n = \frac{-q(1-\bar{R})I_0\beta}{h\nu D} e^{-\rho^2/R^2} e^{-\beta z} , \qquad (1)$$

[*]   The Shockley-Read model for recombination through monovalent flaws is assumed to be valid for the samples discussed in this chapter. In the experiments to be discussed, the injected photocarrier density is significantly larger than the equilibrium free electron and hole densities in the silicon. It therefore follows that the lifetime that is measured is $\tau = \tau_\infty = \tau_{no} + \tau_{po}$, in the notation of, for example: J.S. Blakemore, (1962), *Semiconductor Statistics*, Pergamon Press, New York, p. 270.

subject to the boundary conditions

$$D\frac{d(\Delta n)}{dz}\bigg|_{z=0} = s_o \,\Delta n\,|_{z=0} \tag{2}$$

and

$$D\frac{d(\Delta n)}{dz}\bigg|_{z=d} = s_d \,\Delta n\,|_{z=d}, \tag{3}$$

where

$$\mu_{el}^2 = \frac{2D}{\omega + i/\tau} \;. \tag{4}$$

and $D = 2D_n D_p/(D_n + D_p)$ is the ambipolar diffusion coefficient. $\bar{R}$ is the optical reflectivity of the semiconductor, the parameter β is the optical absorption coefficient, while s_o and s_d are the surface recombination velocities at the surfaces $z=0$ and $z=d$, respectively. The photocarrier lifetime within the bulk material is denoted by τ, and q is the quantum efficiency of the semiconductor. In our experiments on silicon, an Argon laser was employed, with a wavelength $\lambda = 5145$ Å and $h\nu = 2.41$ eV. Therefore, it will henceforth be assumed that $q = 1$. The excess free hole density in the semiconductor at the modulation frequency ω is, as already stated, equal to $\Delta n(\mathbf{r})$.

Under steady state excitation, ($\omega=0$), the parameter μ_{el} is the $1/e$ electronic diffusion length and is, in general, complex. When $\omega \gg 1/\tau$, the decay length of the diffusion wave becomes $\mu_{el} \approx \sqrt{2D/\omega}$, and is much smaller than the diffusion length at low frequencies. From Eq. (1), it is seen that the quantity μ_{el} defines the spatial extent of the photocarrier diffusion into the semiconductor.

The periodic heating in the semiconductor $\theta(\mathbf{r})$ at the modulation frequency ω is governed by a diffusion equation of the form:

$$\nabla^2\theta + \left[\frac{2i}{\mu_{th}^2}\right]\theta = \frac{1(1-\bar{R})I_o\beta}{h\nu k_{th}}(h\nu - E_g - 3k_B T)e^{-\rho^2/R^2}e^{-\beta z}$$
$$-\Delta n\left[\frac{(E_g + 3k_B T)}{k_{th}\tau}\right] \tag{5}$$

subject to the boundary conditions:

$$k_{th}\frac{d\theta}{dz}\bigg|_{z=0} = -s_o\Delta n\,|_{z=0}(E_g + 3k_B T) \tag{6}$$

and

$$k_{th}\frac{d\theta}{dz}\bigg|_{z=d} = -s_d\Delta n\,|_{z=d}\,(E_g+3k_BT)\,,\qquad(7)$$

where

$$\mu_{th}^2 = \frac{2k_{th}}{\rho C\omega} = \frac{2\kappa}{\omega}\qquad(8)$$

Here k_{th} denotes the thermal conductivity, κ the thermal diffusivity, μ_{th} the thermal diffusion length, and C the specific heat of the sample. The parameters E_g, k_B, and T denote, respectively, the band gap energy of the semiconductor, the Boltzmann constant, and the ambient temperature.

The sources of the periodic heating in the semiconductor are of interest. The first term on the right hand side of Eq. (5) corresponds to the heat released in the semiconductor due to rapid de-excitation of the photogenerated electrons and holes to the conduction and valence band edges, respectively. The de-excitation, occurring on a time scale of picoseconds, may be considered instantaneous for the purposes of the present work. This source term may therefore be considered to represent a direct conversion of a fraction of the incident absorbed laser energy to heat in the semiconductor, that fraction being equal to $(h\nu-E_g-3k_BT)/h\nu$.

The second term on the right hand side of Eq. (5) represents the heat generated in the semiconductor due to the recombination of the photogenerated electrons and holes. This term is thus linearly related to the photocarrier density. It is assumed in writing the term, as represented in Eq. (5), that the recombination process is predominantly nonradiative. This is certainly the case for silicon at room temperature.

Finally, the boundary conditions of Eqs. (6) and (7) represent the heat generated at the semiconductor surfaces due to surface recombination.

The homogeneous differential equations associated with Eqs. (1) and (5) are similar in form. The difference between the homogeneous periodic free carrier and thermal diffusion equations has its origin in the finite lifetime of the photocarriers, as can be seen from a comparison of the two decay lengths μ_{el} and μ_{th}. For $\omega\gg1/\tau$, the two equations are isomorphic, and the periodic photocarrier distribution can be described generally by an exponentially damped wave, exactly analogous to a thermal wave. For $\omega\ll1/\tau$, the carrier diffusion is seen to become independent of ω, and therefore we do not expect the periodic photocarrier distribution to exhibit a wavelike nature in this regime, but rather expect it to be described by a purely exponential decay in space.

Both Eqs. (1) and (5) may be solved exactly, for example by the use of Hankel transform techniques; however the solutions are not amenable to direct physical insight. To obtain a general idea of the periodic photocarrier distribution in the illuminated semiconductor, we will solve Eq. (1) under the simplifying conditions that $|\beta\mu_{el}|\gg1$, $R\gg|\mu_{el}|$, $d\gg|\mu_{el}|$, and that s_o and s_d are negligible. Under these conditions, Eq. (1) may be readily solved to obtain the result:

$$\Delta n(\mathbf{r}, \omega) = \frac{I_O(1-\bar{R})}{h\nu\, D} \frac{\mu_{el}}{(1-i)} e^{(i-1)z/\mu_{el}} \tag{9}$$

The general form of Eq. (9) confirms the above predictions: for $\omega \gg 1/\tau$ the periodic carrier distribution propagates as an exponentially damped wave, while for $\omega \ll 1/\tau$, $\Delta n(\mathbf{r})$ decays exponentially. In all cases, the characteristic decay length is equal to $|\mu_{el}|$. It is also seen in Eq. (9) that for $\omega \gg 1/\tau$, the periodic photocarrier density is proportional to $\omega^{-\frac{1}{2}}$, while the value of $\Delta n(\mathbf{r})$ is integrated throughout the semiconductor $\propto \omega^{-1}$. For values of $\omega \ll 1/\tau$, the periodic photocarrier density is independent of ω, as expected. These general features of the periodic photocarrier distribution in the irradiated semiconductor will be observed to be evidenced clearly in the experiment to be discussed in the following section.

III. The Acoustic Phase Perturbation in Silicon

The first technique which we will discuss involves the measurement of the phase perturbation of an externally-generated acoustic wave propagating through a region of silicon irradiated by a modulated Argon-ion laser beam (Fig. 2). The acoustic wave in this case is a Rayleigh type surface acoustic wave. In the region of the optical absorption, a periodic photocarrier population and temperature variation is generated in the semiconductor. As with any insulator or metal, this periodic heating produces a periodic phase perturbation of the acoustic wave, due largely to the dependence of the elastic constants of the sample on temperature.

In silicon, as in most multivalley semiconductors, it is found that in addition to the local heating, the periodic photocarrier population also contributes directly to the acoustic phase perturbation. This contribution results from the dependence of the elastic constants of the semiconductor on the free carrier density. Hence, a measurement of the acoustic phase perturbation that results from laser absorption in silicon yields information concerning the photoinjected free carrier population, as well as the periodic heating in the sample.

The effect of the free electron density on the elastic constants was first described by Keyes (1961,1967). Subsequently, the dependence of the elastic constants of silicon on the free electron density has been thoroughly investigated experimentally and, with the exception of the elastic constant c_{44}, is well understood theoretically (Keyes, 1961; 1967;Einspruch and Csavinsky, 1963; Hall, 1967). The dependence of the elastic constants on the free hole density is more complex in theory, due to the nature of the valence bands in silicon; experimental results are quite scarce. However, it is believed that the effect of free holes on the elastic properties of silicon is comparable to that of free electrons (Table I).

This electronic contribution to the elastic constants is small. For silicon in the nondegenerate regime, a variation in the free carrier density of Δn cm^{-3} will

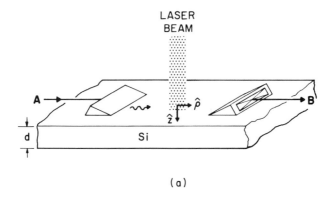

(a)

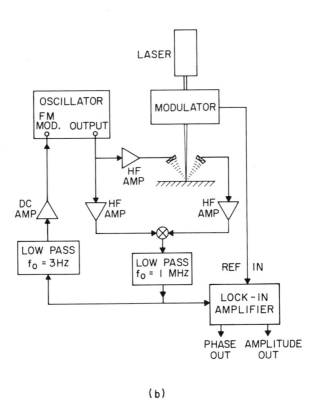

(b)

Fig. 2 (a) The transducer configuration used in the phase perturbation measurement of a 50 MHz Rayleigh wave on silicon. (b) The experimental system used to detect the acoustic phase perturbation.

Table I

The Dependences of the Elastic Constants of Silicon on the Free Electron
and Free Hole Densities, as Measured Experimentally
(Units: $N{-}cm^3/m^3{\times}10^{11}$)

	c_{11}	c_{12}	c_{44}
d /dn	$-16^{a,b}$	$7.7^{a,b}$	$-3.0^{a,b}$
d /dp	$-2.2^{c,d}$	$1.1^{c,d}$	-3.2^{e}

[a] Einspruch N.G. and Csavinsky P. (1963). Appl. Phys. Lett. **2**, 1
[b] Hall J.J. (1967). Phys. Rev. **161**, 756
[c] Csavinsky P. and Einspruch N.G. (1963). Phys. Rev. **132**, 2434
[d] Mason W.P. and Bateman T.B. (1964). Phys. Rev. **A134**, 1387
[e] Kim C.K., Cardona M. and Rodriguez S. (1976). Phys. Rev. **B13**, 5429

produce a relative change in the elastic constants of the order of 10^{-21} Δn (note that the variation in the elastic constants is linear with Δn for nondegenerate statistics). In previous studies, which involved the measurement of the absolute acoustic velocity in silicon, electronic effects in the elastic properties have been very difficult to observe for free carrier concentrations below $\sim 10^{18}$ cm^{-3}. With the present technique, which employs the detection of a periodic acoustic phase perturbation, it will be seen that periodic photocarrier densities of the order of $\sim 1{\times}10^{13}$ cm^{-3} may be detected. Hence, even though the governing effects are small, measurement of the acoustic phase perturbation proves to be quite a sensitive means of detecting photocarrier densities.

A schematic of the acoustic phase measurement system is shown in Fig. 2. Two surface wave transducers are placed on the sample to be investigated. One transducer is driven by an external oscillator, the other receives the transmitted acoustic wave. A modulated laser beam is directed onto the semiconductor sample in the path of the acoustic wave.

We define the output of the external oscillator to be $A \cos\xi t$. For sinusoidal modulation of the phase of the acoustic wave, the output of the receiving transducer is of the form:

$$V_1(t) = B \cos[\xi t + \delta \cos(\omega t + \psi) - \phi_{dc}], \qquad (10)$$

where ω is again the modulation frequency of the laser, ϕ_{dc} is the phase shift associated with the propagation of the acoustic wave between the transducers, and is therefore dependent on ω, as well as on the D.C. heating and D.C. photocarrier population in the sample. The magnitude of the periodic phase perturbation δ is assumed to be small. The outputs of the oscillator at frequency ξ and of the receiving transducer [i.e., $V_1(t)$] are mixed. The resulting waveform is then low-pass filtered to yield an output of the form

$$V_2(t) = \cos[\delta \cos (\omega t + \psi) - \phi_{dc}] \qquad (11)$$

Since the phase perturbation δ is small in magnitude, the signal $V_2(t)$ can be made to be directly proportional to the phase perturbation of the acoustic wave, $\Delta\phi = \delta \cos (\omega t + \psi)$, by adjusting the oscillator frequency slightly so that $\phi_{dc} = \pi/2$. The output signal is then fed into a two-phase lock-in amplifier.

The time-averaged voltage level from the output of the mixer is proportional to $\cos \phi_{dc}$. Hence, the mixer output must be zero for the signal frequency ω to be truly proportional to the acoustic phase perturbation. To assure that this condition is satisfied, the negative feedback system shown in Fig. 2b is employed. Slow drifts in the time-averaged voltage are amplified and fed into the FM modulation port on the external oscillator. This causes the oscillation frequency to shift in such a way as to maintain zero average voltage at the mixer output. In this way, the measurement is stable to long term drift.

In the present experiment, the frequency of the acoustic waves is 50 MHz, and the radius of the incident laser beam is $R \approx 2$ mm. Under these conditions, it has been determined that the minimum acoustic phase perturbation that may be detected is equal to $\sim 1.2 \times 10^{-7}$ rad (rms) in a 1 Hz bandwidth (Stearns, 1985). This corresponds to a relative change in the elastic constants of the order of 5×10^{-9}. From the magnitude of the electronic dependence of the elastic constants indicated earlier, we should thus expect to be able to measure a periodic photocarrier density of the order of 10^{13} cm^{-3} in the silicon.

A particularly important aspect of the measurement technique is the range of modulation frequencies obtainable. The modulation frequency ω may be any value within the bandwidth of the transducer. For a 50 MHz Rayleigh wave transducer, this means that $\omega/2\pi$ may vary from a few hertz to perhaps 10 MHz. The practical significance of this large bandwidth will be evident in the results to be discussed below.

To study the electronic contribution to the acoustic phase perturbation in silicon, an experiment is performed on a well-characterized sample of n-type silicon, doped with 3×10^{13} P atoms/cm^3. The sample is thermally oxidized to reduce surface recombination, and has been measured independently to have a bulk carrier lifetime of 153 $\mu s^{(*)}$. A 50 MHz Rayleigh wave is launched and received on the silicon wafer by a pair of wedge transducers. The beamwidth of the Rayleigh wave is approximately 3 mm. An Argon-ion laser beam (514.5 nm) of 80 mW CW power is acousto-optically modulated and directed onto the oxidized silicon surface in the path of the Rayleigh wave, midway between the two transducers. The laser beam diameter at the silicon surface is approximately 4 mm.

The acoustic phase perturbation of the Rayleigh wave may be shown to be of the form (Kino, 1978):

(*)   The measurement of the lifetime τ was determined from the decay of the photoconductivity of the sample following transient optical excitation.

$$\Delta\phi = \frac{-\omega}{4P_O} \int_V (\Delta\rho \, v_i \cdot v_i^* + S_{ij} \, \Delta c_{ijkl} \, S_{kl}^*) dV \qquad (12)$$

Here $\Delta\rho$ and Δc_{ijkl} denote the change in the mass density and elastic constant of the silicon, S_{ij} and v_i are the strain and particle velocity due to the Rayleigh wave, V is the volume of the perturbed region, and P_O is the total acoustic beam power. It is assumed in Eq. (12) that the changes in mass density and elastic constant act as a weak perturbation on the surface acoustic wave. Furthermore, diffraction of the acoustic beam between transducers is neglected. This is valid in the present experiment, where the transducers are only some 1.5 cm apart.

In the experiment, the Rayleigh wave is made to propagate along the [100] direction of the (001) cut silicon wafer. For this geometry, the particle velocity and strain fields of the Rayleigh wave are well known (*Microwave Acoustics Handbook*, 1973). In Eq. (12), the changes in the elastic constants and mass density result not only from periodic heating, but also from the periodic photocarrier density in the silicon. It is therefore necessary to determine both the modulated free carrier density and temperature fields within the silicon.

To determine the photogenerated free carrier density in the illuminated silicon, Eq. (1) must be evaluated. The ambipolar diffusivity is taken to be $D = 15$ cm^2/s, a typical value. It follows then that the largest value of the electronic diffusion length in the silicon, which occurs when $\omega \ll 1/\tau$, will be $|\mu_{el}| = 0.479$ mm. Hence, the photocarrier diffusion may be considered to be one-dimensional, and Eqs. (1)-(4) may be readily solved to obtain the result:

$$\Delta n = \frac{I_0(1-\bar{R})}{2h\nu} [\{(\xi+s_d)\exp[(i-1)\frac{z-d}{\mu_{el}}]+(\xi+s_d)\exp[(1-i)\frac{z-d}{\mu_{el}}]\}*$$

$$\{(\xi+s_o)(\xi+s_d)\exp[(1-i)d/\mu_{el}]-(\xi-s_o)(\xi-s_d)\exp[(i-1)d/\mu_{el}]\}^{-1}+c.c] \qquad (13)$$

where

$$\xi = D\frac{(1-i)}{\mu_{el}} . \qquad (14)$$

In the experiment, the optical reflectivity of the oxidized silicon is measured to be $\bar{R}=0.12$, and the wafer thickness to be $d = 0.33$ mm. Evaluating Eq. (13) in the regime $\omega \ll 1/\tau$, with $I_0=0.41$ W/cm^2, $h\nu=4.0\times10^{-19}$ J, and assuming temporarily that $s_o=s_d=0$, we predict that $\Delta n \approx 4.9\times10^{15}$ cm^{-3} at the illuminated surface of the silicon. This clearly justifies the assumption of ambipolar diffusion.

The periodic heating in the silicon may be predicted by solving Eq. (5), using Δn, given by the expression shown in Eq. (13). It is reasonable to assume one-dimensional heating for the geometry of the present experiment; for the modulation frequencies of interest, one may take $\beta\mu_{th}\gg1$. Even with these simplifications, the form of the periodic heating in the silicon is algebraically very complex, and will not be explicitly shown here.

Having solved for the fields $\Delta n(z)$, $\Delta p(z)$, and $\theta(z)$ in the silicon, in order next to predict the phase perturbation of the surface wave, the dependences of the elastic constants and mass density on the temperature and free carrier density in the sample must be known. The changes in the elastic constants with temperature are taken to be: $dc_{11}/dT = -1.4 \times 10^7 \ N/m^2\text{-}K$, $dc_{12}/dT = -5.8 \times 10^6 \ N/m^2\text{-}K$, and $dc_{44}/dT = -6.2 \times 10^6 \ N/m^2\text{-}K$ (Burenkov and Nikanorov, 1974). The dependence of the mass density on temperature is $d\rho/dT = -2.1 \times 10^{-2} \ kg/m^3\text{-}K$ (*Handbook of Chemistry and Physics*, 1973). The dependences of the elastic constants on the free electron and free hole density are shown in Table I. Finally, the dependence of the mass density on the electron-hole pair concentration may be written as $d\rho/d(n-p) = 6.3 \times 10^{-27} \ kg-m^3/m^3$, where $n-p$ denotes the density of photogenerated electron-hole pairs in the silicon (Figielski, 1961).

Eq. (12) is then evaluated to predict the phase perturbation of the Rayleigh wave due to the absorption of the incident modulated laser beam, assuming a bulk carrier lifetime of $\tau = 155 \ \mu s$, and front and back surface recombination velocities of $s_o = 15 \ cm/s$ and $s_d = 30 \ cm/s$, respectively. The predicted phase perturbation is shown in Fig. 3. Shown separately are the thermal and electronic contributions to the acoustic perturbations. The magnitude of the thermal contribution is seen to vary as ω^{-1} for modulation frequencies $\omega/2\pi < 500 Hz$. The inflection in both the magnitude and phase of the thermal contribution at $\omega/2\pi \sim 1$ kHz is due to a damped resonance of the thermal wave in the thickness of the silicon wafer.

The electronic contribution to the acoustic phase perturbation in the silicon is seen to be independent of ω for modulation frequencies less than ~1 kHz. This is to be expected from the earlier discussion of Eq. (9): for values of $\omega < 1/\tau$, the photocarrier densities Δn and Δp do not vary with frequency. The phase of the electronic contribution to the acoustic perturbation, with respect to the laser beam modulation, is $0°$ for values of $\omega < 1/\tau$, as expected. At $\omega \sim 1/\tau$, the magnitude of the electronic contribution begins to roll off, while the phase becomes less than $0°$. This roll-off occurs at $\omega/2\pi \sim 1$ kHz in the present case. The inflection in both the magnitude and phase of the electronic contribution at $\omega/2\pi \sim 15$ kHz may be interpreted as arising from a damped resonance of the photocarrier population in the thickness of the silicon. This inflection reflects the low surface recombination velocities present at the surfaces of the silicon wafer.

The fact that, for modulation frequencies larger than ~100 Hz, the thermal contribution to the total acoustic phase perturbation is negligible, is of primary importance. This implies that for values of $\omega/2\pi > 100$ Hz, the acoustic phase perturbation measurement is essentially a probe of the photocarrier density in the silicon. Information directly concerning the electronic properties of the silicon may therefore be obtained with the measurement. For example, the bulk photocarrier lifetime may be determined from the value of ω at which the magnitude of the acoustic perturbation rolls off.

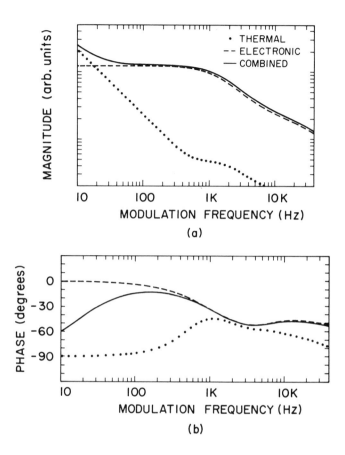

Fig. 3 Theoretical curves showing the thermal and electronic contributions to the phase perturbation of a 50 MHz Rayleigh wave on silicon. (a) Magnitude of the acoustic perturbation as a function of modulation frequency. (b) Phase of the perturbation relative to the incident laser beam modulation.

Comparison of the predicted acoustic phase perturbation with the experimental results is shown in Fig. 4. It will be observed that the data is recorded over a very large range of modulation frequencies. The value of being able to work over such a range of frequencies is immediately evident, for we may clearly see from the data the decay of the thermal contribution to the phase perturbation for $\omega/2\pi > 100$ Hz, as well as the roll-off in amplitude of the overall perturbation at $\omega \sim 1/\tau$. The agreement between experiment and theory is seen to be excellent, indicating that our understanding of the electronic effects on the propagation of acoustic waves in silicon is both qualitatively and quantitatively correct.

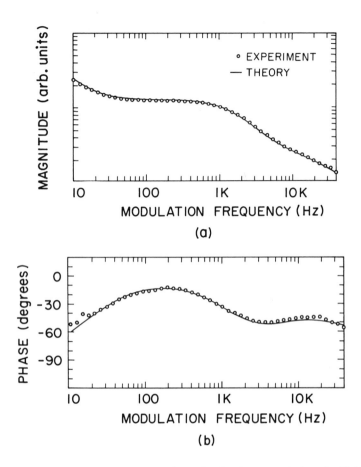

Fig. 4 Comparison of theory with experiment for the phase perturbation of a 50 MHz Rayleigh wave on Silicon. (a) Magnitude of the acoustic perturbation as function of modulation frequency. (b) Phase of the acoustic perturbation.

Because of the wide range of modulation frequencies obtainable in the experiment, it should be possible with this phase perturbation technique to measure bulk carrier lifetimes in the silicon over a large range of values, from perhaps 0.01 μs to 1 ms. It should also be possible to focus the incident laser beam, and hence to obtain information on recombination effects and local photocarrier density with high spatial resolution. An image of the magnitude of the acoustic phase perturbation at a modulation frequency where the thermal contribution has become negligible would then directly reflect such variations in the electronic properties of the sample.

The technique should be applicable to most multivalley semiconductors, and to any other semiconductors that exhibit an electronic contribution to the elastic constants. Furthermore, the technique should be sensitive to any method of periodic carrier injection. Other types of acoustic waves may, of course, be used, for example bulk shear and longitudinal waves in the semiconductor. For such simple acoustic modes, only one elastic constant may be involved in the acoustic phase perturbation [i.e., in Eq. (12)]. In that case, it may be possible to use the perturbation technique to investigate, with high sensitivity, the fundamental interaction between the acoustic and free carrier fields in the semiconductor.

IV. The Electronic Contribution to Photoacoustic Generation in Silicon

We have shown in the preceding section that the effect of photoinjected free carrier population on the elastic constants of a semiconductor may be detected acoustically, and that the results yield useful information concerning the photocarrier lifetime. In general, an injected free carrier population will not only perturb the elastic constants of a semiconductor, but will also introduce a local strain in the sample. This strain in turn may produce acoustic waves in the semiconductor, in a manner wholly analogous to acoustic wave generation by local periodic thermal expansion.

It is important to understand the relative sizes of the thermal and electronic contributions, especially in light of several recent and interesting results on photoacoustic imaging of semiconductor devices (Rosencwaig and White, 1981; Rosencwaig, 1982). For example, if in a given experiment it is the case that electronic strain is dominant in the production of acoustic waves, then a photoacoustic image of the semiconductor may reflect directly the electronic properties of that semiconductor. We therefore investigate, in this section, the contribution of the strain produced by the periodic photoinjection of free carriers to the photoacoustic generation in silicon.

The strain produced in a cubic semiconductor, due to the injection of Δn electron-hole pairs, may be shown to be of the form (Figielski, 1961)

$$S_{ij}^{el} = \frac{1}{3} \frac{dE_g}{dP} (\Delta n) \delta_{ij} \tag{15}$$

Here δ_{ij} denotes the unit tensor; the parameter dE_g/dP denotes the (hydrostatic) pressure dependence of the energy gap of the semiconductor at constant temperature. In silicon, this strain has the numerical value $S_{ij}^{el} = -9.0\times10^{-25}(\Delta n)\,\delta_{ij}$, where Δn is expressed in units of cm^{-3} (*Handbook of Chemistry and Physics*, 1973). This electronic strain may be compared with the thermal strain that arises from the periodic heating of the semiconductor, $S_{ij}^{th} = \alpha\,\theta\,\delta_{ij}$, where α is the linear coefficient of thermal expansion of the semiconductor.

It is clear from the above equations that the mechanical strain due to photocarrier injection and local heating are very similar in form for semiconductors of cubic symmetry. The strains vary linearly with Δn and θ, respectively, and

both are isotropic. In a measurement of photoacoustic generation in a semiconductor, we would therefore expect that, as long as the incident radiation is of an
energy greater than the band gap energy, the electronic strain will produce
acoustic waves in the semiconductor in a manner analogous to the thermal
strain. Furthermore, in the case of silicon, where $\alpha = 3.0 \times 10^{-6} \ C^{-1}$ the electronic and thermal strains will be of opposite sign (*Handbook of Chemistry and
Physics*, 1973). Hence, we would expect acoustic waves generated from the
photoinjection of free carriers in the silicon to be opposite in phase to acoustic
waves generated from the periodic heating of the silicon.

In general, it is difficult to determine *a priori* whether the thermal or the
electronic contribution to the photoacoustic generation in the semiconductor is
dominant. By making use of the difference in sign between the two contributions, we have performed an experiment to specifically investigate the impor-

Fig. 5 Schematic of the photoacoustic measurement. (a) The system used to detect the
1.10 MHz photoacoustic signal. (b) The configuration of the silicon wafer, aluminum
buffer rod, and PZT transducer. The silicon wafer is mounted with wax onto the buffer
rod. The PZT transducer is 1.0" in diameter.

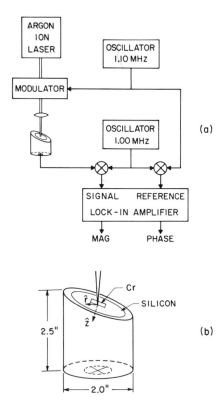

tance of the electronic strain on photoacoustic generation in silicon. A strip of chromium, 1000 Å thick, is evaporated onto a thermally-oxidized wafer of <100>-cut silicon, doped with 3×10^{13} P atoms/cm^3. The wafer is then mounted onto an aluminum buffer rod, onto the back of which is bonded a 1.0 MHz lead zirconium titanate (PZT) longitudinal bulk-wave transducer (Fig. 5). It will be noted that, in the experiment, the plane of the transducer is fixed at an angle of 30° with respect to the semiconductor surface. This is done to enhance the detection of the generated longitudinal wave which is known to exhibit strong directivity with very weak generation normal to the illuminated surface (Scruby, Dewhurst, Hutchins and Palmer, 1981). An Argon-ion laser beam (514.5 nm) of 20 mW CW power is acousto-optically modulated at 1.10 MHz and focused onto the surface of the silicon wafer. The signal from the acoustic transducer is electronically mixed down to ~100 kHz and fed into a two-phase lock-in amplifier. The output of the lock-in amplifier is then recorded, as the laser beam is scanned across the strip of metallizaiton and onto the oxidized silicon surface.

The photoacoustic signal that results when the laser beam is incident on the chromium film is entirely thermal in origin, as the incident light is absorbed within the chrome, and no excess carriers are generated in the underlying silicon. When the laser beam is focused onto the unmetallized silicon surface, both thermal and electronic contributions are present in the photoacoustic generation. Hence, by comparing the photoacoustic signal that results when the laser beam is directed onto the metallized and unmetallized silicon, the relative sizes of the thermal and electronic contributions to the photoacoustic generation in the silicon may be inferred.

When the laser beam is incident on the oxidized silicon surface, Eqs. (1) through (8) govern the periodic generation and diffusion of the photocarriers and heating, respectively, in the silicon. Several simplifications of these equations may be made. The bulk carrier lifetime in the silicon is on the order of 50 μsec, and hence in the experiment $\omega \gg 1/\tau$. This implies that $\beta |\mu_{el}| \gg 1$, as $\beta \approx 2.4 \ \mu m^{-1}$ and $|\mu_{el}| \sim 21 \mu m$ (assuming D = 15 cm^2/sec (*American Institute of Physics Handbook*, 1972). Furthermore, $d/|\mu_{el}| \gg 1$. Under these conditions, Eq. (1) may be written in the simplified form:

$$\nabla^2(\Delta n) + \left[\frac{i\omega}{D} \right] \Delta n = 0 , \qquad (16)$$

subject to the boundary condition:

$$D \left. \frac{d(\Delta n)}{dz} \right|_{z=0} = \frac{(1-\bar{R}_{Si})}{h\nu} I_0 e^{-p^2/R^2} + s_o \Delta n \left. \right|_{z=0} \qquad (17)$$

and the condition that $\Delta n \to 0$ as $z \to \infty$. Furthermore, as $D/|\mu_{el}| \gg s_o$ ($s_o \approx 20$ cm/s), the boundary term involving s_o in Eq. (17) may be neglected.

When the laser beam is incident on the metallized portion of the silicon, assuming all of the light is absorbed in the 1000 Å layer of chromium, there will be no photogeneration of carriers, and hence $\Delta n=0$ throughout the silicon.

In Eq. (5), for the periodic heating in the silicon, the source term proportional to Δn may be neglected, as $\omega \gg 1/\tau$. The thermal diffusion length in the silicon at 1.1 MHz is $\mu_{th} = 5.1$ μm, so we may write that $d/\mu_{th} \gg 1$; it also follows that $\mu_{th}\beta \gg 1$. Under these conditions, Eq. (5) may be simplified to the form:

$$\nabla^2\theta + \left[\frac{i\omega\rho C}{k_{th}}\right]\theta = 0 \tag{18}$$

subject to the boundary condition:

$$k_{th}\frac{d\theta}{dz}\bigg|_{z=0} = \frac{(1-\bar{R}_{Si})}{h\nu}(h\nu - E_g - 3k_B T)I_0 e^{-p^2/R^2} \tag{19}$$

and the condition that $\theta \rightarrow 0$ as $z \rightarrow \infty$.

When the laser beam is incident on the metallized portion of the sample, the boundary condition of Eq. (19) becomes:

$$k_{th}\frac{d\theta}{dz}\bigg|_{z=0} = (1-\bar{R}_{Cr})I_0 e^{-p^2/R^2}, \tag{20}$$

where $\bar{R}_{Cr}$ is the optical reflectivity of the chromium metallization. Here the 1000 Å layer of chrome is treated merely as a source of heating at the surface of the silicon. Thermal diffusion within the layer itself may be ignored, as the thermal wavelength in chromium at 1.1 MHz is ~17 µm.

To derive the excitation of the acoustic transducer in the experiment, we will follow closely the theory of Kino and Stearns (1985), which makes use of the reciprocity theorem to derive the generation of acoustic waves due to thermal excitation. Because, in their work, only thermal strains were assumed to be present, a slight modification of the formalism must be made to include the effects of the electronic strain on the photoacoustic generation. This may be achieved in a straightforward manner by writing the stress in the illuminated silicon to be of the form:

$$T_{ij}^1 = c_{ijkl}\{S_{kl} - \alpha\theta\delta_{kl} - \frac{1}{3}\frac{dE_g}{dP}(\Delta n)\,\delta_{kl}\}\,. \tag{21}$$

Here we have included the electronic strain in the stress-strain relation in a manner analogous to the inclusion of thermal strain.

With this definition of the stress tensor, it follows from the work of Kino and Stearns (1985) that the excitation of the receiving transducer by the longitudinal acoustic wave generated in the silicon will be of the form

$$V = K \int_\Gamma \Delta^o \left[\alpha\theta + \frac{1}{3} \frac{dE_g}{dP} (\Delta n) \right] d^3r \; . \tag{22}$$

The parameter V is the output voltage of the acoustic transducer, K is a proportionality constant whose precise value will not be important here, and Δ^o is the volume dilatation associated with the acoustic field that would be excited by the receiving transducer, if it were externally driven, with no illumination present. Hence, in the present experiment, Δ^o is the volume dilatation associated with a quasi-plane wave incident on the free silicon surface at an angle of 30° to the surface normal. The integral is evaluated over the entire volume Γ of the silicon wafer and aluminum buffer rod used in the experiment.

As has been discussed above, the fields Δn and θ in Eq. (22) extend at most only some tens of microns into the bulk of the silicon. This is much less than the acoustic wavelength in silicon at 1.1 MHz ($\lambda_{ac} \approx 8$ mm). As discussed by Kino and Stearns (1985), in this case Δ^o may be taken to be constant in the above integral, and equal to $\Delta^o|_{\rho,z=0}$. Substituting Eqs. (16) and (18) into Eq. (22), we obtain the following expression for the transducer output:

$$V = \left[\frac{K}{-i\omega} \right] \Delta^o \bigg|_{\rho,z=0} \int_\Gamma \left[\frac{k_{th}\alpha}{\rho C} \nabla \cdot \nabla\theta + \frac{D}{3} \frac{dE_g}{dP} \nabla \cdot \nabla(\Delta n) \right] d^3r \tag{23}$$

Applying Gauss' theorem and the boundary conditions of Eqs. (17), (19), and (20), it follows that for illumination of the oxidized silicon,

$$V_{Si} = \left[\frac{K\pi R^2}{-i\omega} \right] (1-\bar{R}_{Si}) \frac{I_o}{h\nu} \Delta^o|_{\rho,z=0} \left[\frac{\alpha}{\rho C} (h\nu - E_g - 3k_BT) + \frac{1}{3} \frac{dE_g}{dP} \right] \tag{24}$$

For illumination of the metallized silicon, we find

$$V_{Cr} = \left[\frac{K\pi R^2}{-i\omega} \right] (1-\bar{R}_{Cr}) I_o \Delta^o|_{\rho,z=0} \left[\frac{\alpha}{\rho C} \right] \tag{25}$$

The bracketed expression in Eq. (24) consists of two terms, one involving α, the other involving dE_g/dP. These terms correspond to the contributions of the thermal and electronic strains, respectively, to the total photoacoustic signal. In the experiment, $h\nu = 3.9\times10^{-12}$ erg, $dE_g/dP = -2.9\times10^{-24}$ cm³, $E_g = 1.78\times10^{-12}$ erg, $\rho = 2.33$ gr/cm³, and $C = 7.12\times10^6$ erg/gr - K (*Handbook of Chemistry and Physics*, 1973; *American Institute of Physics Handbook*, 1972). Substituting these values into the bracketed expression in Eq. (24), it is found that the contribution of the electronic strain to the total photoacoustic signal is approximately -2.6 times that of the thermal strain.

The ratio of the total transducer response, when the laser beam is incident on the oxidized silicon, to the response when incident on the metallized silicon surface, is simply equal to the ratio of Eq. (24) to Eq. (25). In the experiment, $\bar{R}_{Si} = 0.18$ and $\bar{R}_{Cr} = 0.54$. Thus, from Eqs. (24) and (25), we predict that $V_{Si}/V_{Cr} = -1.3$.

Fig. 6 shows the magnitude and phase relative to the incident modulated laser beam of the experimental photoacoustic signal, as the laser beam is scanned across the edge of the chromium metallization. As predicted, there is a clear 180° phase shift in the photoacoustic signal across the edge of the metallization. Furthermore, the magnitude of the photoacoustic signal is approximately 40% larger when the laser beam is incident on the oxidized silicon than when incident on the chromium, in good agreement with the theoretical prediction.

Fig. 6 The photoacoustic signal, as the laser beam is scanned across the edge of the Cr metallization. (a) Magnitude of the transducer output. Positive values of the horizontal axis refer to the region of metallized silicon; negative numbers correspond to unmetallized silicon. (b) Phase of the output relative to the modulation of the incident laser beam. A 180° phase shift is clearly present at the chromium edge.

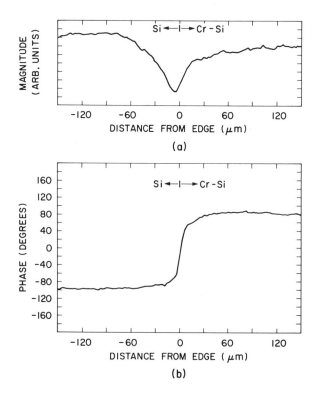

We therefore have a good understanding of the electronic contribution to photoacoustic generation in silicon. It is evident that, for laser excitation at 514.5 nm, the 1.1 MHz photoacoustic signal is actually dominated by the electronically-induced component of the generation. We may note, in fact, that for an incident laser beam of energy nearer E_g, the electronic contribution would dominate even more thoroughly, as the term $(h\nu - E_g - 3k_B T)$ in Eq. (24) would become smaller.

In the derivation above, no mention has been made of the bond between the silicon wafer and the aluminum buffer rod. This bond will presumably affect the value of $\Delta^o \vert_{\rho,z=0}$, but as long as the bond is uniform over the distance scanned experimentally (i.e., over the 100 microns centered on the edge of the metallization), the predicted ratio of V_{Si}/V_{Cr} will be independent of $\Delta^o \vert_{\rho,z=0}$, and hence will be unaffected by the bond.

Furthermore, it has tacitly been assumed that in the above derivation, the injected photocarrier density is sufficiently low to ignore the effects of Auger recombination in the silicon. To be convinced of this assumption, Eqs. (16) and (17) may be solved directly, to obtain the result:

$$\Delta n(\rho,z) = \frac{(1-\bar{R}_{Si})I_O \pi R^2}{2\pi\, h\nu} \int_0^\infty \frac{\exp(-bz)}{Db+s_o}\, e^{-\lambda^2 R^2/4} J_O(\lambda\rho)\lambda d\lambda \qquad (26)$$

where

$$b = (\lambda^2 - i\omega/D)^{1/2} . \qquad (27)$$

In the experiment, $R \approx 10\ \mu m$ and $I_O \pi R^2 (1-\bar{R}_{Si}) = 12.7$mW. Eq. (26) may be evaluated for $z=\rho=O$, and it is found that $\Delta n(\rho,z=O) \approx 5.3 \times 10^{17} cm^{-3}$. Although this value is near that of the free carrier density at which Auger effects become important, it is still a fair assumption to neglect such effects here.

As a final check of the measurement, an experiment identical to that described above has been performed, except instead of silicon, a thin slice (~1 mm thick) of the ceramic silicon nitride (Si_3N_4) has been used as the absorbing substrate. Here, because silicon nitride is an insulator, no phase shift is predicted in the photoacoustic signal, as the laser beam is scanned across the edge of a 1000 Å strip of chromium on the Si_3N_4. Fig. 7 shows the photoacoustic signal for the experiment on Si_3N_4. There is a clear change in the magnitude of the signal as the laser beam is scanned over the edge of the metallizaiton, associated with the difference in optical reflectivity between the chromium and Si_3N_4. As expected, however, there is essentially no change in the phase of the photoacoustic signal across the edge of the chromium metallization. The few degrees of phase shift that are present presumably result from the additional propagation of the thermal wave within the thin layer of chromium.

It therefore appears that we have unambiguously presented evidence, both experimental and theoretical, that for the parameters of the present experiment,

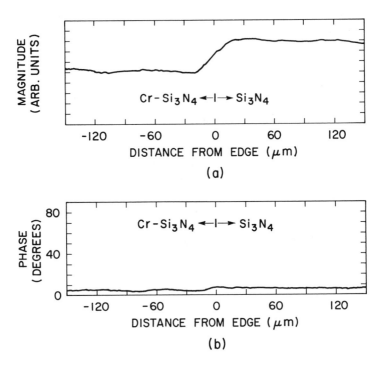

Fig. 7 The photoacoustic signal, as the laser beam is scanned across the edge of the Cr metallization on Si_3N_4. (a) Magnitude of the photoacoustic signal. (b) Phase of the transducer output.

the electronically-induced strain is the dominant source of the photoacoustic generation in the silicon. This fact may be of some practical interest, for example in the interpretation of photoacoustic images of inhomogeneous semiconductors. It may be noted that most semiconductors may be expected to exhibit a strain proportional to the density of injected free carriers, so that an electronic contribution to photoacoustic generation will be a common property to many semiconductors.

V. Photoacoustic Imaging of Surface Recombination in Silicon

The experiments of the two previous sections have been founded on the direct interaction between the photogenerated free carrier density and the elastic properties of silicon. The final experiment to be addressed in this Chapter involves the measurement of the periodic surface heating of an illuminated silicon sample at 2 MHz, and the contribution to the periodic heating arising from the surface recombination of photoinjected free carriers in the silicon.

The surface heating is measured by detection of the 2 MHz acoustic wave generated in the air above the silicon. In order to detect the acoustic wave, a novel high-frequency air transducer is used. The transducer has been described in a previous publication (Fox, Khuri-Yakub and Kino, 1983). Briefly, it consists of a spherical shell of PZT-5H, one-half acoustic wavelength in thickness (in the PZT-5H), with a one inch radius of curvature. A film of RTV silicone rubber, one-quarter acoustic wavelength in thickness, is employed as a matching layer from the ceramic into the air. The diameter of the PZT shell is one inch; the active area of the transducer is one-half inch in diameter. The acoustic transducer therefore forms an f/2 lens, with a one inch focal distance. The transducer is measured to have a one-way insertion loss of 23 dB at 2.0 MHz, neglecting acoustic losses in air. The 3 dB fractional bandwidth of the transducer is approximately 4%.

Fig. 8 Measurement of 2 MHz photoacoustic generation in air. (a) Configuration of the sample and the 2 MHz air transducer. (b) Schematic of the system used to detect the 2 MHz photoacoustic signal.

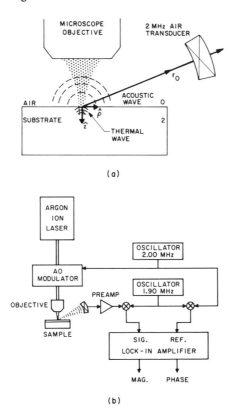

The photoacoustic measurement in air is performed as shown in Fig. 8. An Argon-ion laser beam is acousto-optically modulated at 2.00 MHz, and is focused through a microscope objective onto the sample to be investigated. The 2 MHz focused acoustic air transducer is secured in an optical gimbal mount, and is aligned so that the geometrical focus of the transducer is coincident with the focused laser beam at the sample surface. For high power microscope objectives, the distance between the sample surface and the front of the objective may be only a millimeter or less. In this case, the transducer must be mounted so that its axis makes a large angle with the normal to the sample surface. This is not inherently a problem; when the laser beam is highly focused, as will be discussed, the acoustic radiation is nearly isotropic in the air. Hence, even when placed at a large angle to the sample normal, the photoacoustic receiver is an efficient collector of the acoustic radiation. In order to obtain a photoacoustic image of the periodic heating at the silicon surface, the sample may then be scanned beneath the laser beam and acoustic transducer, while the transducer output is monitored.

The acoustic generation in the air above the illuminated solid, which results from the periodic heating of the sample, may be predicted in a straight-forward manner. The wave equation that governs the generation of acoustic waves in the air may be written in the form:

$$\nabla^2 \psi + (k+i\gamma)^2 \psi = 3\alpha\theta(r), \quad \text{with} \quad \mathbf{u}(r) = \nabla\psi(r) \tag{28}$$

subject to the boundary conditions that:

$$u_z|_{z=0} = 0 \quad \text{and} \quad u(r\rightarrow\infty) = 0. \tag{29}$$

Here α is the linear coefficient of thermal expansion in the air, $\mathbf{u}(r)$ is the acoustic particle displacement of the generated wave, and $\theta(r)$ is the periodic temperature variation in the air at the modulation frequency ω. The parameter k is the real part of the acoustic wavevector; γ is the imaginary part of the acoustic wavevector (the attenuation in air). The boundary at a $z=0$ is assumed to be perfectly rigid, which is an excellent approximation in the present case. In Eq. (29), a spherical coordinate system is assumed, with radial coordinate r.

It will be assumed that the radius of the laser spot at the sample surface R is much smaller than an acoustic wavelength in air at 2.00 MHz. This condition is easily fulfilled; at 2 MHz the acoustic wavelength in air is $\lambda_{ac} = 116\,\mu m$, while the typical spot size of the focused laser beam is 10 μm or less. It is also the case that at 2 MHz both the thermal and electronic diffusion lengths in silicon are much less than λ_{ac}. It therefore follows that in predicting the acoustic generation, the region of periodic heating in the air may be assumed to be much smaller than an acoustic wavelength is extent. Under these conditions, in predicting the far field acoustic radiation, the Green's function which satisfies the boundary conditions of Eq. (29) may be taken to be uniform over the region of periodic heating in the air, and equal to $(2\pi r)^{-1} \exp[(ik-\gamma)r]$. Eq. (28) may then be readily solved to obtain the following result for the photoacoustic generation in the far field:

$$\mathbf{u}(\mathbf{r}) = \hat{\mathbf{r}}\frac{-3\alpha(ik-\lambda)}{2\pi r}e^{(ik-\gamma)r}\int\theta(\mathbf{r})dV \qquad (30)$$

where the integral in Eq. (30) is taken over the heated volume of air above the illuminated sample surface, and $\hat{\mathbf{r}}$ is the unit vector in the radial direction. It is assumed that the thermal wavelength is much less than the acoustic wavelength, or the radius r. A time dependence of $\exp(-i\omega t)$ is assumed in this derivation.

It is interesting to note that the 2 MHz photoacoustic measurement in air is expected to be some 10-50 times more sensitive in detecting surface heating than is infrared detection at the same modulation frequency. This is primarily due to the fact that the efficiency of photoacoustic generation increases with modulation frequency, while the efficiency of blackbody radiation does not. A brief discussion of the relative sensitivities of photoacoustic and infrared detection techniques at high thermal modulation frequencies is presented in the Appendix.

We have carried out an experiment where a silicon sample containing regions of varying surface recombination (i.e., variations is s_o) is studied. The values of s_o along the sample surface have previously been determined through an infrared absorption measurement[*]. In the experiment, it is evident from Eq. (5) that there are three dominant sources of periodic heating in the silicon. The first source is associated with the immediate de-excitation of the photoinjected carriers to the band edges of the silicon. This source of heating will, in general, be only weakly dependent on the electronic properties of the sample. The second source of heating in the sample arises from the recombination of the photocarriers in the bulk of the silicon. In general this term is of interest, as it is dependent on both Δn and τ. However, in the present experiment, this term yields a negligible contribution to the periodic surface heating of the silicon. The third source of heating at the silicon surface corresponds to the surface recombination of the photogenerated free carriers. Because, in the sample to be investigated, s_o is known to vary along the silicon surface, it is expected that the periodic heating will likewise vary, as dictated by the boundary condition at $z=0$.

Before describing the experimental results, we will address some subtleties and approximations associated with the prediction of the photoacoustic signal due to the illumination of the silicon. In order to solve Eq. (5) to predict the periodic heating at the silicon surface, we must first solve for the injected periodic photocarrier density in the silicon Δn. It may be assumed in the present experiment that $\beta|\mu_{el}|$, $d/|\mu_{el}| \gg 1$. Under these conditions, Eq. (26) is the correct solution to the photocarrier diffusion equation, Eq. (1), if we write that in Eq. (27) $b = [\lambda^2 + (1/\tau - i\omega)/D]^{\frac{1}{2}}$. We shall estimate the value of $\Delta n(\rho, z=0)$ for the present experiment. In the experiment, we take the radius of the incident laser beam to be $R \approx 2\,\mu m$, and $I_o\pi R^2(1-\bar{R}_{Si}) = 15.9$ mW (here

[*] This measurement, which involves the detection of the absorption of a CO_2 laser beam (10.6 μm) due to periodic photocarrier injection by an Argon-ion laser beam (514 nm), is described in Appendix A of R.G. Stearns (1985).

$\bar{R}_{Si} = 0.3$). The bulk carrier lifetime in the silicon has been measured to be $\tau = 17$ μs (Stearns, 1985). Eq. (26) is then evaluated to obtain the value $\Delta n\,|_{\rho,z=0} \approx 8\times10^{18}\ cm^{-3}$. This is a large photocarrier density. The effective lifetime corresponding to Auger recombination in silicon may be written in the form: $\tau_A = \gamma^{-1} n^{-2}\,[s]$, where $\gamma = 2\times10^{-31}\ cm^6/s$, and n is the free electron (or free hole) density in cm^{-3} (Passari and Sosi, 1983; Nilsson and Svantesson, 1972). For $\Delta n = 8\times10^{18}\ cm^{-3}$, $\tau_A = 0.08$ μsec.

It would therefore seem that Auger recombination should play an important role in the photocarrier diffusion in the present experiment. This would imply that Eqs. (1) and (26), as written, should not be expected to correctly predict the photogenerated free carrier density in the silicon. The proper photocarrier diffusion equation for the present experiment should include a term on the left hand side of Eq. (1) equal to $-(3\gamma/D)(\Delta n_{DC})^2 \Delta n$, where Δn_{DC} is the D.C. density of photogenerated free carriers in the silicon. This additional term, which accounts for Auger recombination, makes the diffusion equation very difficult to solve. Fortunately, and perhaps somewhat counter to initial expectation, the inclusion of this extra term in Eq. (1) does not greatly affect the actual photocarrier distribution in the silicon. The reason for this may be seen from a study of Fig. 9.

Fig. 9 Theoretical periodic photoinjected free carrier density at the silicon surface ($z=0$). The absorbed laser power at a modulation frequency of 2.00 MHz is assumed to be 15.9 mW; R=2 μm. The front surface recombination velocity is assumed to be zero. The bulk carrier lifetime for the curves is: (a) 10 μs (b) 1.0 μs, and (c) 0.1 μs.

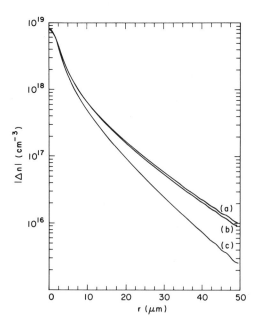

Curves of Δn as a function of ρ, produced from the numerical evaluation of Eq. (26) for $z=0$, are shown in Fig. 9. All parameters are kept constant and set equal to the values associated with the experiment, with the exception of the bulk carrier lifetime, which is varied from curve to curve. It is seen that the surface concentration of photocarriers exhibits little dependence of τ, for $\Delta n > 2 \times 10^{18} cm^{-3}$, corresponding to $\rho < 5\mu m$. Hence, in this region, the photocarrier diffusion is essentially quasistatic, and is dominated by the Laplacian term in the photocarrier diffusion equation. Because this is the case, we expect the Auger recombination term to have little effect on the carrier diffusion within this region (remember the minimum value of $\tau_A \sim 0.1 \mu s$). For values of $\rho > 5\mu m$, the photocarrier concentration is $< 2 \times 10^{19} cm^{-3}$, and hence Auger recombination is no longer of much importance to the carrier diffusion (at $2 \times 10^{18} cm^{-3}$, $\tau_A^{-1} \approx 8 \times 10^5 s^{-1}$ compared with $\omega = 1.3 \times 10^7 s^{-1}$).

Hence, in the present experiment, due to the quasistatic nature of the photocarrier diffusion in the region of high photocarrier density, Auger recombination effects do not markedly affect the overall diffusion of the carriers, and we will henceforth ignore Auger recombination in the present interpretation of the experiment.

Eq. (26) may then be directly substituted into Eq. (5) to solve for the periodic heating in the silicon. We note that the 2 MHz photoacoustic signal, from Eq. (30), is directly proportional to the volume integral of θ in the air above the silicon. Solving Eq. (5) for the periodic heating in the silicon, and noting that θ must be continuous across the silicon-air interface at $z=0$, it is found that:

$$V_{PA} \propto \int_{air} \theta \, d^3r = \frac{I_O \pi R^2 (1-\bar{R}_{Si}) \mu_g \mu_{th}}{2i \, h\nu \, k_{th}}$$

$$\times \left\{ \frac{\mu_{th}\beta(h\nu-E_g-3k_BT)}{\mu_{th}\beta+(1-i)} + \frac{s_o\mu_{el}(E_g + 3k_BT)}{D(1-i)+s_o\mu_{el}} + \frac{(1-i)\mu_{el}^2\mu_{th}(E_g+3k_BT)}{2\pi \, (\mu_{el}-\mu_{th})[D(1-i)+s_o\mu_{el}]} \right\} \quad (31)$$

Here V_{PA} is the output voltage of the 2 MHz air transducer, and μ_g is the thermal diffusion length in the air. It has been assumed in Eq. (31) that $d/\mu_{th} \gg 1$.

The bracketed expression in Eq. (31) contains three terms. The first term corresponds to the heating in the silicon due to immediate de-excitation of the photogenerated carriers to the band edges. The second and third terms correspond to the heating that arises from the surface and bulk recombination of the photocarriers, respectively. In the present experiment at 2 MHz, $\mu_{el} = (2D/\omega)^{1/2} = 15.5 \, \mu m$ and $\mu_{th} = 3.75 \, \mu m$. Fig. 10 shows the predicted dependence of the 2 MHz photoacoustic signal on the front surface recombination velocity of the silicon sample for the incident Argon-ion laser beam of energy $h\nu = 2.14 \, eV$. Also shown are the individual contributions to the total signal from the first and second terms in the bracketed expression of Eq. (30). The contribution of the third term is negligible.

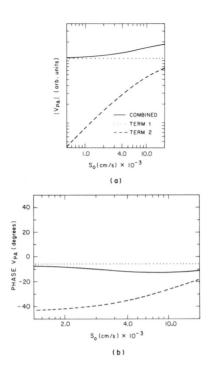

Fig. 10 The predicted dependence of the 2.0 MHz photoacoustic signal in air on the front surface recombination velocity, s_o. Shown also are the individual contributions to the total signal from the first and second terms in the bracketed expression of Eq. (31).

It is clear that the dependence of the photoacoustic signal on the front surface recombination velocity of the silicon is measurable, for values of s_o between 500 cm/s and 20,000 cm/s. For $s_o < 500$ cm/s, the heating due to rapid photocarrier de-excitation dominates. It will be noted that this contribution to the heating could be made less significant by utilizing an incident light beam of energy more nearly equal to the band gap energy of the silicon. In theory, as $h\nu \rightarrow E_g$, the contribution to the periodic heating from the rapid photocarrier de-excitation will become relatively small. It must be noted, however, that under this condition, β also becomes small. The effect of this is to multiply the second term in the bracketed expression of Eq. (31) by the factor $\beta\mu_{el}/(\beta\mu_{el}+1-i)$ [the third term in Eq. (31) is still negligible]. This results in a decrease of the photoacoustic signal as $h\nu \rightarrow E_g$, with an increased sensitivity of the measurement to the surface recombination velocity s_o.

For large values of s_o, it can be seen from Eq. (31) that V_{PA} will become insensitive to the surface recombination velocity; in the present experiment this occurs at $s_o \sim 50,000$ cm/s. Sensitivity of the measurement to s_o for values larger than 50,000 cm/s may be increased by raising the modulation frequency ω, hence decreasing μ_{el}.

The results of the present experiment are shown in Fig. 11. The sample measured consists of a wafer of <100>-cut silicon, doped with ~2×10^{15} P-atoms/cm^3. The sample has been prepared such that there exist square regions along its surface, 30 μm × 30μm in size, in which the surface recombination velocity s_o = 4700 cm/s (Stearns, 1985). The background surface recombination velocity of the sample is s_o = 2400 cm/s. Fig. 11(a) is an optical photograph of the region scanned in the photoacoustic measurement. The three dark areas in the photograph correspond to regions of heavy doping; the contrast is due to topography associated with the doping. It will be noted that in the optical image there is no contrast between regions of different surface recombination velocities. Fig. 11(c) shows the image of the phase of the photoacoustic signal in the scanned region. The areas of higher surface recombination velocity are barely visible in the figure, indicating a change in the phase of V_{PA}< 3° between the regions of different surface recombination velocity. The predicted phase shift of the photoacoustic signal between the regions of s_o = 2400 cm/s and s_o = 4700 cm/s, from Eq. (31), is 1.9°.

Therefore, the 2 MHz photoacoustic measurement in air may be used to image the variations in the surface recombination velocity of the silicon, at least in the range of the values of s_o encompassed in the present experiment. As indicated earlier, greater sensitivity to s_o for values <1000 cm/sec may be obtained by using lower incident laser energies. The ability to image changes in the surface recombination velocity of the silicon, in a noncontacting manner, may be quite useful in the detection and study of surface defects in the silicon.

VI. Conclusions

In this chapter we have discussed three photoacoustic measurements, each of which indicates some promise for the photoacoustic study of semiconductors. Two of the measurements are based on the dependence of the elastic properties of the semiconductor on free carrier concentration. In the first experiment, the perturbation of the elastic constants of the semiconductor is measured, due to the periodic photoinjection of excess free carriers. In the second experiment, the direct effect of the modulated photocarrier population on photoacoustic generation within the semiconductor is both predicted and detected.

In both of these measurements, the effects of periodic heating in the sample are found to compete with the effects produced by the modulated photocarrier density. This is an inherent complication. In the acoustic phase perturbation measurement (Section 3), it is found that at sufficiently large modulation frequencies, the thermal contribution to the acoustic perturbation is negligible, and the perturbation is dominated by the modulated free carrier population. Hence, at high enough modulation frequencies (in the experiment, at frequencies of $\omega/2\pi$>100 Hz) the acoustic phase perturbation measurement becomes essentially a measure of the photoinjected carrier density. From this measure-

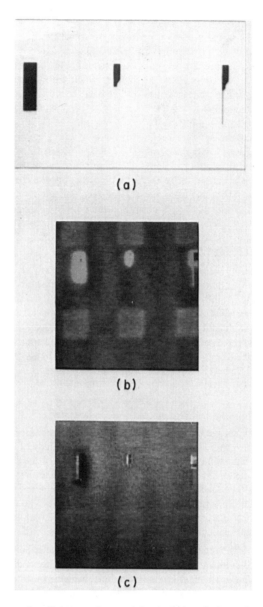

(a)

(b)

(c)

Fig. 11 Images of a silicon sample containing variations in its surface recombination velocity. (a) Optical image of a region of the silicon wafer. (b) Magnitude of the photoacoustic image of the same region. Light squares correspond to regions of surface recombination velocity $s_o = 4700$ cm/s; dark background in image corresponds to $s_o = 2400$ cm/s. (c) Phase of the photoacoustic image.

ment, it has been shown how certain electronic properties, such as the photocarrier lifetime, may be determined.

In the measurement of the direct photoacoustic generation within the semiconductor, it is not generally simple to differentiate between the thermal and electronic contributions. In the experiment described in this Chapter (Section IV), it was seen that the electronic contribution to the photoacoustic generation is of the same order as the thermal contribution. Which of the two contributions dominates in a particular photoacoustic measurement will depend on the details of the measurement (i.e., the energy of the incident light or electron beam, the modulation frequency, and the relative values of the thermal and electronic diffusion lengths). Perhaps the most important conclusion to be drawn from the experiment is that in general an electronic contribution does exist in the photoacoustic generation process within semiconductors, and that this contribution should be appreciated when, for example, interpreting photoacoustic images of semiconductors.

In the final experiment, the photoacoustic generation in the air above the irradiated semiconductor was investigated. In this case, the photoacoustic measurement is sensitive only to the periodic heating at the semiconductor surface. The novel feature of the measurement is the high acoustic (or modulation) frequency employed. The chief advantages of using 2 MHz photoacoustic detection are the increased spatial resolution of the measurement and the increased efficiency of photoacoustic generation at the higher frequency. Furthermore, use of a relatively small acoustic lens, which may be mounted in the open environment without need for the use of a resonant gas cell, makes the technique attractive. The lens may be mounted conveniently in conjunction with an optical microscope, allowing the possibility of simultaneous optical and photoacoustic imaging. Finally, the photoacoustic detection in air is noncontacting and typically noncontaminating.

It has been shown that variations in surface recombination velocity along a silicon sample may be detected using the 2 MHz photoacoustic measurement in air. In this case the contrast mechanism is the difference in local heating associated with the different surface recombination velocities of the photocarriers. At 2 MHz, the measurement is seen to be much more sensitive to surface recombination than to bulk recombination.

We have concentrated in this Chapter on electronic measurements. The same techniques, however, are also useful for measurements of the change in temperature caused by the poor thermal conductivity of a bad bond or of a poorly adhering thin film. Measurements using the air transducer show the change in temperature due to these effects. Measurements carried out over a range of frequencies, of the phase of the photoacoustically-generated or perturbed signal relative to the phase of the laser beam modulation, have also given a great deal of information. We have used these techniques to determine the thickness of transparent and opaque films, and to measure the thermal conductivity of a silver epoxy bond between two different materials (Stearns, 1985; Stearns and Kino, 1986).

Finally, it may be noted that all of the measurements discussed in this Chapter are directly amenable to imaging, as they involve the use of an incident laser beam to irradiate the sample which may be highly focused.

VII. Appendix
Relative Sensitivities of an Acoustic Transducer in Air and of an Infrared Detector

Our aim here is to compare the sensitivities of a high-frequency acoustic transducer operating in air with an infrared detector. Assume a modulated laser beam to be incident on a plane surface. Absorption of the laser beam produces a periodically heated region of area A at the sample surface. We assume that the heated area A is much larger in radial extent than a thermal wavelength in air. In this case, the acoustic power radiates into a solid angle $d\Omega$ in the air, a distance r above the heated region of the surface is:

$$dP = \frac{1}{2}|v|^2 Z_A r^2 d\omega \tag{A.1}$$

where v is the acoustic particle velocity and Z_A is the acoustic impedance of air. It follows from Eq. (30) that

$$|v| = \frac{3\alpha\omega^2\mu_A}{2\sqrt{2}\,\pi V_A}\frac{e^{-\gamma r}}{r}|\bar{\theta}|A \tag{A.2}$$

or

$$dP = \frac{9}{16\pi^2}a^2\omega^2\mu_A^2\frac{\rho_A}{V_A}e^{-2\gamma r}|\bar{\theta}|^2 A^2 d\Omega \tag{A.3}$$

We have assumed here that $\gamma \ll k$. The parameter $\bar{\theta}$ is the spatial average of the peak periodic temperature variations over the area A, V_A is the acoustic velocity in air, μ_A is the thermal diffusion length in air, and ρ_A is the mass density of air.

Since the spot size is larger than the thermal diffusion length, and $\mu_A \propto 1/\omega^{1/2}$, the volume penetrated by the thermal diffusion wave in air is proportional to $1/\omega^{1/2}$. Hence, as $\bar{\theta} \propto 1/\omega^{1/2}$, it follows that the particle velocity varies linearly with the frequency ω and the generated acoustic power varies as the square of the frequency ω. The efficiency increases as the square of the frequency to the point where the acoustic losses in air associated with the $e^{-2\gamma r}$ term become dominant.

The noise equivalent power (NEP) of the acoustic transducer-preamplifier combination may be expressed by the relation:

$$(NEP)_{ac} = \frac{k_B T\, B\, F}{\eta_p}, \tag{A.4}$$

where η_p is the conversion efficiency of the transducer from acoustic to electrical power; k_B is Boltzmann's constant, T is the ambient temperature, B is the bandwidth in hertz of the detection system, and F is the noise figure of the preamplifier. For our 2 MHz focused acoustic transducer, η_p is measured to be 5.3×10^{-3}; for the preamplifier used, $F = 1.35$. The NEP of the 2 MHz acoustic transducer and preamplifier combination at T = 300° is therefore

$$(NEP)_{ac} = 1.05\times10^{-18}B \quad [W] \qquad (A.5)$$

Thus, the signal-to-noise ratio of the acoustic transducer system is

$$(S/N)^s_{ac} = 5.4\times10^{16}\alpha^2\omega^4\mu_A^2\left[\frac{\rho_A}{V_A}\right] e^{-2\gamma r}|\bar\theta|^2A^2\Omega_{ac}B^{-1} \qquad (A.6)$$

where Ω_{ac} is the solid angle of acceptance of the acoustic transducer.

An infrared experiment similar to the photoacoustic measurement at 2 MHz may be imagined, where the focused acoustic transducer is replaced with an IR lens and detector. The electromagnetic power P_{IR} radiated into a solid angle by an illuminated sample, modulated at the frequency ω, and integrated over the entire energy spectrum of the radiation, may be written in the form (Stearns, 1985):

$$P_{IR} \approx \left[\frac{4\sigma_B}{\pi}\right] T^3 \cos\psi\,|\bar\theta|\,A\Omega_{IR} \qquad (A.7)$$

Here σ_B is the Stefan-Boltzmann constant ($\sigma_B = 5.67\times10^{-8}$ W/m^2–k^4), and ψ is the angle between the central axis of the detector of solid angle Ω_{IR} and the sample normal (we assume $\Omega_{IR}\ll2\pi$ steradians). In deriving Eq. (A.7), it has been assumed, for simplicity, that the emissivity of the sample is unity at all wavelengths of the black body radiation.

To compare directly the sensitivities of the photoacoustic and infrared detection methods, the detectivity of the IR detector must be known. A typical value for the NEP of a liquid nitrogen cooled IR detector is (*Laser Focus Buyer's Guide*, 1984).

$$(NEP)_{IR} \approx 2.5\times10^{-12}B^{\frac12} \quad [W], \qquad (A.8)$$

where the bandwidth B of the measurement is again expressed in units of hertz. Combining Eqs. (A.7) and (A.8), the signal-to-noise ratio of the infrared measurement for a sample at room temperature may be expressed in the approximate form:

$$(S/N)_{IR} \approx \frac{7.8 \times 10^{11} \, |\bar{\theta}| A \Omega_{IR}}{B^{\frac{1}{2}}} \qquad \text{(A.9)}$$

It is apparent that the sensitivity of the infrared detection system falls off as $1/\omega^{\frac{1}{2}}$ since $\bar{\theta}$ varies as $1/\omega^{\frac{1}{2}}$.

Comparing Eqs. (A.6) and (A.9), it is clear that the sensitivities of the photoacoustic and infrared techniques show markedly different dependences on a number of parameters. Foremost among these parameters is the thermal modulation frequency. As indicated above, photoacoustic detection becomes relatively more sensitive as ω is increased (this is ultimately limited by acoustic losses is the air). Infrared detection, on the other hand, becomes relatively less sensitive at higher values of ω.

From Eqs. (A.6) and (A.9), it is also evident that the photoacoustic and infrared detection sensitivities depend differently on the absolute periodic heating of the sample $|\bar{\theta}| \, A$, as well as on the bandwidth B of the measurement. Thus, for smaller temperature fluctuations at the sample surface (i.e., lower incident laser power), infrared detection becomes relatively more sensitive than photoacoustic detection. As the bandwidth of the measurement is decreased, photoacoustic detection becomes relatively more sensitive.

Finally, we see that the sensitivities of the photoacoustic and infrared detection techniques vary directly with the solid angle subtended by the acoustic and infrared detector, respectively. Because of the different dependences of the two detection techniques on such a large number of parameters, it is difficult to claim in general that one technique is more sensitive than another. The relative sensitivities of the photoacoustic and infrared detection methods must be compared for each specific set of experimental parameters employed.

We have evaluated the signal-to-noise ratio of both the photoacoustic and infrared detection techniques, assuming the experimental values of ω, B, and $|\bar{\theta}| A$ employed in Section V. That is, we have assumed an incident modulated laser power of 0.121 W where ψ is taken to be 45° and the solid angle of the photoacoustic and infrared detectors is assumed to be 0.19 steradians. The irradiated sample was assumed to be silicon and the thermal modulation frequency was assumed to be 2 MHz. Under these assumptions, the predicted signal-to-noise ratio of the infrared measurement, in a 100 kHz detection bandwidth, is $(S/N)_{IR} = 0.75$, whereas, for the photoacoustic sensor $(S/N)_{ac} = 9.9$. Hence, under these experimental conditions, we expect the photoacoustic measurement to be ~13 times more sensitive than infrared detection. The estimates we have made here are crude; they assume the same solid angle for both the photoacoustic and infrared devices, whereas an infrared device is slightly easier to make with a larger solid angle. On the other hand, we have assumed that the sample is a perfect black body which is not normally the case. Finally, the comparison between the two techniques depends on the nature of the device being observed. With transparent optical films, the infrared device measures the temperature of the underlying absorbing substrate while the acoustic device measures the surface temperature of the top surface.

VIII. References

American Institute of Physics Handbook, 3rd Edition, (1972) (Coordinating Editor: D.E. Gray), McGraw-Hill, New York, pp. 6-147.

Burenkov Y.A. and Nikanorov S.P., (1974) Sov. Phys. Solid State. Engl. Transl. **16**, 963.

Einspruch N.G. and Csavinsky P., (1963), Appl. Phys. Lett. **2**, 1.

Figielski T., (1961), Phys. Stat. Sol. **1**, 306.

Fox J.D., Khuri-Yakub B.T., and Kino G.S., (1983), *Proc. Ultrasonics Symp.*, 581.

Hall J.J., (1967), Phys. Rev. **161**, 756.

Handbook of Chemistry and Physics, 59th Edition, (1973), (Weast Robert C., Ed.), CRC Press, Inc., West Palm Beach, Florida.

Keys R.W., (1961), IBM J. Res. Develop. **5**, 266.

Keys R.W., (1967), *"Electronic Effects in the Elastic Properties of Semiconductors"*, Solid State Phys. **20**, 37.

Kino G.S., (1978), J. Appl. Phys. **49**, 3190.

Kino G.S. and Stearns R.G., (1985), Appl. Phys. Lett. **49**, 926.

Laser Focus Buyer's Guide, (1984), (M.R. Levitt, Ed.), Penn Well Publ., Newton, Mass.

Microwave Acoustics Handbook, (1973), (A.J. Slobodnik, Jr., E.D. Conway, and R.T. Delmonico, Eds.), Vol. 1A, Air Force Cambridge Research Laboratories, Bedford, Mass., 540.

Nilsson N.G. and Svantesson K.G., (1972), Solid State Commun. **11**, 155.

Passari L. and Sosi E., (1983), J. Appl. Phys. **54**, 3935.

Rosencwaig A. (1977) in *Optoacoustic Spectroscopy and Detection*, (Pao Y-H., Ed.), Academic Press, New York.

Rosencwaig A., (1982), Science **218**, 223.

Rosencwaig A. and White R.M., (1981), Appl. Phys. Lett. **38**, 165.

Scruby C.B., Dewhurst R.J., Hutchins D.A., and Palmer S.B., (May 1981), *Laser Generation of Ultrasound in Metals*, AERE Harwell Report No. 10099, Oxfordshire, England Department of Appl. Phys., University of Hull.

Stearns R.G., Khuri-Yakub B.T., and Kino G.S., (1983), Appl. Phys. Lett. **43**, 748.

Stearns R.G., (1985), *Photoacoustic Techniques for the Quantitative Characterization of Materials*, Ph.D. Thesis, Stanford University.

Stearns R.G. and Kino G.S., (1986), IEEE Trans. UFFC, Special Issue on Acoustic Sensors, (submitted).

PHOTOTHERMAL DEFLECTION SPECTROSCOPY OF SEMICONDUCTORS

Danielle Fournier and A. Claude Boccara

Laboratoire d'Optique Physique – ER 5 CNRS
Ecole Supérieure de Physique et de Chimie
10, rue Vauquelin, 75231
Paris Cedex 05 – FRANCE

I. Introduction

Photothermal methods have achieved great success in a few years, because they have allowed the solution of "difficult" cases in the area of solid state spectroscopy. The combination of thermal and optical investigations of solids leads to the determination of optical absorption coefficients of opaque, or scattering samples, of thin films and multilayered structures. Photothermal spectroscopy has been successfully applied to semiconductor samples in the case of powders, non-polished samples, crystalline, polycrystalline or amorphous thin films, multilayered structures, etc.

In Section II of this Chapter we will describe the principles of a photothermal experiment and will give the outline of the calculations which lead to the absolute determination of the optical absorption coefficient. In Section III we will describe some characteristic examples which reveal interesting features of photothermal spectroscopy.

II. Photothermal Spectroscopy of Semiconductors: Instrumentation and Calculations

1.   EXPERIMENTAL APPARATUS

When a sample is irradiated by a light flux, usually there is conversion of the absorbed light into thermal energy resulting in heating of the sample. If the impinging beam is modulated, the sample surface exhibits a modulated surface temperature which is the parameter that has to be measured in a photothermal scheme. There are several solutions in implementing this measurement directly or indirectly: the photoacoustic scheme (with a microphonic cell), and the optical scheme (I.R. detection or "mirage" detection) to name a few. The aim of this Chapter is to describe in detail the photothermal deflection spectroscopy scheme only, as it was found to be very suitable for thermal-wave probing as well as a most sensitive technique. We refer the reader to other chapters in this volume and other references in order to find experimental details of other detection schemes (e.g. Rosencwaig, 1980).

Fig. 1 "Mirage" detection principle.

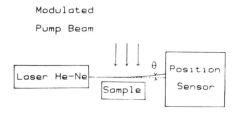

In order to implement a photothermal deflection spectroscopy experiment we must join a light source such as a dye laser or a xenon arc with a monochromator, a FTIR spectrometer or a visible FT spectrometer (Débarre, Boccara, and Fournier, 1981) with a "mirage" detection scheme. Fig. 1 describes the principle of this "mirage" detection (Boccara, Fournier, and Badoz, 1980). The periodically heated sample surface (at frequency f = ω/2π) is a heat source for the surrounding medium, which exhibits a periodic temperature gradient near the sample surface. This periodic temperature gradient gives rise to a refractive index gradient suitable for periodically deflecting a probe beam propagating along the surface of the solid. The amplitude of this periodic deflection is given by:

$$\Theta = \frac{l}{n} \left[\frac{dn}{dT} \right] < \frac{dT(x,t)}{dx} > \qquad (1)$$

where l is the interaction pathlength between the probe beam and the temperature gradient dT/dx and n is the refractive index of the medium. We will show below how dT/dx can be calculated and related to the optical properties of the sample. The brackets in Eq. (1) denote a spatial average over both the pathlength l and the finite waist size of the probe beam. This method is particularly sensitive if the sample is immersed in a transparent liquid, such as CCl_4 for instance ($dn/dT = 5 \times 10^{-4} {}^{\circ}K^{-1}$). The smallest measured sample surface temperature variations correspond to

$$\Delta T_s = 10^{-4} \, {}^{\circ}/\sqrt{Hz} \text{ in air and}$$
$$\Delta T_s = 10^{-7} \, {}^{\circ}/\sqrt{Hz} \text{ in } CCl_4$$

Fig. 2 Experimental arrangement of the compact mirage cell.

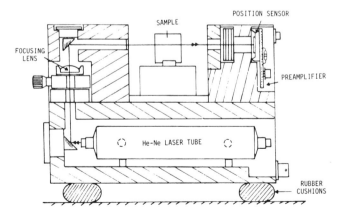

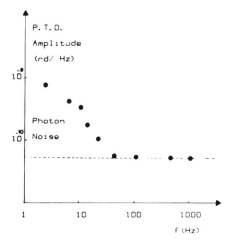

Fig. 3 Spectral density of the noise exhibited by the compact "mirage" cell.

Fig. 2 describes a new geometry for the "mirage" apparatus (Charbonnier and Fournier, 1986). This cell includes the three blocks: the laser compartment, the focusing optics, and the position sensor with its filter in order to avoid the stray scattered light. This structure is sufficiently small for setting in the sample compartment of a commercial Fourier transform I.R. spectrometer. Moreover, its sensitivity is more than an order of magnitude larger than the previous set-up. The theoretical noise (i.e., photon noise) has been reached for frequencies above 30 Hz (Fig. 3). Note that a semiconductor laser can be used instead of a He-Ne laser for probing the index gradient. This laser source presents two advantages: its small size and a strong reduction of the pointing and intensity noises above 50 kHz in comparison with He-Ne lasers.

2.   CALCULATION OF THE PERIODIC TEMPERATURE VARIATION AT THE SAMPLE SURFACE

In order to estimate the surface temperature of a sample, we have to solve the heat diffusion equation in all the regions where heat can propagate. These calculations were performed by Rosencwaig and Gersho (1976) for the case of photoacoustic detection and have been extended to the case of "mirage" detection by several authors (see for instance Murphy and Aamodt, 1980; Mandelis, 1983).

a. Sample on Transparent Substrate

Fig. 4 describes the 1-D geometry of the system. This geometry is the most appropriate for spectroscopic applications, but it is sometimes more efficient to use tight focusing of the pump beam to increase the signal. In that case a 3-D calculation is necessary (Jackson, Amer, Boccara, and Fournier 1981).

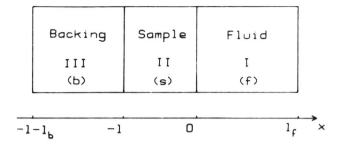

Fig. 4 Geometry of a 1-D system.

The sample is irradiated by a modulated light beam whose amplitude at the frequency $\omega/2\pi$ is I_o. Thus heat diffusion equations can be written in the three regions shown in Fig. 4, i.e. the transparent fluid (f), sample (s), and transparent backing (b):

$$\frac{\partial^2 T_f}{\partial x^2} = \frac{1}{D_f} \frac{\partial T_f}{\partial t} \; ; \qquad \text{for } 0 < x < l_f \tag{2}$$

$$\frac{\partial^2 T_s}{\partial x^2} = \frac{1}{D_s} \frac{\partial T_s}{\partial t} - A \exp(\alpha x) e^{j\omega t} \; ; \tag{3}$$

$$\text{for } -l < x < 0$$

$$\frac{\partial^2 T_b}{\partial x^2} = \frac{1}{D_b} \frac{\partial T_b}{\partial T} \; ; \qquad \text{for } -l-l_b < x < -l \tag{4}$$

where T_i ($i=f,s,b$) is the modulated temperature in the region (i) and D_i the thermal diffusivity. The thermal diffusivity $D_i = k_i/\rho_i c_i$, where k_i is the thermal conductivity, c_i the specific heat and ρ_i the density.

The term $A \exp(\alpha) \exp(j\omega t)$ represents the modulated heat source due to the light absorbed by the sample, α being the optical absorption coefficient. In Eq. (3):

$$A = \frac{\alpha I_o \eta}{2k_e} \; ,$$

where η is the optical-to-thermal energy conversion efficiency by nonradiative de-excitations. We shall take $\eta = 1$ in the following calculations. The general steady state solutions of Eqs. (2), (3), and (4) are given by:

region I : $T_f(x,t) = T_s \exp(-\sigma_f x + j\omega t)$ \tag{5}

region II : $T_s(x,t) = [U \exp(\sigma_s x) + V \exp(-\sigma_s x)$ \tag{6}

$\qquad\qquad - E \exp(\alpha x)] \exp(j\omega t)$

region III : $T_b(x,t) = W \exp[\sigma_b(x+l) + j\omega t]$ \tag{7} where T_s, U, V, E and W are complex constants; $\sigma_i = (1+j)/\mu_i$ where μ_i is the thermal diffusion length; $\mu_i = \sqrt{\dfrac{D_i}{\pi f}}$ in region (i)

These constants are found when applying the boundary conditions for heat flux and temperature continuity at the interfaces.

We can thus write the modulated sample surface temperature at x=0 (Mandelis, 1983):

$$T_s = \frac{\alpha I_o}{2k_e(\alpha^2-\sigma_s^2)} \frac{(r-1)(b+1)\exp(\sigma_s l)-(r+1)(b-1)\exp(-\sigma_s l)+2(b-r)\exp(-al)}{(g+1)(b+1)\exp(\sigma_s l)-(g-1)(b-1)\exp(-\sigma_s l)}$$

$$\text{with } \quad b = \frac{k_b\sigma_b}{k_s\sigma_s} = \frac{k_b}{k_s}\left[\frac{D_s}{D_b}\right]^{\frac{1}{2}} \qquad (8)$$

$$g = k_f\sigma_f/k_s\sigma_s$$

$$r = (1-j)\frac{\alpha}{2a_s}$$

This complex temperature can be written as: $T_s = |T_s| \exp(-j\phi)$. We will now examine two cases of practical interest for spectroscopic measurements.

a.1 Thermally thick sample

For a thermally thick sample, the thermal diffusion length μ_s is much smaller than the sample thickness l. (i.e. bulk sample). Under these conditions Eq. (8) becomes:

$$T_s = \frac{I_o}{2k_s}\left[\frac{1}{1+g}\right]\frac{(r-1)\alpha}{(a^2-\sigma_s^2)} \qquad (9)$$

where $|T_s|$ and ϕ can be written as an explicit function of the product $(\alpha\mu_s)$.

$$|T_s| = \frac{I_o}{2k_s}\left[\frac{1}{1+g}\right]\alpha\mu_s^2\left[\frac{((\alpha\mu_s/2)-1)^2+\alpha^2\mu_s^2/4}{\alpha^4\mu_s^4+4}\right]^{\frac{1}{2}} \qquad (10)$$

$$\text{and } \quad \phi = \tan^{-1}(2/\alpha^2\mu_s^2)+\tan^{-1}[\alpha\mu/(2-\alpha\mu)] \qquad (11)$$

We see that $|T_s|$ and ϕ are independent of b, which accounts for the thermal contact between the sample and the backing. The phase variation between very small and large absorption coefficients is $\pi/4$. We will show in the next section that a careful analysis of the phase signal will allow an absolute determination of α.

a.2 Thermally thin sample

We examine here the case of an absorbing layer deposited on a transparent substrate.

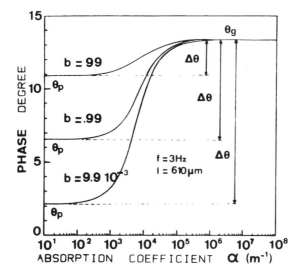

Fig. 5 Phase variation of the surface temperature of a GaAs 610 μm thick sample with the optical absorption coefficient α for different values of b.

When $\sigma_s l < l$, then $\exp(\pm\sigma_s l)\approx 1\pm\sigma_s l$ and Eq. (8) reduces to:

$$T_s = \frac{\alpha I_o[(r-b)(1-\exp(-\alpha l)+\sigma_s l(rb-1)]}{2k_s(\alpha^2-\sigma_s^2)(b+g)} \tag{12}$$

For very thin samples ($\alpha_s l \ll 1$) such as micron thick semiconductor films the phase of Eq. (12) reduces to $\pi/4$ independently of α (Rosencwaig, 1980). Fig. 5 shows the influence on the phase of T_s of the thermal mismatch between the film and the substrate for a thermally thin GaAs sample. One can see that the phase variation when going from $\alpha l \ll 1$ to $\alpha l \gg 1$ is indeed strongly dependent on the parameter b.

Fig. 6 Geometry of the 1-D one layer system.

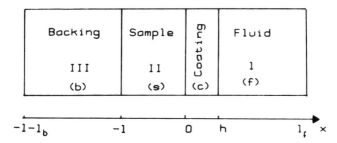

b. Multilayered Structures

Mandelis, Teng, and Royce (1979) and Fernelius (1980) have extended the Rosencwaig and Gersho (1976) model to the case of an absorbing layer (thickness h, absorption coefficient α_c) deposited on an absorbing substrate (thickness l, absorption coefficient α_s) (Fig. (6)).

The sample surface temperature T_s can be obtained by solving heat equations in the four media with suitable boundary conditions. The resulting expression is rather complicated (Fernelius, 1980):

$$
T_s \times \{(1-b)\exp(-\sigma_s l)\left[(1-c)\left[1+\frac{g}{c}\right]\exp(\sigma_c h)+(1+c)\left[1-\frac{g}{c}\right]\exp(-\sigma_c h)\right]
$$

$$
- (1+b)\exp(\sigma_s l)\left[(1+c)\left[1+\frac{g}{c}\right]\exp(\sigma_c h)+(1-c)\left[1-\frac{g}{c}\right]\exp(-\sigma_c h)\right]
$$

$$
= 2E\left[2(r_s-b)\exp(-\alpha_s l)+(1+b)(1-r_s)\exp(\sigma_s l)\right.
$$

$$
-(1-b)(1+r_s)\exp(-\sigma_s l)\right] + Z\left[2(1-b)(1+r_c)\exp(-\sigma_s l-\alpha_c h)\right.
$$

$$
-2(1+b)(1-r_c)\exp(\sigma_s l-\alpha_c h)-(1-b)(1-c)\left[1-\frac{r_c}{c}\right]\exp(-\sigma_s l+\sigma_c h)
$$

$$
-(1-b)(1+c)\left[1+\frac{r_c}{c}\right]\exp(-\sigma_s l-\sigma_c h)+(1+b)(1+c)\left[1-\frac{r_c}{c}\right]\exp(\sigma_s l+\sigma_c h)
$$

$$
+(1+b)(1-c)\left[1+\frac{r_c}{c}\right]\exp(\sigma_s l-\sigma_c h)\right] \tag{13}
$$

where:

$$
c = \frac{k_c\sigma_c}{k_s\sigma_s} \quad , \quad b = \frac{k_b\sigma_b}{k_s\sigma_s} \quad ,
$$

$$
g = \frac{k_f\sigma_f}{k_s\sigma_s} \quad , \quad r_s = (1-j)\frac{\alpha_s\mu_s}{2} \quad , \quad r_c = (1-j)\frac{\alpha_c k_c\mu_s}{2k_s}
$$

$$
E = \frac{\alpha_s I_o\exp(-\alpha_c h)}{2k_s(\alpha_c^2-\sigma_s^2)} \quad , \quad Z = \frac{\alpha I_o}{2k_c(\alpha_c^2-\sigma_c^2)}
$$

$$
\text{and } \sigma_i = \frac{(1+j)}{\mu_i} \quad (i=f,c,s,b)
$$

Eq. (13) will be used below with semiconductors in two cases of interest: i) when the bandgap of the substrate is smaller than the bandgap of the coating (e.g., GaAlAs/GaAs), and ii) when the bandgap of the substrate is larger than the gap of the coating (e.g., GaAsSb/GaAs).

3. SPECIFIC PROBLEMS WITH SEMICONDUCTORS

The heat diffusion equations that we have used up to now do not account for the photoinduced carriers which can diffuse through the sample before they recombine. Indeed it has been shown theoretically that the photothermal signals are affected by carrier diffusion processes. Nevertheless these effects are usually small, especially in the low frequency regime which is used in spectroscopic experiments (See Chapter 12). However, when the sample under study is a thin film for which a careful analysis of the signal phase must be done, we expect errors on the calculation of the absorption coefficient, especially for indirect gap semiconductors with large carrier lifetimes.

III. Photothermal Spectroscopy of Semiconductors: Illustrative Examples

1. POWDERED SEMICONDUCTORS

In the early days of photoacoustics, photothermal techniques seemed very convenient for studying light scattering materials. Indeed, the signal being proportional to the energy deposited into the sample, one can believe that, to first approximation, it is independent of light scattering processes. A more careful analysis has been performed by Yasa, Jackson, and Amer (1982) who have shown that for low scattering coefficients (α_s) the photothermal signal does not effectively vary with α_s whereas for larger values, strong discrepancies are observed (Fig. (7)).

Fig. 7 Variation of the photoacoustic signal magnitude with the light scattering factor $\alpha_s l$ for several values of αl [From Tam (1986)].

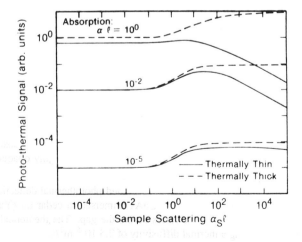

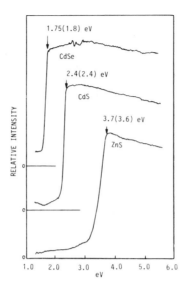

Fig. 8 Photoacoustic spectra of three direct-band semiconductors in powder form at 300 K (Rosencwaig, 1975). The band gaps are compared to the values derived from specular reflectance measurements (in parentheses). See also Chapter 15.

Despite the strong light scattering, Rosencwaig (1975) has recorded three direct-band gap semiconductor spectra all in powder form. The gap nature and the position of the band gap E_g have not been studied by the classical energy dependence of the absorption coefficient as a function of the photon energy, $\alpha(h\nu) \sim (h\nu - E_g)^{1/2}$, (Pankove, 1971), however, the band edges as measured by the position of the "knee", agree very well with the values usually found in the literature (Fig. 8). A detailed discussion of this feature is presented in Chapter 15. The same kind of measurements has also been applied to indirect bandgap semiconductors (GaP). We wish to emphasize that these experiments on powdered materials are very easy to perform and require only a few milligrams of unprepared sample.

2.   BULK SEMICONDUCTORS

Although such experiments using photothermal techniques may appear useless because they can be performed easily with classical spectrometers, let us point out that they allow the determination of absolute optical absorption coefficients when thermal parameters are known, even for highly opaque samples.

Fig. 9 shows the experimental and computed photothermal deflection signal phase variation for thermally thick GaAs immersed in cedar oil (Yacoubi, 1986) as the absorption coefficient varies across the gap. The theoretical curve has been calculated using a thermal diffusivity of $2.5 \ 10^{-5} \ \text{m}^2/\text{s}$.

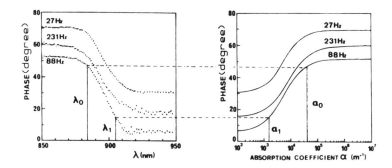

Fig. 9 Experimental phase variation of the photothermal deflection signal versus the wavelength for various modulation frequencies; and theoretical phase variation of the photothermal deflection signal versus the optical absorption coefficient α for various modulation frequencies for a thermally thick GaAs sample.

From the analysis of the two curves of Fig. 9, one can deduce the absolute absorption coefficient as a function of the wavelength as shown in Fig. 10.

Fig. 10 Optical absorption spectrum of GaAs deduced from Fig. 9 (curve at 231 Hz) and compared to data obtained with ellipsometry (E.S.).

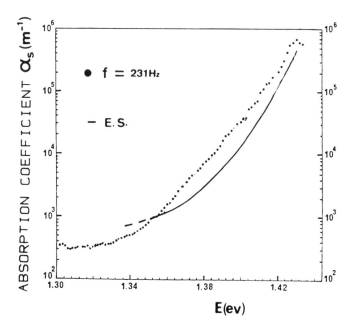

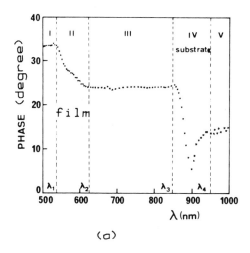

(a)

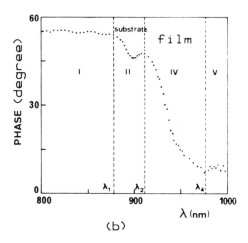

(b)

Fig. 11 Experimental phase variation of the photothermal deflection signal versus wavelength. a) GaAlAs deposited on GaAs, b) GaAsSb deposited on GaAs.

Furthermore, the good sensitivity of the mirage detection and its ability to operate with light diffusing samples has led us (Fournier and Boccara *"Oxygen Concentration Measurement in Unpolished Si Wafer"*, to be published) to construct an apparatus for monitoring impurity concentration in rough silicon wafers. For the measurement of oxygen concentration, we have used the SiO vibration at 1,106 cm^{-1} which can be excited with the IR light of a CO_2 laser.

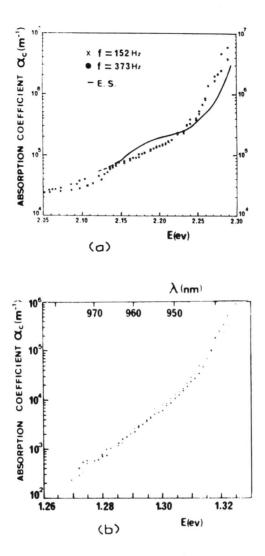

Fig. 12 Optical absorption spectrum of GaAlAs deduced from data of Fig. 11.
a) GaAlAs, b) GaAsSb.

3.   MULTILAYERED CRYSTALLINE SEMICONDUCTORS

This case is of particular interest because the use of photothermal investigation leads easily to the determination of the optical absorption coefficients of both the substrate *and* the film. The main rival method is ellipsometry which is very convenient for monitoring the growth process of the film, as polarized light is mainly sensitive to both the film thickness and the real part of the index of refraction. Nevertheless, for low absorption coefficient measurements photothermal methods exhibit a larger sensitivity.

Here we mention, for instance, the very interesting results obtained by Yacoubi (1986) with GaAlAs and GaAsSb films on GaAs. These two examples are characteristic of semiconductor stacks whose bandgap energy is located in the visible or very near infrared. Using the model by Fernelius (1980), (see Section II), this author has determined the optical spectra of all the compounds. Fig. 11 illustrates the phase variation of the mirage signal in various spectral regions where the substrate and the coating are successively opaque and transparent. A careful analysis of region II for GaAlAs/GaAs and of region IV for GaAsSb/GaAs leads to the absolute determination of the absorption coefficients of GaAlAs and GaAsSb (Fig. 12). The GaAs spectrum is deduced from region IV or II.

The extremely high sensitivity of the mirage detection coupled with a color center laser excitation has allowed the measurements of optical absorption in single quantum wells (SQW) (Penna, Shah, DiGiovanni, Cho, and Gossard, 1985). The room temperature spectra are reported for quantum wells of GaAs (104 Å thick) and InGaAs (75 Å thick) where structures due to heavy and light excitons are distinguishable (Fig. 13).

In such experiments the absorbing quantum well structure (thermally ultrathin) is surrounded by two transparent complex semiconducting structures.

Fig. 13 Absorption spectrum of the 104 Å thick single quantum well of GaAs at room temperature [From Penna *et al.* (1985)].

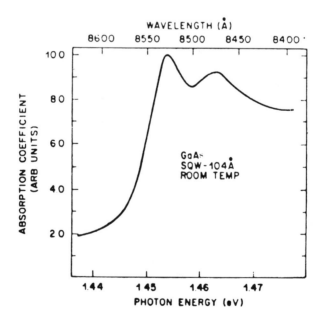

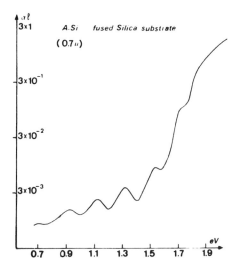

Fig. 14 Optical absorption of a 0.7 μm thick amorphous silicon sample deposited on fused silica (photothermal deflection experiment) [From Debarre *et al.* (1981)].

In the spectral region of interest, only the amplitude of the signal (and not the phase which does not vary; Mandelis, Siu and Ho, 1984) allows the determination of the (relative) absorption of the exciton.

4.   AMORPHOUS AND POLYCRYSTALLINE FILMS

Photothermal deflection spectroscopy of amorphous semiconductors deposited on non-absorbing substrates has been very successful since it has been proposed (Fournier, Boccara and Badoz 1982) as a sensitive technique for measurements of low absorptance associated with gap-and pseudo-gap states. Indeed, nowadays more than twenty laboratories are using these techniques for materials characterization (Amer and Jackson, 1984).

The a-silicon spectrum shown on Fig. 14 has been recorded with a 0.7 μm thick sample deposited on fused silica and immersed in CCl_4 in order to increase the deflection signal. The sample illumination has been carried out by using a visible and near infrared Fourier transform spectrometer (Debarre *et al.* 1981) constructed in our laboratory. With this instrument αl as low as 10^{-5} can be measured, but a reasonable sensitivity has also been achieved by using xenon arc or quartz halogen tungsten sources associated with conventional monochromators.

The amplitude of the deflection signal shows two characteristic features: 1. In the visible region for strong αl the deflection signal varies as $1-\exp(-\alpha l)$ (Eq. (12) in the limit of very small l). In order to obtain αl, as shown in Fig. 14, the signal has to be "desaturated". 2. In the near infrared, for $\alpha l < 1$ we can observe a signal modulation due to optical interference effect in the film

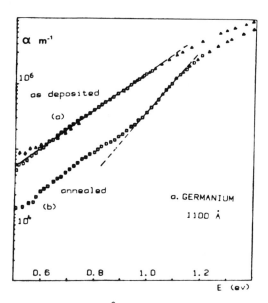

Fig. 15 Optical absorption of 1100 Å thick a-germanium before and after annealing [From Theye *et al.* (1985)].

(Buckley and Beaglehole, 1977; Mandelis *et al.*, 1984). This modulation, which is due to the variation of optical energy "stored" in the film as a function of the wavelength, is much less important than the large modulation of the transmission obtained with semiconductor samples whose refractive index is large. One can get rid of this modulation by a computation which accounts for the optical properties of the film (Theye, 1985).

The phase of the signal for such thin films ($l < 1\mu m$) is constant all along the spectrum at low frequency (10-100 Hz) experiments and thus does not carry any useful information.

Fig. 15 shows the photothermal deflection spectra of a 1,100 Å thick amorphous germanium sample before and after annealing. The shoulder at 0.7-0.8 eV of the annealed sample is related to pseudogap states due to dangling bonds (Theye, Georghiu, Driss and Boccara, 1985). A more detailed analysis of the density of states in the pseudogap which has been initiated by Amer and Jackson (1984) will be found in this book in the chapter by Tanaka.

Finally, we would like to recall that the photothermal techniques allow the determination of thin film optical properties even when they are deposited on light scattering or opaque substrates. Indeed, Roger Fournier, Boccara, Noufi and Cahen (1985) have shown that it is possible to characterize a few micron-thick polycrystalline $Cu_x In_{1-x} Se$ layers deposited on various substrates (Al_2O_3

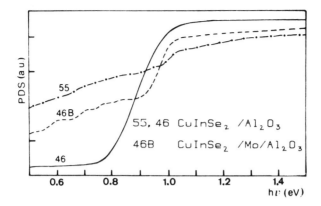

Fig. 16 Photothermal deflection amplitude signal of $Cu_xIn_{1-x}Se$ films deposited on Al_2O_3 and Mo/Al_2O_3 [From Roger *et al.* (1985)].

or Mo/Al_2O_3, Fig. 16). Several experiments have been performed on these samples. Spectroscopic data obtained with low frequency modulation through the signal amplitude, as well as more sophisticated analysis carried out at high frequencies (kHz range) on both amplitude and phase have allowed the calculation of thermal parameters and have revealed the spatial distribution of the absorbing centers (Roger, 1987).

5. SURFACE STUDIES

The same kind of experimental apparatus as described earlier can be used for surface studies. However, for these studies the samples are often placed in an ultra-high vacuum chamber and the experimental detection scheme must be

Fig. 17 Experimental apparatus for probing the local deformation induced by a focused pump beam [From Olmstead *et al.* (1983)].

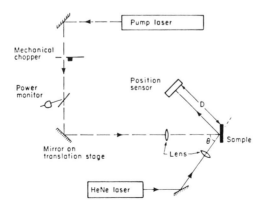

modified. One can probe the temperature gradient within the sample itself using either a collinear of a transverse geometry (Jackson, Amer, Boccara and Fournier, 1981) or measure the local surface thermal deformation induced by a focused laser beam (Amer, Olmstead, Fournier, and Boccara, 1983; Olmstead, Amer, Kohn, Fournier, and Boccara, 1983) (Fig. (17)). Surface and bulk absorption can be distinguished by this latter technique and surface absorptance as low as 10^{-6} W of incident power can be measured. The theoretical basis of the signal generation implies first the determination of the temperature field distribution and then the solution of the relevant Navier-Stokes equations. A nice application of this method has been presented by Olmstead and Amer (1984) who have studied the polarization dependence of midgap absorption by Si(III) 2 × 1 single domain surfaces. Their results supported the π-bonded chain model of the surface reconstruction.

IV. Conclusion

In this Chapter, we have shown that photothermal deflection spectroscopy is a unique technique for the determination of very low absolute optical absorption coefficients in cases of thin films deposited on transparent or light scattering substrates (a-silicon, a-germanium, quantum wells, etc). Furthermore, we have illustrated the ability of this method to determine optical parameters of each layer of a multilayered structure such as GaAlAs/GaAs or GaAsSb/GaAs. It will soon be possible to extend these quantitative studies on single layer structures to multilayers and, if there is a thermal mismatch between the layers, to determine the optical coefficient of each individual layer of a large stack.

Unfortunately, this method needs the simultaneous determination of all the thermal parameters of the system to obtain its absolute optical characterization. Thus, it is often necessary to combine thermal measurements and spectroscopic experiments. Nevertheless, photothermal deflection spectroscopy seems to be a very fruitful method for the characterization of microelectronic materials (amorphous silicon, MOS transistors, etc) and for many optoelectronic devices (laser diodes, detectors, etc) whose layered structure is relevant to this technique. We strongly believe that future techniques for integrated circuit inspection will be a combination of photothermal spectroscopy and imaging on a micrometer scale.

V. References

Amer, N., Olmstead, M., Fournier, D., and Baccara, A.C., (1983). J. Phys. (Paris) Vol. **44**, C6-317

Amer, N., and Jackson, W., (1984). in *"Semiconductors and Semimetals"*, Vol. **21 B**, (J.I. Pankove, ed.), Academic Press, New York, N.Y.

Boccara, A.C., Fournier, D., and Badoz, J., (1980). Appl. Phys. Lett., **36**, 130.

Buckley, R. and Beaglehole, D. (1977). Appl. Opt. **16**, 2495.

Charbonner, F., and Fournier, D., (1986). Rev. Sci. Instrum., **57**, 1126.

Debarre, D., Boccara, A.C., and Fournier, D., (1981). Appl. Opt., **20**, 4281.

Fernelius, N.C., (1980). J. Appl. Phys., **51**, 650.

Fournier, D., Boccara, A.C., and Badoz, J., (1981). *Proc. 2nd International Topical Meeting on Photoacoustic Spectroscopy.* Technical Digest, Berkeley, CA.

Fournier, D., Boccara, A.C., and Badoz, J., (1982). App. Opt., **21**, 74.

Jackson, W., Amer, N., Fournier, D., and Boccara, A.C., (1981). *Proc. 2nd International Topical Meeting on Photoacoustic Spectroscopy.* Technical Digest, Berkeley, CA.

Jackson, W., Amer, N., Boccara, A.C., and Fournier, D., (1981). Appl. Opt. **20**, 1333.

Mandelis, A., (1983). J. Appl. Phys. **54**, 3404.

Mandelis, A., Teng, Y.C., and Royce, B.S.H., (1979). J. Appl. Phys. **50**, 7138.

Mandelis, A., Siu, E., and Ho, S. (1984). Appl. Phys. **A33**, 153.

Murphy, J. and Aamodt, L., (1980). J. Appl. Phys. **51**, 4580.

Olmstead, M., Amer, N., Kohn, S., Fournier, D., and Boccara, A.C., (1983). Appl. Phys. **A32**, 141.

Olmstead, M. and Amer, N., (1984). Phys. Rev. Lett. **52**, 1148.

Pankove, J., (1971). *"Optical processes in semiconductors"* Dover publications Inc., New York.

Penna, A.F.S., Shah, J., DiGiovanni, A.E., Cho, A., and Gossard, A.C. (1985). Appl. Phys. Lett. **47**, 591.

Roger, J.P., Fournier, D., Boccara, A.C., Noufi, R., and Cahen, D., (1985). Thin Solid Films, **128**, 11.

Roger, J.P., (1987). (To be published).

Rosencwaig, A., (1975). Physics Today **28**, 23; and Anal. Chem. **47**, 592A.

Rosencwaig, A., (1980). *"Photoacoustics and Photoacoustic Spectroscopy".* in *Chemical Analysis.* (P.J. Elving and J.D. Winefordner, eds). Vol. **57**. J. Wiley, New York, N.Y.

Rosencwaig, A., and Gersho, A., (1976). J. Appl. Phys. **47**, 64.

Tam, A., (1986). Rev. Mod. Phys. **58**, 381.

Theye, M.L., (1985) (Private communication).

Theye, M.L., Georghiu, A., Driss-Khodja, K., and Boccara, A.C., (1985). *11th Int. Conf. on amorphous and liquid semiconductors.* Rome.

Yacoubi, (1986). Ph.D. Thesis, University of Montpellier (France).

Yasa, Z., Jackson, W., and Amer, N., (1982). Appl. Opt., **21**, 21.

FOURIER TRANSFORM PHOTOACOUSTIC
SPECTROSCOPY OF SOLIDS

Barrie S.H. Royce

Applied Physics and Materials Laboratory
Princeton University
Princeton, NJ 08544, U.S.A.

I. Introduction

The excitation of a material by the absorption of modulated or pulsed radiation, and the detection of the thermal or displacive response of the material or its immediate environment that follows this event, are common features of a group of non-destructive testing techniques that have been called photoacoustic spectroscopy (PAS). The radiation sources used for this type of measurement have had wavelengths extending from the microwave to the x-ray region of the electromagnetic spectrum, and samples have been studied at temperatures ranging from a few degrees Kelvin to about 800 K. The photoacoustic technique is normally detector noise limited making the use of high power light sources desirable, with laser sources being used extensively for studies over narrow spectral regions. When a wider spectral survey is required either dispersive or interferometric spectrometers may be employed. Because of their multiplexing and throughput performance, interferometric spectrometers offer advantages over dispersive instruments, and are particularly useful in the infra-red spectral region.

This Chapter will review the application of Fourier transform spectrometers to photoacoustic studies in both the visible and infra-red spectral regions. Even though no Fourier Transform PA studies of semiconducting materials have been reported to the author's best knowledge, it is felt that exposure of workers in the semiconductor field to the capabilities and potential of the Fourier transform photoacoustic technique will be a positive catalyst to the development of such applications in the future. Two approaches will be discussed: the constant scan spectrometers that are available from many commercial vendors, and the step-and-integrate spectrometers that have been developed in some laboratories for visible and near infra-red studies. The general principles of interferometric spectroscopy, and many details of the experimental technique, are covered in books by Chamberlain (1979) and by Griffiths and de Haseth (1986). The two interferometric approaches under review were introduced as PAS methods at about the same time. Farrow, Burnham and Eyring (1978) used a step-and-integrate spectrometer for visible spectroscopy, and a similar apparatus covering the visible and near infra-red range was developed by Debarre *et al.* (1981). Constant scanning FTIR-PA interferometry was initially applied to gas phase samples by Busse and Bullemer (1978). The extension of the method to solid samples was pioneered by Rockley (1979, 1980), Vidrine (1980), and Laufer *et al.* (1980). The general features of Fourier transform PAS have been discussed in reviews by Vidrine (1982), Tam (1983, 1986), and Royce and Benzinger (1986). Yasa *et al.* (1982) have presented a compact general discussion of photoacoustic signal generation in light scattering media.

In Fourier transform PAS, the radiation from a Michelson interferometer is incident on a sample that is to be studied using photoacoustic detection. This radiation is modulated at audio frequencies, and the nonradiative de-excitation processes that follow the absorption of the radiation give rise to a periodic heating of the absorbing sample. For a condensed phase sample, the PAS signal is obtained by measuring a change in a property of a fluid in contact with the sample, or of the sample itself, that results from this periodic heating. Three detection modes have been favored in interferometric PAS: the measurement of the expansion of the solid using piezoelectric detectors, the measurement of the change in pressure of a gas in contact with the sample in a closed cell using a sensitive microphone, and the detection of the refractive index gradient near the sample surface using the deflection of a probe laser beam. The measured signal is recorded over a range of interferometer mirror positions and used to construct an interferogram which contains information about the optical properties of the sample. The Fourier transform of this interferogram is used to obtain the sample's absorption spectrum, however, because of the complex signal path in PAS, the data will also involve the thermal properties of the sample and the ambient fluid.

II. Signal Generation

1.   THE IDEAL INTERFEROMETER

The central component of a Fourier transform photoacoustic spectrometer is a Michelson interferometer. The optical configuration of this device is shown schematically in Fig. 1, and indicates that the ideal interferometer is symmetric with respect to the input and output ports. Radiation of wavenumber, W, enters the interferometer from the source and is divided into two beams by a non-absorbing beam splitter with 50% reflectivity. The divided beam is sent along two different optical paths. Mirrors at the end of each interferometer arm return the beams to the beam splitter where they recombine to generate the exit beam. One of the interferometer arms has a moving mirror which may be displaced in a controlled way by an amount x, with respect to the zero path difference (ZPD) location of the two mirrors. This introduces a retardation (phase delay), $2\pi Wx$, between the two beams when they are recombined at the beam splitter.

Fig. 1 Schematic diagram of the Michelson Interferometer used for FTIR-PAS

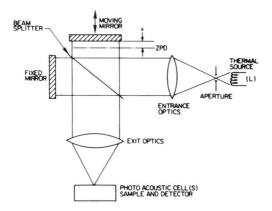

The optical components of the spectrometer are assumed to be non-dispersive, and spectral sampling of the output is assumed to occur at precisely known values of x, including x = 0, so that no phase terms other than those associated with the spectrometer retardation are present. The power transmitted by the interferometer is then a function of the retardation. For a source with a spectral power, B(W)dW, in the wavenumber interval between W and (W+dW) the transmitted power has the form (Chamberlain, 1979):

$$I(x) = \int\limits_0^\infty B(W)dW + \int\limits_0^\infty B(W) \cos(2\pi Wx)\, dW \tag{1}$$

I(x) contains a contribution from each spectral element in the source, and depends upon the path difference between the two arms of the interferometer. When there is no path difference, i.e. at the ZPD location for which x=0, all of the radiation entering the interferometer from the source is transmitted and

$$I(0) = 2 \int\limits_0^\infty B(W)dW \tag{2}$$

Eq. (1) can therefore be written as:

$$I(x) = 0.5I(0) + \int\limits_0^\infty B(W) \cos(2\pi Wx)dW \tag{3}$$

The first term of Eq. (3) is independent of the mirror position, x, and represents the average power transmitted by the interferometer. The second term is the interferogram which will be used to obtain information about the optical properties of a sample in connection with a photoacoustic detector. For this ideal case, the interferogram is perfectly symmetric about x = 0. Instrumental and photoacoustic effects to be discussed below may modify this situation.

If the source is monochromatic with a spectral power at wavenumber W of $B(W)$, the interferogram extends to infinity with constant amplitude and has the form:

$$I(x) = B(W) \cos(2\pi Wx) \tag{4}$$

For a source of finite spectral width, components of different wavenumbers interfere destructively for non-zero values of x. The interferogram is, therefore, attenuated as x increases, so that for large values of x, the interferometer throughput tends to a value of 0.5I(0).

The expression for the interferogram given in Eq. (3) displays the multiplexing (Fellget's) advantage that is important in detector noise limited systems such as PAS. Radiation components from the whole spectral band are simultaneously incident upon the sample, producing an increased signal strength from the PAS detector and improving the signal to noise ratio as the square root of the spectral collection time. The interferometer also has a larger angular aperture than a dispersive instrument for which the resolution is determined by the slit width. For the same resolving power the interferometer, therefore, achieves more energy throughput. This property is known as Jacquinot's advantage. Griffiths *et al.* (1977) have shown that for the Digilab FTS-14 this throughput advantage may be about two orders of magnitude when compared to a Beckman 4240 dispersive spectrometer. Griffiths and de Haseth (1986) point out that the combined effect of Jacquinot's and Fellget's advantages may be as much as 2000, although the full gain is seldom realized in practice. Despite this, it is these two advantages of interferometric spectroscopy that make the method of particular interest in photoacoustic applications in the infra-red spectral region where high intensity radiation sources covering a wide spectral range are not available.

2.   THE REAL INTERFEROMETER AND PHOTOACOUSTIC DETECTION

A real spectrometer differs from the ideal device in a number of ways. In general, the spectrometer will contain dispersive elements, such as a beam splitter, that will introduce a wavenumber dependent phase delay into the spectral throughput. The reflectivity of the beam splitter will also be wavenumber dependent, and so its efficiency, $E(W) = 4RT$ (where R is the reflectivity and T the transmission of the beam splitter), will vary with wavenumber. This will cause the modulation depth of the components of the interferogram to be wavenumber dependent. The measured signal is normally digitally sampled at discrete values of x, and a systematic error in sampling position will add a contribution to the phase of the interferogram. The sampling is also restricted to a finite range of x on either side of the ZPD (eg. $-L < x < D$). This data truncation, together with the apodization function, $A(W,D)$, used to modify its abruptness at $x = D$, will determine the spectral resolution that can be obtained when $I(x)$ is Fourier transformed to yield spectral information about the sample. In addition to these instrument dependent effects, the photoacoustic method introduces an additional wavenumber and modulation frequency dependent phase delay, $\phi(W)$, due to thermal transport processes involved in the signal generation path (Choquet *et al.*, 1986). For a homogeneous sample of low optical absorption coefficient, the centroid of the energy deposition profile is at some depth below the sample surface and the periodic heat transfer to the ambient exhibits a phase shift with respect to the exciting radiation. As the optical absorption coefficient increases, the centroid of energy deposition moves towards the sample surface, decreasing the phase delay associated with the heat transfer processes.

The sample in the PAS cell has a wavenumber dependent photoacoustic response, S(W), that depends upon the sample's optical absorption coefficient, β, and the radiation modulation frequency, f, together with the thermal properties of the sample and the transducer gas in the PAS cell. The interferogram generated by the PAS signal therefore has the form:

$$F(x) = \int_0^\infty S(W)B(W)E(W) \cos[2\pi Wx - \phi(W)]dW \qquad (5)$$

The photoacoustic interferogram contains only contribution from those wavenumbers for which S(W) is non-zero. This implies that for a broad band source, the PAS interferogram has the form associated with an emission interferogram, with a reduced intensity at the ZPD and more information in the wings. This behavior is seen in Fig. 2, where the PAS interferograms for a black graphite absorber and a silica sample are presented. If additional phase components due to dispersive elements in the interferometer and the sampling

Fig. 2 Photoacoustic Interferograms measured with a Gas-Microphone Cell. a) Black Graphite Powder Reference b) Powdered Silica Sample

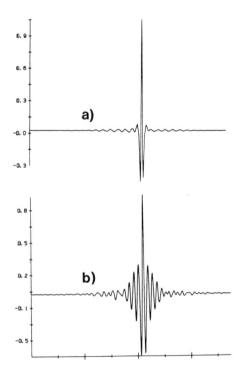

process are present, the argument of the cosine term in Eq. 5 must be modified to include them and becomes: $2\pi Wx - \Phi(x)$. If accurate spectral data are to be obtained, it is necessary to correct the interferogram for these phase terms. This is normally done by collecting a double sided interferogram over a range of path differences extending between $-L < x < +L$, where $L < D$, and using this low resolution interferogram to compute a phase correction function. For many solid state samples, a resolution of between 4 and 8 cm^{-1} is all that is required in the final spectrum, and the phase correction function can be obtained at essentially the needed resolution, i.e. $L \sim D$. Since both the amplitude and phase of the PA response are sample and cell condition dependent, the phase correction must be determined for each sample using its own PAS response under conditions identical to those that will be used for the spectral measurements. Convolving the computed phase correction function with the measured interferogram will then yield a single sided cosine Fourier integral. Fourier inversion of this integral yields the product $S(W)B(W)E(W)*A(W,D)$ from which the sample photoacoustic response may be determined, if the optical throughput of the spectrometer $B(W)E(W)*A(W,D)$ is measured independently. This is normally done by measuring the photoacoustic response of the system with the sample replaced by a black absorber, such as a graphite powder. The response, $S(W)$, of a black absorber corresponds to photoacoustic saturation and will be unity across the full spectral range. In the near infrared, graphite is a reasonable approximation to such a sample, having an essentially constant value for $S(W)$ over the 4000 to 400 cm^{-1} region. As pointed out above, these optical throughput data must also be corrected by their own phase function, and must be measured under the same photoacoustic conditions of cell gas and mirror velocity as those employed for the sample. Taking the ratio of the spectral data collected with a sample in the PAS cell to that obtained with the black absorber will yield a sample spectrum that is corrected for the optical throughput of the interferometer.

Because the PAS signal depends on both the modulation frequency and the thermal properties of the sample, it is necessary to further correct the sample spectrum for effects due to these parameters. For the step and integrate spectrometers employing either amplitude or phase modulation of their throughput, the modulation frequency is the same at each wavenumber, and the thermal diffusion length in the sample is, therefore, independent of wavenumber. The photoacoustic regime appropriate to the sample must be understood in order to interpret relative peak heights across the spectral range, but this interferometric measuring technique does not introduce any new spectral distortions. For a constant scan rate spectrometer, the modulation frequency is both wavenumber and mirror velocity dependent. The thermal diffusion length will, therefore, vary with wavenumber and influence the measured spectral amplitudes, but not the peak positions.

At a given wavenumber, the PAS signal is inversely proportional to the modulation frequency, f, of the radiation as a result of the reduced number of photons reaching the sample in each modulation period. The signal is proportional to the energy deposited in the sample by the absorption of the radiation and has the form (Teng and Royce, 1982):

$$q = (B/f)\cos\phi \tag{6}$$

where B is independent of the optical absorption coefficient and the modulation frequency, but depends on both the number of photons and their energy. For a constant scan rate spectrometer operating at a given mirror velocity, the modulation frequency increases in proportion to the wavenumber of the radiation, that is, in proportion to the photon energy. The increased energy per photon, therefore, compensates for the decreased number of photons reaching the sample, and the PAS signal does not exhibit a (1/f) behavior over the measured spectral range at a given mirror velocity. If the mirror velocity is changed, the modulation frequency at each wavenumber changes, with the photon energy remaining constant, and the expected (1/f) dependence of the signal is observed at each wavenumber. The photoacoustic amplitude is wavenumber dependent because the PA phase $\phi(W)$ depends upon the wavenumber through both $\beta(W)$ and the frequency dependence of the thermal diffusion length.

The spectral dependence of the modulation frequency causes the thermal diffusion length in the sample to be wavenumber dependent, and decreases the PAS signal as the wavenumber increases. For a homogeneous sample, the thermal diffusion length is inversely proportional to the square root of the modulation frequency. As has been shown by Teng and Royce (1982), the corrected peak heights for a thermally thick sample that is not in photoacoustic saturation may be obtained to within a multiplicative constant by multiplying the amplitude spectrum, ratioed to that of a black absorber, by the square root of the modulation frequency at each wavenumber.

The photoacoustic phase is expected to show a general frequency dependence in addition to excursions in those spectral regions in which absorption features occur. This can be understood by considering the expected phase response of a homogeneous, thermally thick, grey body sample. For such a sample the phase is given by the expression:

$$\tan\phi = (1+2/\beta\mu) \tag{7}$$

where β is independent of wavenumber. The thermal diffusion length in the sample, μ, depends on the inverse square root of the frequency and, therefore, decreases as the wavenumber of the radiation increases. Substituting for the wavenumber dependence of the modulation frequency at a constant mirror velocity, V, gives:

$$\tan\phi = \left[1+2\left[\frac{V\rho c}{k}\right]^{\frac{1}{2}}\sqrt{W/\beta}\right] \tag{8}$$

where ρ is the density, c the specific heat, and k the thermal conductivity of the material. The phase of the grey body sample therefore increases as the inverse

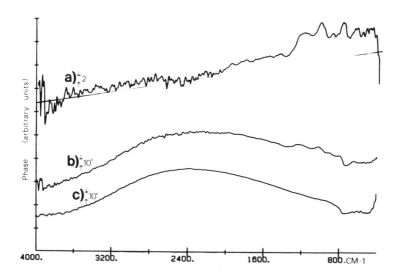

Fig. 3 Photoacoustic Phase as a function of Wavenumber. a) Graphite Reference
b) Silica Sample   c) Phase Difference (b - a)

tangent of a quantity that depends upon the square root of the wavenumber. For a black body, β is infinite and the phase is constant at all wavenumbers for normal modulation frequencies. Differencing or ratioing the grey body phase spectrum with that of a black body will, therefore, display the wavenumber (frequency) dependence of the sample's phase. Fig. 3 shows phase data for both a graphite reference powder and a silica sample. It is seen that the phase response of the black body varies slowly across the spectral range due to beam splitter and other instrument effects. The silica spectrum shows a similar general trend, but with local phase excursions due to the presence of absorption features. The difference between these two phase spectra is plotted on the same diagram, and the overall linear frequency trend of the photoacoustic phase for an absorbing sample is clearly seen. In addition, the phase shifts associated with the spectral features due to lattice absorption are superimposed on this general trend.

Although all rapid scan interferometers apply a phase correction to the collected interferograms, this may not always have the desired spectral resolution. In addition, the photoacoustic signal is not a direct measure of the absorption coefficient of a homogeneous sample, yielding the product of the absorption coefficient and the thermal diffusion length. The absence of a range of interferometer mirror velocities may also make the correction method proposed by Teng and Royce impracticable. Recently, Choquet *et al.* (1986) have suggested a method in which the interferometer's optical configuration and data acquisition algorithm are modified to permit the real and imaginary components of a double sided FTIR-PAS interferogram to be collected at the desired spectral resolution. These modifications enable quantitative spectra to be obtained from measurements made at a single mirror velocity. Considerable data manipulation is still required, including a manual phase correction procedure. The method therefore seems to be unready for general application at this stage of its development.

3.   EFFECTS DUE TO SAMPLE CHARACTERISTICS

Under some experimental conditions the thermal properties of the sample may change as a function of time. An example of this type of behavior (Royce *et al.*, 1981) might be the curing of a polymer with the material changing from a liquid to a solid during the reaction. In such an experiment the degree of polymerization can be followed by observing peak heights associated with the breaking of monomer double bonds and the formation of the polymer chains. In order to compensate for the changing thermal properties of the sample during this reaction it is desirable to use an "internal standard" that is known not to change during the time the reaction is being followed. The chromophore being used as a standard should have a spectral feature in approximately the same wavenumber region as the peaks that are to be used to follow the reaction, and this feature should have approximately the same optical absorption coefficient as those that are being used to monitor the reaction. Under these conditions the signal to noise ratio and the optical and thermal depths associated with all the chromophores will be similar. The peak height of the reacting chromophore can then be normalized to that of the reference peak to compensate for changes in the cell environment and the physical properties of the sample. An example of this use of an internal standard is shown in Fig. 4 where the Bis-phenol-A peak at 1509 cm^{-1} in a photocuring acrylate system is left unchanged by the exposure to UV-light and is used to normalize the change in magnitude of the peak at 1407 cm^{-1} associated with the CH bending vibration of the CH$_2$ chromophore of the acrylate group that is participating in the polymerization reaction.

Fig. 4 PA-Spectrum of acrylated diglycidyl-Epoxy-Bis-phenol A during cure. a) 0 sec, b) 10 sec, c) 20 sec, d) 30 sec, of UV exposure

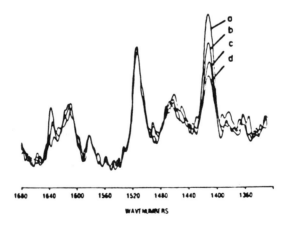

WAVENUMBERS

For powder samples the signal path is more complex than for homogeneous materials. If the PAS response is being measured with a microphone in a closed cell, both amplitude and phase contributions arise due to the expansion of the interparticle gas. McGovern *et al.* (1985a) have suggested that a powder sample for which the particle size is small compared to a thermal diffusion length may be regarded as a composite material with a thermal conductivity governed by the interparticle gas, and a heat capacity dominated by the solid phase. The thermal diffusion length in the powder depends upon both of these quantities, and will be greater for the powder than for a homogeneous solid of the same material. The interparticle gas has a much larger thermal expansion coefficient than the solid phase, and heat transferred to it from the solid phase within an optical absorption depth of the illuminated surface will cause a pressure change that will be transmitted to the gas above the composite sample at the velocity of sound. For high porosity materials, this pressure wave suffers little attenuation, and will be additive to the cell pressure produced by heat transfer from the powder-gas composite to the cell gas above its surface. This model is essentially an extension of the "composite piston model" of McDonald and Wetsel (1978), developed for homogeneous materials, to a two phase system with a

Fig. 5 Computer PA response for an aerosol powder sample in nitrogen gas. A light scattering coefficient of 100 cm^{-1} is assumed. a) Signal Magnitude b) Signal Phase.
_____ Total Signal, ... Pressure Contribution, ----- Thermal Contribution

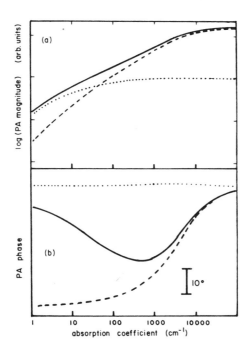

significant thermal expansion coefficient. The pressure and thermal terms of the composite material model are characterized by different length scales. The thermal term is associated with the thermal diffusion length of the composite sample, whereas the pressure term due to the interparticle gas is associated with the optical deposition depth for the absorption process. At low values of the optical absorption coefficient the deposition length is large and the pressure term may dominate the total signal. For high sample optical absorption coefficients the energy deposition length is short and the main contribution to the PAS signal comes from the thermal term. This behavior is illustrated by the computed magnitude curves as a function of sample absorption coefficient shown in Fig. 5 together with the corresponding photoacoustic phase contribution. It can be seen that the pressure term increases the PAS sensitivity at low absorption coefficients, and because of its essentially instantaneous response, significantly modifies the phase.

The magnitude of the pressure term is also dependent upon the porosity of the powder and the particle size. As a powder of a given particle size is compacted the attenuation of the pressure component of the total signal is increased. Particle-particle contact is still poor and the thermal contribution to the total signal does not increase significantly, consequently the overall signal tends to decrease. Yang and Fateley (1986) have pointed out that only the heat generated within a thermal diffusion length of a particles' surface will contribute to the periodic component of the PAS pressure signal. For a powder composed of particles of low optical absorption coefficient, optical energy will be deposited throughout the volume of a particle. If the particle size is less than a thermal diffusion length at the relevant modulation frequency, all of the absorbed energy will contribute to the periodic signal. For a larger particle, only that fraction of the absorbed energy within approximately two thermal diffusion lengths of the particle surface will contribute to the periodic heat transfer to the interparticle gas, and hence to the pressure signal. For particles of irregular shape and low optical absorption coefficient it is the smallest linear dimension that will determine the volume of the particle able to participate in the periodic heat transfer process.

If the particles have a high specific surface area, or are capable of gas adsorption in internal pores, periodic, thermally driven, desorption processes may also contribute to the pressure signal by increasing the number density of gas molecules in the PAS cell. Ganguly and Somasundaram (1983) have discussed this process for condensible gases, and Korpiun *et al.* (1986) have extended Korpiun's earlier work on liquids (Korpiun, 1984) to microporous materials such as zeolites.

The presence of these additional pressure contributions complicates the normalization of peak heights across the spectral range, and makes quantitative measurements difficult. In a limited spectral range, one mechanism may dominate the PAS response. Under this condition, changes in peak height will be indicative of the changes in the number of chromophores, however, it must always be remembered that the response is only linearly dependent upon the optical absorption coefficient for a restricted set of experimental conditions and for constant sample thermal properties.

For some experiments it is desirable to have a sample at a temperature either above or below that of the spectrometer. Fig. 1 indicates that the ideal interferometer is symmetric with respect to the sample and the source, and an expression similar to Eq. (3) describes the intensity of the radiation leaving the sample that exits the interferometer towards the light source (Tanner and McCall, 1984). Energy conservation considerations show that radiation from the source or sample that does not leave the interferometer by the opposite arm is returned to its point of origin. The interferometer is, therefore, a device that switches energy between its entrance and exit ports, either permitting energy at a given wavenumber to pass through and exit at the opposite arm, or returning the photons to the sample and the source for the same mirror position. McGovern *et al.* (1985b) have shown that the PAS response of a sample, which results from the change in sample temperature produced by the net transfer of radiation at a given mirror position, must include both sample absorption and emission processes. These processes occur simultaneously, however, the sample temperature increases as a result of the absorption of energy from the light source, and decreases as a result of emission to the spectrometer so that their photoacoustic phases differ by 180 degrees. For a room temperature sample, emission effects are normally unimportant over the mid-IR spectral region. As the sample temperature is raised above about 100°C, an emission contribution to the signal begins to modify measured spectral amplitudes at the low wavenumber (400 cm^{-1}) end of such a spectrum, and influences spectral information at progressively higher wavenumber as the sample temperature is increased. Low (1984) has reported similar sample emission effects which were detected using a photothermal beam deflection technique in conjunction with a liquid nitrogen temperature source.

III. Experimental Methods

1. TYPES OF INTERFEROMETER

Two methods have been employed for the use of the interferogram generated by the Michelson interferometer in a photoacoustic experiment. In the first, which is the basis for most commercial FTIR-spectrometers, the moving mirror is translated at a constant velocity, V, over the distance needed to achieve a desired spectral resolution. This constant mirror velocity encodes each wavenumber at an audio frequency given by the expression: f=2VW. The modulated radiation from the spectrometer falls on the sample and produces a PAS signal that is measured at known positions of the moving mirror to generate the desired interferogram. Cosine Fourier transformation of the interferogram obtained by this method, after correction for the spectral output of the source and the phase delays associated with the instrument and signal path, provides optical absorption data for the sample. In the second method, the step-and-integrate mode, the moving mirror of the interferometer is stepped over the range of mirror displacement needed to provide the desired spectral resolution. At each step the PAS signal is measured by modulating the incident radiation at

**BOMEM DA3 SERIES
OPTICAL CONFIGURATION**

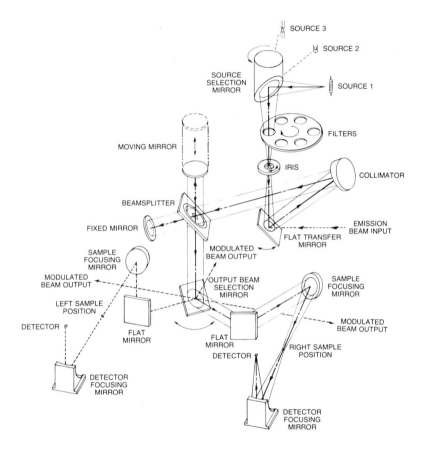

Fig. 6 Optical Configuration of the Bomem DA3 FTIR Spectrometer. The PAS cell is normally located at the right sample position

a fixed frequency, and the PAS response of the sample is measured at this modulation frequency using lock-in detection techniques.

a)   Constant Scan Spectrometers

The optical configuration of the Bomem DA3 spectrometer is shown in Fig. 6. This spectrometer has selectable mirror velocities, which is an advantage for PAS measurements as it permits control of the modulation frequency, and hence, the signal amplitude and the thermal diffusion length in the sample. The sample compartment of this instrument is large, which permits considerable flexibility in the design of a PAS cell and its acoustic isolation mounts. Using this spectrometer, Benziger *et al.* (1984a) have followed chemical reactions taking place in a PAS cell. The necessary flow tubes for the reactants, thermocouple leads, and heater power lines are easily brought to the cell through one of the removable side plates of the sample compartment.

Most PAS cells require the radiation to enter the cell from the top. A plane or focusing front surface mirror is used to direct the horizontal beam from the interferometer onto the sample. Provided all of the radiation from the interferometer is incident on the sample, its exact distribution is not important if a gas-microphone cell is being used. For a photothermal deflection detector, care should be taken to focus the modulated radiation along the path followed by the probe laser beam. In the mid-IR spectral region black body sources, such as glow bars, are normally used. Because of their higher emissivity, silicon carbide glow bars have an intensity advantage as compared to a nichrome wire wound bar and are, therefore, desirable for PAS applications.

In order to minimize the electronic noise due to ground loops, the PAS signal should be taken to the spectrometer detector inputs via shielded leads that are connected to the instrument ground at only one location. A low noise amplifier between the PAS detector and the spectrometer is useful for signal conditioning, and for ensuring that the signal amplitude matches the range of the analog-to-digital converter of the spectrometer. This amplifier should be used at constant gain across the interferogram and the internal amplifier of the spectrometer should also be used in a constant gain mode.

The Bomem spectrometer has the additional advantage of providing access to the phase of the interferogram. This may be acquired at 8 or 16 cm^{-1} resolution, and clearly shows both the photoacoustic and instrumental contributions. In general, the resolution of the phase file will be adequate for making the required corrections to the interferogram which is transformed using the algorithms provided with the spectrometer to yield uncorrected spectral data. This then must be normalized for the optical transfer function of the spectrometer in the manner discussed above.

b)   Step-and-Integrate Spectrometers

In a step-and-integrate spectrometer, intensity modulation may be produced by means of a mechanical chopper which interrupts the light source, or by periodically modulating the position of one of the interferometer mirrors about its equilibrium position. The position variation of the mirror introduces a small additional retardation in the optical path which modulates the output intensity from the interferometer in a manner that is dependent upon the local slope of the interferogram. The absorption spectrum can be obtained by Fourier transforming the measured signal, which for the phase modulation technique has the form of a sine Fourier integral, whereas for amplitude modulation the signal yields a cosine Fourier integral. The spectra so obtained must then be normalized by the optical transfer function of the spectrometer.

Fig. 7 shows the optical, mechanical, and electronic configuration employed by Debarre *et al.* (1981) for their step-and-integrate PAS spectrometer. The moving mirror is driven by a linear motor and serves to both control the retardation and introduce a phase modulation at a fixed frequency, f. A He-Ne laser is used to measure the location of the moving mirror and the interferogram is sampled at each L/4 location of this wavelength, L. By using active control to stop the mirror at zero crossings of the He-Ne modulation signals measured alternately at f and 2f, the interferometer mirror can be positioned to an accuracy of L/50 at these locations. In their initial measurements, Debarre *et al.* (1981) focused the modulated white light from the interferometer into a non-resonant gas-microphone PAS cell, and detected the PA response using a lock-in amplifier tuned to f, the white light modulation frequency. This signal was digitized using a 12 bit analog-to-digital converter, and transformed to give spectral data using a Cooley-Tukey fast Fourier transform program. In later measurements (Fournier *et al.*, 1982), this interferometer was used in conjunction with a PDS detector, the sample being either immersed in a liquid or in a gaseous ambient. To reduce the effects of convection currents on the throughput intensity, the interferometer was contained in a heated, isothermal enclosure. The use of phase modulation for the collection of the PAS response,

Fig. 7 The Step-and-Integrate interferometer developed by Debarre *et al.* (1981) for Visible and Near-IR Photoacoustic Spectroscopy

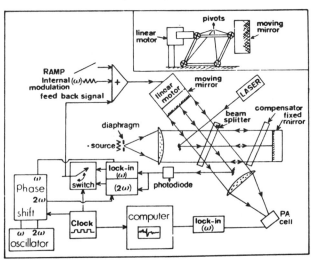

gives a signal which looks like the first derivative of the interferogram and which, therefore, has a zero mean value.

2.   PHOTOACOUSTIC CONSIDERATIONS

From the PAS standpoint, these two interferometric approaches differ in several important ways. The constant modulation frequency of the step-and-integrate technique implies that the thermal diffusion length in the sample, which depends on the inverse square root of the modulation frequency, is the same for each wavenumber. The photoacoustic phase will then only vary due to changes in the optical absorption coefficient of the sample. For the constant scan rate method, the modulation frequency is wavenumber dependent, and consequently, the thermal diffusion length will be different for each wavenumber. Since both the magnitude and phase of the PAS signal depend upon the product of the optical absorption coefficient and the thermal diffusion length, care must be taken to determine the phase correction function at the same mirror velocity as is used for the spectral data. The relationship between the optical deposition depth and the thermal diffusion length is also important in photoacoustic saturation. For the variable scan rate method both of these quantities will be wavenumber dependent and the saturation condition may change across a spectral region of interest. Varying the spectrometer mirror velocity, and hence the modulation frequency at a given wavenumber, permits the state of saturation to be checked.

Long-term source stability is an important factor in step-and-integrate spectrometers used with amplitude modulation because the interferogram is built-up in steps of equal retardation with the data collected at significantly different times. The data obtained photoacoustically gives the amplitude of the interferogram at each step, and source instability contributes amplitude noise which can lead to spurious spectral features. To minimize this effect, the source output should be independently monitored and the PAS signal should be normalized to the incident intensity. When phase modulation is employed the method is less sensitive to source stability as the interferogram has a zero mean value. This has the advantage of avoiding amplitude fluctuations in the interferogram (IGM) due to light source instabilities. However, the signal to noise ratio at each step will be dependent upon the stability of the source.

3.   PHOTOACOUSTIC DETECTORS

a)   The Gas-Microphone Cell

The most frequently employed mode of photoacoustic detection involves a gas-microphone photoacoustic cell. In this cell, the sample and the microphone occupy an acoustically isolated volume. The microphone detects the pressure fluctuations that result from the periodic heating of the sample by radiation from the interferometer, and heat transfer to the cell gas. The cell will have a resonant frequency that is associated with its geometry, and which will also depend upon the sample volume. Many commercial cells separate the microphone from the sample compartment by a small diameter tube and exhibit a Helmholtz resonance at a frequency that corresponds to one of the spectral multiplexing frequencies of the constant scan spectrometers. Under these conditions, care must be taken to ensure that the modulation frequency of the incident radiation is stable, so that the cell resonance does not move to another spectral location. This requires stable and repeatable mirror velocities for each scan. It is also important to normalize spectra obtained with a cell of this type by a reference spectrum obtained under identical conditions of mirror velocity, cell gas, and cell resonance frequency. If this is done, both the amplitude and phase contributions of the cell resonance can be properly taken into account. For some commercial cells, the lowest resonance frequency is about 1kHz, and for the slowest mirror velocities of many FTIR-spectrometers, this places the resonance in the 400 to 4000 cm^{-1} spectral region. Spectrometers, such as the Bomem DA3, which have both low and variable mirror speeds, permit PAS operation below this resonant frequency range with modulation frequencies between 40 and 400 Hz corresponding to the 400 to 4000 cm^{-1} spectral range.

The variation of the radiation modulation frequency with wavenumber in a rapid scan spectrometer causes the photoacoustic sampling depth to be wavenumber dependent. For a homogeneous material, the photoacoustic signal is obtained from about the first two thermal diffusion lengths in the sample. In the mid-infrared spectral region, between 4000 and 400 cm^{-1}, the signal depth varies by about a factor of three across the spectral range. By varying the mirror velocity of the interferometer all modulation frequencies can be changed, and the spectral information will come from a different range of depths in the sample. Information about the near surface distribution of chromophores may be obtained by systematically varying the mirror velocity, and differencing spectral data collected at different velocities. For this data to be easily interpreted, it is an advantage to make these measurements with a cell that does not have acoustic resonances at any of the frequencies encountered. Donini and Michaelian (1984) have discussed the effects of cell resonances on depth profiling of chromophores. Donini and Michaelian (1986) have also reported spectral measurements using a PARC photoacoustic cell in conjunction with a Bruker 113 vacuum spectrometer over the 10,000 to 4,000 cm^{-1} range. When used with nitrogen gas, this cell has a resonant frequency of about 1.3kHz which, at the lowest mirror velocity of the unmodified Bruker spectrometer, places the resonance at approximately 5,500 cm^{-1}. By modifying the mirror velocity control electronics, it was possible to reduce the mirror velocity of the double-sided moving mirror to 0.032 cm/sec and raise the wavenumber at which the cell resonance occurred to 10,200 cm^{-1}, thus avoiding amplitude and phase complications associated with the resonance.

In a gas-microphone cell, the gas must be non-absorbing in the spectral region being studied if contributions to the PAS signal from this source are to be avoided. Carbon dioxide and water vapor both have strong IR absorption features and must be removed from the cell if clean solid phase spectra are to be obtained in this spectral region. Because of the very direct heat transfer path from a vibrationally excited gas molecule to the translational kinetic energy of the gas, the gas phase signals may be two orders of magnitude larger than those

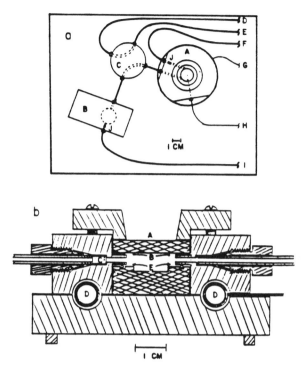

Fig. 8 Variable Temperature Gas-Microphone Cell with a gas flow capability. a) Overall configuration; A:Sample Cell, B:Microphone Chamber, C:Gas Flow Control Valve, D:Gas Outlet, E:Sample Chamber Inlet, F:Sample Chamber Outlet, G:Heater Power, H:Thermocouple, I:Microphone Chamber Inlet, J:5 Micron Acoustic Isolation Frits. b) Detail of Sample Compartment; A:KBr window, B:Graphite Gasket, C:5 Micron Frit, D:Heater, E:Metal O-Ring.

from the condensed phase, even with the short optical absorption path in the cell. The complexity of the gas phase water spectrum also makes this hard to remove from the solid state spectrum by spectral subtraction techniques. It is, therefore, desirable to be able to purge the PAS cell with dry nitrogen, helium, or some other transparent gas before making the spectral measurements. Rockley and Devlin (1980) suggested that undesirable gas phase species could be removed by placing some molecular sieve (zeolite) in the PAS cell sample compartment, but out of the region illuminated by the radiation. This method works well if in-cell reactions are not being followed.

Benziger *et al.* (1984b) have shown that a gas flow can be maintained through the cell during data collection provided the acoustic volume of the cell is defined with frits of the type used in gas chromatography. A diagram of their cell is presented in Fig. 8. For pore sizes between about 0.5 and 50 microns, these frits permit the gas in the cell to be changed in about one second while still providing acoustic isolation over the multiplexing frequency range. This technique is particularly useful when gas is being evolved from either the sample being studied or the cell walls, or when the gas is being used in a chemical reaction taking place in the cell. The velocity of sound, and hence the cell resonance frequency, is dependent upon the cell gas. The gas also influences the signal amplitude through its thermal properties. It is therefore important to measure the sample and reference spectra using the same ambient in order to properly normalize the spectral data for the optical throughput of the spectrometer.

A major disadvantage of the gas-microphone cell is its sensitivity to ambient noise and low frequency room vibrations. In order to obtain the best signal-to-noise ratio in a spectrum, it is necessary to pay attention to the acoustic isolation of the spectrometer and the cell. Royce *et al.* (1985) have mounted their Bomem DA3 interferometer on air suspension legs of the type used to isolate laser tables. This removes the low frequency acoustic noise resulting from building vibrations and improves the signal to noise ratio of the PAS signal by about a factor of five. The PA-cell is contained in the sealed vacuum compartment of the spectrometer, and is isolated by a double anti-vibration mount from the noise generated by the mirror scanning and alignment system. Although data are normally collected with the spectrometer used in the purged mode, the rigid walls of the vacuum sample compartment reflect airborne room noise.

Electronic noise must also be minimized, and it is important to have the microphone preamplifier as close to the microphone as possible. For their studies of catalytic surface reactions, McGovern *et al.* (1985a) used a B&K microphone with a sensitivity of 50 mV/Pa in conjunction with a low noise B&K preamplifier that screwed into the back of the microphone. These components were isolated from external acoustic and electrical noise sources by being housed in an aluminum cylinder. The amplified signal was then taken to a PARC 113 preamplifier for further conditioning before being returned to the spectrometer for digitizing and processing. Care was taken to avoid ground loops in the signal path.

The importance of minimizing noise sources can easily be understood in terms of the associated reduction of data acquisition time. To reduce the effects of random noise on the PAS interferograms collected in the constant scan rate mode, it is usual to co-add phase corrected IGMs. Typically, between 50 and 1000 IGMs are added together to yield a noise reduction that depends on the square root of this number. An improvement in the signal-to-noise ratio in the IGM of a factor of two decreases the required data collection time for the same quality spectrum by a factor of 4. With the low-mirror velocities used in FTIR-PAS, a single scan at a resolution of 4 cm^{-1} may take 4 sec, and so this noise reduction will reduce the data collection time from about one hour to 15 min.

b)   Photothermal Deflection Detectors

The periodic heat transfer between an absorbing sample and a fluid in contact with it causes a periodic change in the refractive index of the fluid to

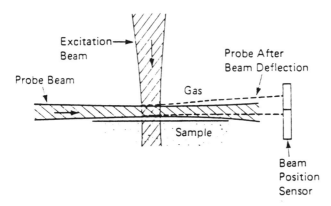

Fig. 9 Optical Configuration of a Photothermal Deflection (Mirage Effect) Detector [From Tam (1982)]

occur. This refractive index change is confined to approximately a thermal diffusion length in the fluid, and may be probed by passing a laser beam tangentially across the sample surface and observing the deflection of the beam. This method was initially developed by Boccara *et al.* (1980) and used in connection with a step-and-integrate spectrometer developed in their laboratory. Murphy and Aamodt (1980), Mandelis (1983), and Mandelis and Royce (1984), extended the theory of the technique as applied to solid samples. Low and thermal diffusion length in the fluid, and may be probed by passing a laser beam tangentially across the sample surface and observing the deflection of the beam. This method was initially developed by Boccara *et al.* (1980) and used in connection with a step-and-integrate spectrometer developed in their laboratory. Murphy and Aamodt (1980), Mandelis (1983), and Mandelis and Royce (1984), extended the theory of the technique as applied to solid samples. Low and Parodi (1982) and Low, Lacroix and Morterra (1982) developed several experimental configurations for use with constant scan spectrometers with the sample in an ambient gas, and Palmer and Smith (1986) have made constant scan measurements in the infra-red using a liquid as the transducer fluid.

The optical configuration of a typical PDS detector is shown in Fig. 9. The sample may be contained in a closed cell or in the ambient gas of the spectrometer. The laser beam passes within a thermal diffusion length of the sample surface, which is illuminated with the modulated radiation from the interferometer. If the sample is transparent at the probe laer wavelength, the probe beam may be used to measure the periodic refractive index change induced in the near surface region of the sample, otherwise this laser detects the change in refractive index of the fluid in contact with the sample region, the beam deflection is enhanced by an optical lever arm and its displacement is measured with a position sensitive detector. The displacement signal depends on i) the refractive index gradient traversed by the beam and involves the temperature dependence of the refractive index of the sample or the fluid with which it is in contact; and ii) the product of the optical absorption coefficient and thermal diffusion length of the sample. After suitable amplification, this signal may be processed by the spectrometer's data acquisition electronics to yield the interferogram, from which the optical absorption of the sample can be determined.

In order to maximize the deflection signal, it is desirable to focus the probe laser beam to a beam waist that can be brought close to the sample surface where the refractive index gradient is a maximum. Seager (private communication) has suggested using a cylindrical lens for this purpose. The sheet of laser radiation produced by this lens will sample a wider path of the illuminated sample surface than will the beam waist of a spherical lens, and will average the effects of sample non-uniformities over this area. The beam deflections are relatively small and refractive index fluctuations along the lever arm may make significant contribution to the noise of the signal. Enclosing the optical path in a constant temperature tube reduces these effects. The optical components must also be rigidly mounted in order to avoid noise due to room vibrations and acoustic sources in the room and the spectrometer. The pointing stability of the laser is also important at low signal levels. Beam wander noise can be partially compensated by splitting the beam prior to passing it over the sample and sending one of the beams over an approximately equivalent optical path to another detector. Using the differential signal from these two paths removes the effects of the laser instability. The effectiveness of the compensation can be tested by observing the difference signal as the angular position of the laser is changed in the absence of a sample.

Johnson *et al.* (1983) suggested that the PDS method may be made more sensitive by placing the sample in an intracavity location in the probe argon laser used for their measurements. Masujima (private communication) has used this configuration for measurements on solid samples, and by tilting one of the cavity mirrors to make the He-Ne laser gain very sensitive to further beam path changes, has further improved the sensitivity of the method. The output intensity of the laser is measured to produce the required interferogram. Neither of these experimenters used an interferometer as the source of modulated radiation for their measurements, but the concept merits further exploration for use with the FTIR configuration.

The sensitivity of the PDS method may by improved by selecting a material with a high temperature coefficient of refractive index change, (dn/dT), as the transducer fluid. In the visible and near infra-red spectral region a particle free liquid such as carbon tetrachloride has been found to give good results. Due to the absorption of most liquids in the infra-red, this method is less satisfactory if a wide range spectral survey is desired in this spectral region. Low *et al.* (1983) showed that increasing the pressure of the ambient gas in contact with the sample also increased the PDS signal. Moore *et al.* (1985) extended the range of pressures and found that the signal reached a maximum value at about 10 atmospheres after which it decreased as the pressure was further increased. The form of the signal is shown in Eq. (7), where A is a constant, α_i is the thermal diffusivity of material i, and μ is the thermal diffusion length of the fluid (Mandelis, 1983)

$$S = A \, (dn/dT)(\alpha_s/\alpha_f)^{1/2} \exp(-x/\mu) \qquad (9)$$

Using a simple ideal gas model to evaluate the pressure and temperature dependence of the various terms in Eq. (9), gives:

$$S = C \, p^{1.5} \, T^{-2.75} \, \exp(-Dp^{0.5} \, T^{-0.75}) \tag{10}$$

This expression shows that the signal initially increases as the pressure increases due to the pressure dependence of the refractive index term, but then decreases due to the pressure dependence of the thermal diffusion length. The temperature dependence of the signal at constant pressure shows a decrease with increasing temperature due to both the thermal diffusion length and the refractive index terms. These considerations indicate that PDS is less interesting as a tool for high-temperature, high-pressure studies than is suggested by the ease with which a laser probe beam may be brought into such an environment.

When used to detect the response of powder samples, PDS differs from the gas-microphone PAS since it only detects the thermal gradient at the sample surface and is insensitive to the thermal expansion of the interparticle gas, and gas desorption processes. Provided the thermal parameters of the composite sample are correctly evaluated, the PDS signal should yield a spectrum of the powder for which the normalization procedures applicable to homogeneous samples will yield correct peak height ratios over the full spectral range.

Photothermal deflection detectors also avoid effects due to cell resonances that are associated with the microphone systems discussed above. In applications such as optical depth profiling, where a range of modulation frequencies at each wavenumber is used to obtain spectral information from thermal diffusion depths of different thicknesses, the absence of spectrum modification by cell resonances can be of importance. Varlashkin and Low (1986) have discussed the use of PDS in conjunction with an FTIR spectrometer having variable mirror speeds for the study of bilayer polymeric samples. These authors concluded that the method is only suitable for probing a restricted range of depth in the sample. For a polyethylene material, this range may extend between 10 and 60 micrometers at 2000 cm^{-1}, if the modulation frequency is restricted to the range between 30 Hz and 3kHz by noise and sensitivity considerations. Another important factor in this type of measurement is the relative optical absorption of the two layers. If the underlayer is the strong absorber and the overlayer is transparent, radiation can reach the underlayer chromophores. The effect of the overlayer is to attenuate the thermal signal at the surface of the sample and to introduce a phase shift with respect to the exciting radiation (Adams and Kirkbright, 1977; Mandelis *et al.*, 1979). Provided the overlayer is less than about two thermal diffusion lengths in thickness, a PDS signal should still be measurable. If the overlayer is the strong absorber, the surface temperature gradient is dominated by the response of this layer due to the exponential attenuation of the signal from the underlayer with overlayer thickness. In addition, the intensity of the radiation reaching the underlayer is reduced in regions of overlayer spectral absorption, and this also reduces the thermal signal from the underlayer, masking spectral features that occur in this overlayer absorption range. When the sample being studied is porous, PDS is at a disadvantage with respect to PAS in signal strength due to the absence of the "pressure" component of the signal. This can significantly reduce the detectability of surface species, however, the spectral magnitude correction procedures required with FTIR-PAS are more easily applied in this case.

c)   Piezoelectric Detectors

The displacement of a sample surface resulting from the thermal expansion which follows the optical absorption and nonradiative de-excitation process, may be detected directly using a piezoelectric transducer in contact with the sample or its support. Most piezoelectric detectors are also light sensitive and care must be taken to prevent either direct or scattered radiation from falling on the detector. Tam (1986) has discussed several configurations for use in the visible and near infra-red spectral regions. Farrow, Burnham and Eyring (1978) employed piezoelectric detection in conjunction with their step-and-integrate FTIR spectrometer. In these measurements, the sample was mounted on the reflecting surface of a front surface mirror and the PZT detector was fixed to the rear surface. This configuration had been previously employed (Farrow, Burnham, Auzanneau, Olsen, Purdie and Eyring, 1978) in conjunction with either a dispersive system or a dye laser light source. Ikari *et al.* (1986) have made measurements on single crystal semiconducting samples using sputtered ZnO transducers on the rear surface of their thermally thick samples, as discussed in detail in Chapter 16. The ZnO transducers were used to record PAS spectra (using a dispersive spectrometer) at temperatures down to 116 K. Shaw and Howell (1982) have made spectroscopic studies at temperatures below 10K using a PZT transducer coupled to the sample by a quartz substrate attached to the cold finger of a cryostat. Since this class of detector directly measures the displacement of the sample surface, it does not require an ambient atmosphere and may be used for spectroscopic studies under high vacuum conditions.

If the sample to be studied is thin compared to a thermal diffusion length, a pyroelectric detector may be used to measure the sample heating produced by the nonradiative de-excitation processes that follow optical absorption, as discussed in detail in Chapters 6 and 7. This method can also be employed with the sample in a vacuum environment as the sample is in physical contact with the pyroelectric detector. Thin film pyroelectric calorimetric detectors have been used for photothermal studies by Coufal (1984); Coufal, Trager, Chuang and Tam (1984); and by Mandelis (1984, 1985). The pyroelectric element in all cases was a polyvinylidene difluoride (PVDF) thin film with metal electrodes coating both faces. The sample is applied directly to one of the electrode surfaces and heat transfer between the sample and the calorimeter generates the output signal. The low thermal mass of the system ensures a high frequency response, and with pulsed lasers picosecond delay times associated with heat transport processes can be measured. This response time is more than adequate for the relatively low modulation frequencies encountered in FT-PAS. As has been pointed out by Mandelis and Zver (1985), the electrodes of the pyroelectric need to be optically opaque over the spectral range being studied otherwise optical absorption in the polymeric piezoelectric will complicate the measured spectral data. Although this type of detector has not yet been used for FTIR-PAS, it would seem to offer the advantages of reduced acoustic sensitivity and environmental flexibility.

IV. Conclusions

When care is taken to minimize noise from acoustic and electrical sources, Fourier transform photoacoustic spectroscopy becomes a useful spectroscopic method. Photoacoustic measurements are linear in the product of the optical absorption coefficient and the thermal diffusion length of a thermally thick sample over several orders of magnitude, and saturate logarithmically at high values of this product. Although this behavior complicates quantitative measurements, it permits spectra due to bulk and surface species to be observed simultaneously on suitable samples. Minimal preparation of samples is required before spectral data can be acquired, and in particular powder samples do not need to be incorporated in a matrix. This permits *in situ* chemical reactions to be followed on the same sample, and for the case of catalytic reactions, this may be taken through many reversible cycles in a given measurement sequence. The absence of a matrix also avoids spectral distortion due to the Christiansen effect. The photoacoustic signal is a measure of the energy absorbed by a sample. This is relatively insensitive to light scattering in the infra-red region where the particle size may be smaller than the optical wavelengths being used.

In the visible spectral region light scattering effects may be more pronounced, and result in a modification of the energy deposition profile in the sample. Scattering tends to cause the centroid of the deposition profile to move towards the sample surface, and will, therefore, increase the signal contribution due to heat transfer to the ambient. Another advantage of the reduction of effects due to light scattering results from the clear definition of the baseline for a zero signal. Processes such as changed sample conductivity, which induce absorption changes over a wide spectral range, may then be detected, whereas in transmission measurements they tend to be masked by baseline uncertainties.

Quantitative measurements are difficult because of the complexity of the signal path following the absorption of light. In the Fourier transform mode, this influences both the amplitude and phase of the signal, and will be wavenumber dependent. Provided the interferograms are corrected using a phase file acquired on the sample itself under identical conditions of spectrometer mirror velocity, ambient, and sample thermal parameters, the spectra obtained by Fourier inversion may be directly related to those expected theoretically. Ratioing these spectra to similarly acquired data from a sample in photoacoustic saturation over the complete spectral range permits correction for the optical transfer function of the spectrometer. Because the reference signal depends on the optical and thermal parameters of reference sample and its ambient, it is essential that the reference interferograms be correct by their own phase file.

In constant scan FTIR-PAS, which is the method most commonly employed, both the amplitude and phase of the PAS response are wavenumber dependent. The modulation frequency of the incident light depends upon both the spectrometer mirror velocity and the wavenumber of the radiation. The thermal diffusion length in both the sample and its ambient as well as the number of photons arriving at the sample in a single modulation period will therefore depend on wavenumber. In the signal amplitude, the photon energy compensates for the reduced photon number at higher wavenumbers, and at a given mirror velocity the signal only depends on wavenumber through the frequency dependence of the thermal diffusion length. The phase of the signal is wavenumber dependent as a result of the finite time associated with thermal transport processes, and is expected to show local variations in the presence of absorption features as well as a general increase with increasing wavenumber.

Control of the radiation modulation frequency permits the thermal diffusion length to be varied, and hence the depth from which spectroscopic information is obtained. In practice, the depth range accessible is restricted because the signal magnitude at a given wavenumber is inversely dependent on the modulation frequency. In addition, for a constant scan spectrometer, the depth sampled at a given mirror velocity is wavenumber dependent, varying by a factor of three across a typical spectral range. Taking signal-to-noise into account suggests that depth information is restricted to layers between about 30 and 200 microns in thickness. The situation is also complicated in structures for which the surface region has strong absorption features that overlap those of an interior region. For this case, the signal from the interior is reduced, both because the incident radiation intensity is decreased by the surface layer absorption and because the thermal signal is exponentially attenuated in its transport to the surface. Gas-microphone cells may exhibit acoustic resonance phenomena which will modify the signal amplitude and phase and further complicate data acquisition. Photothermal deflection and photopyroelectric detection methods do not suffer from this additional disadvantage.

Although FTIR-PAS has not been used for the study of semiconducting systems the method would seem to offer some features that might be of utility. The method provides another spectroscopic tool, and one that has a high dynamic range as well as a relative insensitivity to sample condition. Provided reasonable care is taken, a photoacoustic cell can be included among the range of detection methods available and spectra with a high signal-to-noise ratio may be collected in a reasonable time. The minimal sample preparation required often compensates for the longer data acquisition time, and the method should certainly be considered if diffuse reflectance or mull techniques are felt to be necessary to achieve a spectral measurement.

V. References

Adams M.J. and Kirkbright, G.F. (1977) Analyst **102**, 281.

Benziger J.B., McGovern S.J. and Royce B.S.H. (1984a) in *Catalyst Characterization Science* (M.L. Deviney and J.L. Gland, Eds.) ACS Washington, DC, p. 449.

Benziger J.B., McGovern, S.J. and Royce B.S.H. (1984b) Appl. Surf. Sci. **18**, 401.

Boccara A.C., Fournier D. and Badoz J. (1980) Appl. Phys. Lett. **36**, 130.

Busse G. and Bullemer B. (1978) Infrared Phys. **18**, 225.

Chamberlain J. (1979) *The Principles of Interferometric Spectroscopy* (G.W. Chantry and N.W.B. Stone, eds.) Wiley-Interscience, New York.

Choquet M., Rousset G. and Bertrand L. (1986) Can. J. Phys. **64**, 1081.

Coufal H. (1984) Appl. Phys. Lett. **44**, 59.

Coufal H., Träger F., Chuang T. and Tam A. (1984) Surf. Sci. **145**, L504.

Debarre D., Boccara A.C. and Fournier D. (1981) Appl. Opt. **20**, 4281.

Donini J.C. and Michaelian K.H. (1984), Infrared Phys. **24**, 157.

Donini J.C. and Michaelian K.H. (1986) Infrared Phys. **26**, 135.

Farrow M.H., Burnham R.K. and Eyring E.M. (1978) Appl. Phys. Lett. **33**, 735.

Farrow M.M., Burnham R.K., Auzanneau M., Olsen S.L., Purdie N. and Eyring E.M. (1978) Appl. Opt. **17**, 1093.

Fournier D., Boccara A., and Badoz J., (1982) Appl. Opt. **21**, 74.

Ganguly P., and Somasundaram T., (1983) Appl. Phys. Lett. **43**, 160.

Griffiths P.R., Sloane H.J. and Hannah R.W., (1977) Appl. Spectr. **31**, 485.

Griffiths P.R., and de Haseth J.A., (1986), *Fourier Transform Infrared Spectrometry*, (P.J. Elving and J.D. Windefordner, Eds.), Wiley-Interscience, New York.

Ikari T, Shigetomi S., Koga Y. and Shigetomi S., (1986) Rev. Sci. Instr., **57**, 17.

Johnson C., Brundage R.T., Glynn T.J. and Yen W.M., (1983) J. de Phys. **44**, C6-253.

Korpiun P., (1984) Appl. Phys. Lett. **44**, 675.

Korpiun P., Herrmann W., Kindermann A., Rothmeyer M. and Buchner B., (1986) Can. J. Phys. **64**, 1042.

Laufer G., Huneke J.T., Royce B.S.H. and Teng Y.C., (1980) Appl. Phys. Lett. **37**, 517.

Low M.J.D., (1984) Spectr. Lett. **17**, 279.

Low M.J.D. and Parodi G.A., (1982) Chem. Biomed. Environm. Inst. **11**, 57.

Low M.J.D., Lacroix M. and Morterra C., (1982) Appl. Spectr. **36**, 139.

Low M.J.D., Arnold T.H. and Severdia A.G., (1983) Infrared Phys. **23**, 199.

Mandelis A., Teng Y.C., and Royce B.S.H., (1979) J. Appl. Phys. **50**, 7138.

Mandelis A., (1983) J. Appl. Phys. **54**, 3403.

Mandelis A. and Royce B.S.H., (1984) Appl. Opt. **23**, 2892.

Mandelis A., (1984) Chem. Phys. Lett. **108**, 388.

Mandelis A., and Zver, M.M., (1985) J. Appl. Phys. **57**, 4421.

McDonald F.A. and Wetsel G.C., (1978) J. Appl. Phys. **49**, 2313.

McGovern S.J., Royce B.S.H. and Benziger J.B., (1985a) J. Appl. Phys. **57**, 1710.

McGovern S.J., Royce B.S.H. and Benziger J.B., (1985b) Appl. Opt. **24**, 1512.

Moore N., Spear J.D., Benziger J.B. and Royce B.S.H., (1985) *Proc. 4th International Conf. on Photoacoustic, Thermal and Related Sciences*, Quebec, paper MA8.

Murphy J.C. and Aamodt L.C., (1980) J. Appl. Phys., **51**, 4580.

Palmer R.A. and Smith M.J., (1986) Can. J. Phys. **64**, 1086.

Rockley M.G., (1979) Chem. Phys. Lett. **68**, 455.

Rockley M.G., (1980) Appl. Spectr., **34**, 405.

Rockley M.G. and Devlin J.P., (1980) Appl. Spectr., **34**, 407.

Royce B.S.H., Teng Y.C. and Ors J.A., (1981) *IEEE Ultrasonics Symposium Proc.*, (B.R. McAvoy, Ed.), IEEE New York, p. 784.

Royce B.S.H., McGovern S.J. and Benzinger J.B., (March 1985) *American Laboratory*.

Royce B.S.H. and Benzinger J.B., (1986) IEEE Trans. UFFC, **UFFC-33**, 561.

Shaw R.W. and Howell H.E., (1982) Appl. Opt. **21**, 100.

Tam A.C., (1983) in *Ultrasensitive Spectroscopic Techniques*, (D. Klinger, Ed.), Academic Press, New York.

Tam A.C., (1986) Rev. Mod. Phys. **58**, 381.

Tanner D.B. and McCall R.P., (1984) Appl. Opt. **14**, 2383.

Teng Y.C. and Royce B.S.H., (1982) Appl. Opt. **21**, 77.

Varlashkin P.G. and Low M.J.D., (1986) Infrared Phys. **26**, 171.

Vidrine D.W., (1982) in *Fourier Transform Infrared and Raman Spectroscopy*, (J.R. Ferro and L.J. Basil, Eds.), Academic Press, New York.

Yang C.Q. and Fateley W.G., (1986) J. Mol. Struct. **141**, 279.

Yasa Z.A., Jackson W.B. and Amer N.A., (1982) Appl. Opt. **21**, 21.

IV

Electronic Transport and Nonradiative Processes in Semiconductors

THERMAL WAVE INVESTIGATION OF TRANSPORT PROPERTIES OF SEMICONDUCTORS I: METHODOLOGY

A. Claude Boccara and Danielle Fournier

Laboratoire d'Optique Physique - ER 5 CNRS
Ecole Supérieure de Physique et de Chimie
10, rue Vauquelin, 75231
Paris Cedex 05 - FRANCE

I. Introduction

In situ non contact techniques for material and device inspection in the field of semiconductors are very active and "hot" areas for scientists and engineers (Sze, 1969). Among the physical parameters of interest, the electronic transport properties have implications in the efficiency and speed of the devices, and the thermal transport properties are of paramount importance for the integration level and the lifetime of the electronic circuits, laser diodes, etc. Due to the fact that usually carrier diffusion is not taken into account, we will analyze in this chapter the physical principles of a photothermal approach to semiconductor transport parameter determination. We will show that such an approach is not only necessary when the study of transport properties is the purpose of the experiment, but also for other kinds of photothermal experiments such as spectroscopy, electrochemistry, etc.

The physical background given in Section II will underline the influence of the physical parameters on the photothermal signals. Then, we will establish in Section III the conditions under which it is realistic to neglect electronic transport properties, when a thermal characterization is the aim of the experiment. Finally, Section IV will describe experiments in which both thermal and electronic transport properties contribute to the signal, and will show how the transport parameters can be measured.

The physical concepts that we will develop here may be applied to various kinds of photothermal experiments, such as photoacoustic (gas-cell and PZT) (Rosencwaig 1980); photothermal deflection ("mirage" effect) (Boccara, Fournier and Badoz; 1980); and reflectivity modulation (Rosencwaig *et al.*, 1985). The greatest care must be taken when using photothermal radiometry with semiconductor samples, because the heat source is often difficult to define due to the weak IR emissivity (Kanstad and Nordal, 1986). We will not discuss here the experimental apparatus, unless specific applications to transport properties such as the "mirage" deflection induced in the bulk of the semiconductor sample (section III) are described.

II. Optically Generated Thermal Waves in Semiconductors

The main processes which have to be considered when a light beam is absorbed by a doped semiconductor sample, at room temperature, are represented in Fig. 1.

This scheme, although over-simplified, will be very useful for discussing most of the features which will appear in this section. The monochromatic light flux impinging on the sample consists of photons whose energy $h\nu$ is larger than the band gap E_g. The absorbed photons give rise to electron-hole pairs. In order to avoid complicated effects such as non-linear behavior in the absorption process, high density plasma effects, and ambipolar diffusion we will consider a doped sample (e.g. p-doped in Fig. 1) and a low level intensity which maintains the photo-induced carrier population at a level lower than the majority carrier

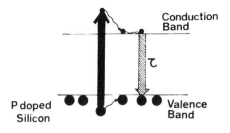

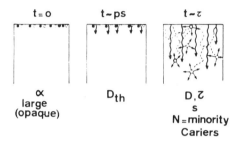

Fig. 1 Optical generation of thermal waves in a p-doped semiconductor (see text).

population (Pankove, 1975; Smith, 1978).

The excess energy $(h\nu - E_g)$ is converted by fast ($\approx$ 1 ps; Shank, 1984) nonradiative processes into thermal energy. Thus the sample exhibits a heat source which reflects the distribution of the absorbed energy: e.g., if the sample is opaque, a thermal wave will be generated on the sample surface.

Then, during their lifetime τ, the photo-induced carriers exhibit a random walk (electronic diffusion) throughout the crystal. During this walk they can cross the sample surface, with a finite probability of recombination there (Surface recombination velocity s).

Finally, after time delay τ, the (mostly nonradiative at room temperature) recombination processes take place, leading to an expanded heat source whose distribution reflects the carrier diffusion processes.

To be more quantitative and obtain analytical results of practical interest we will first develop a calculation which applies to strongly absorbing samples (surface absorption) and leads to the temperature and carrier distribution within the sample volume. Then, we will introduce a more complete calculation taking also into account the light energy distribution within the sample. Following these two approaches obtained for a periodic excitation, we will extend these results to the case of pulsed excitation.

1.   SURFACE ABSORPTION. MODULATED EXCITATION

In order to calculate the carrier density, $N(x,t)$, and temperature, $T(x,t)$, distributions through the sample whose surface is illuminated by a periodic uniform light flux, we have to couple two contributions (Fournier, Boccara, Skumanich, and Amer, 1986; Mikoshiba, Nakamura, and Tsubouchi, 1982):

a. Thermal Contribution

The equation describing the thermal contribution follows from the standard heat diffusion with a source term due to carrier recombination, and a surface source boundary condition:

$$D_{th} \frac{\partial^2 T}{\partial x^2} = \frac{\partial T}{\partial t} - \frac{E_G D_{th}}{k} \frac{N(x,t)}{\tau} , \qquad (1a)$$

with the boundary conditions

$$T(x,0) = 0,$$

$$-k\frac{dT}{dx}(0,t) = sN(0,t)E_G + \frac{hv-E_G}{hv} \phi_o e^{i\omega t} , \qquad (1b)$$

where D_{th} is the thermal diffusivity (cm^2/s), k is the thermal conductivity (W/cm·K), E_G is the semiconductor gap energy (J), hv the photon energy (J), ϕ_o is the light flux (W/cm^2), s is the surface recombination velocity (cm/s), and τ is the minority-carrier lifetime (s).

b. Free-carrier Contribution

The carrier population distribution is given by the non-homogeneous diffusion equation:

$$D\frac{\partial^2 N}{\partial x^2} = \frac{\partial N}{\partial t} + \frac{N}{\tau} \qquad (2a)$$

The boundary conditions are:

$$N(x,0) = 0,$$

$$D\frac{dN}{dx}(0,t) = -\frac{\phi_o}{hv}e^{i\omega t} + sN(0,t), \qquad (2b)$$

where D is the minority-carrier diffusion coefficient (cm^2/s). Here we have assumed for simplification that surface recombination is only localized on the sample surface and we have neglected the influence of a space charge region.

The two differential equations are coupled by the carrier source decay. The steady-state solution for the temperature distribution is given by:

$$T(x,t) = Re\left[-\frac{(\phi_o E_g/\lambda_{el})\left[\dfrac{e^{-x/\lambda_{th}}}{1/\lambda_{th}} - \dfrac{e^{-x/\lambda_{el}}}{1/\lambda_{el}} \right]e^{i\omega t}}{hv\tau Dk(s/D+1/\lambda_{el})[1/\tau D+i(\omega/D)-i(\omega/D_{th})]} \right.$$

$$+ \left[\frac{\phi_o E_G s}{hvDk(s/D+1/\lambda_{el})} + \frac{(hv-E_G)\phi_o}{hvk} \right] \lambda_{th} e^{-x/\lambda_{th}} e^{i\omega t} \right] , \quad (3a)$$

where

$$1/\lambda_{th} = \sqrt{i\omega/D_{th}} \text{ and } 1/\lambda_{el} = \sqrt{(1+i\omega\tau)/D\tau} ,$$

For the minority-carrier distribution, one obtains

$$N(x,t) = Re \left[\frac{\phi_o e^{-x/\lambda_{el}}}{hvD(s/D+1/\lambda_{el})} e^{i\omega t} \right] , \quad (3b)$$

The minority-carrier distribution exhibits a purely electronic term characterized by an exponential dependence on the depth away from the illuminated surface. This decay is characterized by the complex electronic diffusion length λ_{el}. The temperature distribution is a linear combination of two kinds of terms: 1. The thermal wave terms with the exponential dependence characterized by the complex thermal diffusion length λ_{th}; and 2. The electronic wave (sometimes called "plasma wave" (Rosencwaig, 1987)) with the complex diffusion length λ_{el}.

2.   BULK ABSORPTION. MODULATED EXCITATION

This analysis has been performed by quite a few authors for photoacoustic detection (Miranda, 1982; Sablikov and Sandomirskii, 1983; Mikoshiba *et al.*, 1982), the theoretical work of Sablikov and Sandomirskii (1983) being the most complete physical analysis: they account for the optical absorption coefficient α, which governs the light energy density distribution through the sample, and discuss the detailed conditions for which the surface heat generation may be considered to be localized in the space charge region of thickness ω (Band bending lower than ≈ 0.5 eV).

The sample surface periodic temperature induced by a modulated light flux ϕ_o is given by (Sablikov and Sandomirskii, 1983):

$$\Delta T = \left[W_s + \left[\frac{W_1}{\alpha+\eta} \right] e^{-\alpha\bar{\omega}} + \left[\frac{W_2}{\beta+\eta} \right] \right] /\eta k \quad (4)$$

where α is the absorption coefficient, and

$$\eta = (1-i)(\pi f/D_{th}); \ \beta = (1-2\pi i f\tau)^{1/2}/L \ ; \ L = (D\tau)^{1/2}$$

$$W_1 = \alpha\phi_o[1-(E_g/hv)\{1+[L^2(\alpha^2-\beta^2)]\}^{-1}]$$

$$W_2 = \phi_o(E_g/hv)[\alpha^2-\beta^2+[\beta^2+(\alpha S/D)]e^{-\alpha\bar{\omega}}]/\{L^2(\alpha^2-\beta^2)[\beta+(S/D)]\}$$

$$W_S = \phi_o\{1-e^{-\alpha\bar{\omega}}-(E_g/hv)[\beta/(\beta+(S/D))]\}$$

$$+ e^{-\overline{\alpha\omega}}(E_g/h\nu)[\alpha(\beta+(S/D))+\beta^2]/[(\alpha+\beta)(\beta+(S/D))]\}$$

For strong absorption coefficients ($\alpha\overline{\omega}\gg1$) these results are indentical to that obtained in Eq. (3a) above. In the limit of $\alpha\overline{\omega}\ll1$ the authors point out differences between their results and the Rosencwaig and Gersho (RG) theory which are mainly revealed in the frequency dependence of the signal phase (Rosencwaig and Gersho, 1976).

The same kind of approach has been followed by Mikoshiba *et al.* (1982), in order to extend the Jackson-Amer PZT detection theory of the photoacoustic effect (Jackson and Amer, 1980) to the case of semiconductors. This kind of calculation which is performed for a gaussian beam irradiating the semiconductor sample is very useful when the sample is irradiated with a laser beam whose dimensions are in competition with (or smaller than) the thermal diffusion length μ or the electronic diffusion length $L = (D\tau)^{\frac{1}{2}}$. The drawback of such 3-D calculation is that the temperature (or strain and PZT voltage) has to be computed numerically because it implies the (simple) calculation of a Bessel integral and one cannot explicitly distinguish the influence of carrier diffusion on the signal.

An important consideration of this experimental mode is that it can be easily applied to samples of practical interest such as Si wafers without the geometrical size limitation of a photoacoustic cell. The signal is analyzed as a function of the chopping frequency, the influence of the carrier diffusion can be observed through the modification of the amplitude and phase of the signal generated by the PZT when compared with the Jackson-Amer model, or when the physical parameters are changed (Jackson and Amer, 1980).

Fig. 2 Calculated PA signal amplitude (a) and phase angle (b) as a function of chopping frequency using the extended Jackson-Amer model on a 200 μm-thick Si wafer for several values of bulk lifetime τ and constant surface recombination velocity s [From Mikoshiba, Nakamura and Tsubouchi (1982)].

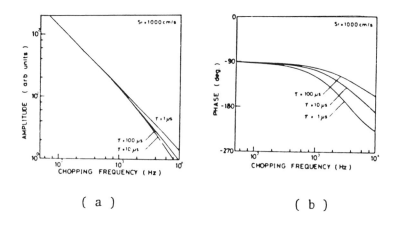

(a)   (b)

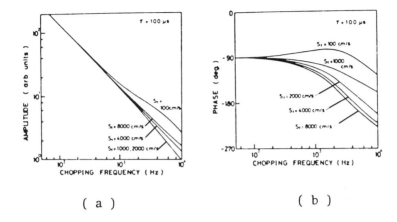

Fig. 3 Calculated PA signal amplitude (a) and phase angle (b) as a function of chopping frequency using the extended Jackson-Amer model on a 200 μm-thick Si wafer for several values of surface recombination velocity s and constant bulk lifetime τ [From Mikoshiba *et al.* (1982)].

Figs. 2 and 3 show the signals computed for a 200 μm thick silicon sample as a function of the chopping frequency with a constant surface recombination velocity s for various carrier lifetimes τ and with a constant τ and various values for s.

3. PULSED EXCITATION

It is relatively easy to solve free carrier population equations with a pulsed excitation using the Laplace transformation. In the case of an opaque sample uniformly irradiated by a short light pulse, the time and spatial dependence of the carriers is given by:

$$N(x,t) = \frac{Q_o E_g}{D} e^{-t/\tau} \left[\sqrt{\frac{D}{\pi t}} e^{-x^2/4Dt} - s e^{(sx/D + s^2 t/D)} \right.$$

$$\left. \times \operatorname{erfc}\left(x/2\sqrt{Dt} + s\sqrt{t/\!\sqrt{D}}\right) \right] \tag{5}$$

where Q_o: pulse energy density; photon energy: $h\nu = E_g$

The temperature distribution is more difficult to obtain in an analytical form of practical use. We have thus simply computed the Fourier transform of expression (3a) with the suitable parameters to get the pulsed response to an excitation.

We will analyze in more detail this kind of signal in section III which is devoted to the local probing of heat and carrier distribution through the sample.

4.   DISCUSSION: ORDERS OF MAGNITUDE

In the proceeding subsection we have considered the illumination of a doped semiconductor and the effect of minority carrier diffusion. Let us examine how this carrier diffusion competes with the thermal diffusion.

For direct gap semiconductors (e.g. GaAs) the recombination time τ is on the order of $10^{-8}-10^{-9}$ seconds. The electronic diffusivity D is in the range 10-400 cm^2/s. During their lifetime the carriers diffuse over a length $\sqrt{D\tau} \approx 10^{-4}-2\times10^{-3}$ cm. Thus, if the thermal diffusion length is much larger than $\sqrt{D\tau}$, the electronic diffusion will not affect the heat distribution. This is the case for frequencies lower than ≈ 0.1 MHz which is reasonable for most photothermal experiments.

For indirect gap semiconductors (Si, Ge) the recombination time τ is much longer (up to 10^{-3} s) and $\sqrt{D\tau}$ may be in the millimeter range. Thus except at low frequencies ($\mu > \sqrt{D\tau}$) the carrier diffusion plays a role in determining the heat distribution.

It must be emphasized that some photothermal experiments are carried out under a high density of optical energy. Thus the photo-induced carrier population is much larger than the majority carrier population. In this latter case ambipolar diffusion takes place which reduces the average carrier expansion (Rosencwaig, 1987).

Finally, let us recall that the carrier effective lifetime is strongly reduced when active surface recombination takes place. Thus, when $s/D \gg (1/D\tau)^{1/2}$ or $s \gg (D/\tau)^{1/2}$ (e.g. 1000 cm/s for Si), the surface thermal wave induced by carrier recombination will be the dominant process of heat generation.

III. Thermal Diffusivity Measurements

As discussed at the end of Section II we will assume here that the diffusion of the photo-induced carriers may be neglected.

To measure the thermal diffusivity D_{th} we have to introduce in the experimental scheme a *geometrical length comparable to the thermal diffusion length* $\mu = \sqrt{D_{th}/\pi f}$ (modulated excitation at a frequency f), or $l_{th} = \sqrt{D_{th}t}$ (pulsed excitation, with t the time of observation after the excitation). From this comparison a phase or an amplitude variation of the signal will lead to the determination of D_{th}.

In this section we will describe some experimental schemes which allow the determination of D_{th} for a large variety of samples using mostly photothermal deflection (Jackson, Amer, Boccara, and Fournier, 1981; Murphy and

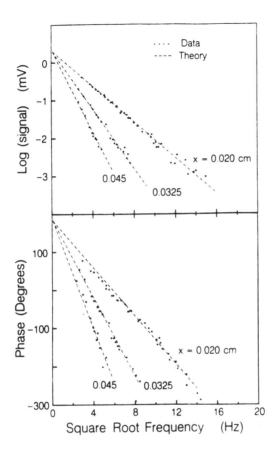

Fig. 4 "Pure" thermal wave probing in a glass sample by mirage effect [From Fournier, Boccara, Skumanich, and Amer (1986)].

Aamodt, 1980).

Indeed the "mirage" detection when used *within* the sample gives a very easy way to characterize quantitatively a thermal wave (Fournier and Boccara, 1980). When the surface temperature is given by $T_o \cos \omega t$, the temperature at depth x in the sample is given by:

$$T(x,t) = T_o \exp(-x/\mu) \cos(\omega t - x/\mu) \tag{6}$$

and the probe laser beam deflection angle can be expressed as

$$\theta(x,t) = \frac{l}{n} \cdot \frac{dn}{dT} \cdot \frac{dT}{dx} = -\frac{l}{n} \cdot \frac{dn}{dT} \left[\frac{2T_o}{\mu} \right] \exp\left[-\frac{x}{\mu} \right] \cos\left[\omega t - \frac{x}{\mu} + \frac{\pi}{4} \right] \tag{7}$$

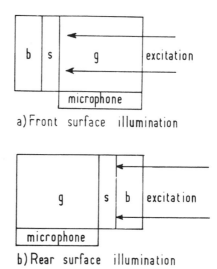

a) Front surface illumination

b) Rear surface illumination

Fig. 5 Front (a) and rear (b) surface illumination of the sample (s) under study in a photoacoustic cell (g = gas, b = backing) [From Charpentier, Lepoutre, and Bertrand (1982)].

Thus, the deflection signal is an ideal probe, which follows the local amplitude and phase of a plane thermal wave with a sensitivity on the order of 10^{-6}–10^{-7} degrees in the bulk of a solid sample. In Eq. (7) l is the interaction length between the probe beam and the heated region. For a fixed frequency θ exhibits an exponential decay, $\exp(-x/\sqrt{D_{th}/\pi f})$, and a linear phase shift $(x/\sqrt{D_{th}/\pi f})$ when varying the distance x between the surface and the probe beam (Fig. 4).

This simple example conveys the fact that to obtain information about the diffusivity, the thermal wave has to propagate along a path whose length (here x) is commensurate with the thermal diffusion length μ. When the sample thickness l is fixed, one must vary the frequency to achieve this condition (Charpentier, Lepourtre, and Bertrand, 1982; Kordeki, Bein, and Pelzl 1986; Pessoa, Cesar, Patel, and Vargas, 1985; Roger, Lepoutre, Fournier, and Boccara, 1987).

Fig. 5 shows the experimental arrangement for a gas-microphone PA cell using front or rear face illumination (Charpentier *et al.* 1982). The amplitudes of the corresponding signals as a function of $f^{1/2}$ are represented in Fig. 6, where it can be seen that they exhibit a slope change for $l \approx \mu$.

Such experiments may provide the experimenter with relatively precise values of the thermal diffusivity, nevertheless great care must be taken to avoid the buckling of the sample which generates spurious photoacoustic signals.

Several authors (Kordecki *et al.* 1986; Pessoa *et al.* 1985) have directly compared the rear and the front surface signals using simultaneous excitation, the main interest of the latter arrangement being that it avoids the buckling of the samples and the spurious signals associated with it. Pessoa *et al.* (1985) have

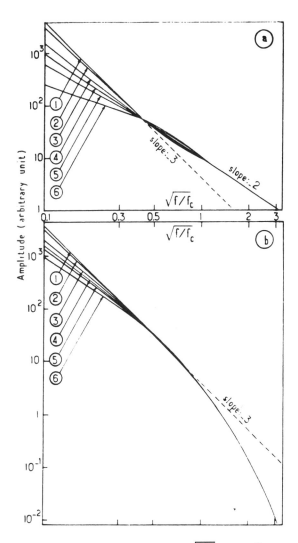

Fig. 6 Amplitude of the photoacoustic signal versus $\sqrt{f/f_c}$ ($f_c=D_{th}/l_s^2$), a: front surface excitation; b: rear surface excitation for various values of the ratio g between the effusivities ($\sqrt{K\rho c}$) of the backing and the sample 1) g = 0; 2) g = 0.1; 3) g = 0.5; 4) g = 1; 5) g = 2. [From Charpentier et al. (1982)].

measured the diffusivity of many semiconductors with a precision of a few percent. Another approach is to use the mirage effect with front or rear surface detection, as the photorefractive signals are much less sensitive to acoustic perturbations than to thermal perturbations (Rousset and Lepoutre, 1982).

One problem which arises when using the mirage detection is that the distance x between the sample surface and the probe beam is difficult to estimate.

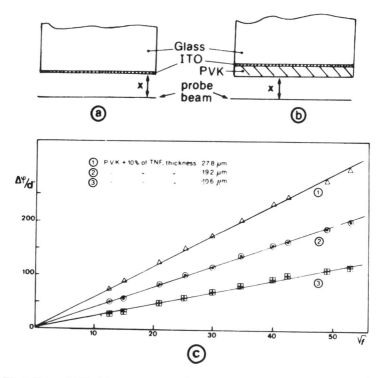

Fig. 7 Thermal diffusivity measurement of doped semiconductor polymer film (PVK). x is kept constant from experiment (a) to experiment (b) by using a reference plane [From Roger, Lepoutre, Fournier, and Boccara (1987)].

In order to avoid this difficulty one has to perform a few measurements at various distances from the sample surface. Another solution which has been used in the measurement of the thermal diffusivity of thin semiconductor films (3 to 30 μm), is to maintain a fixed value for x. In this case when the thickness l of the film is smaller than μ, the difference in phase between experiments a and b of Fig. 7 is simply l/μ. The results obtained are shown in Fig. 7c for semiconducting polymers used in holographic recording that we have characterized as function of the doping. A good check of the validity of the experimental procedure is to verify that the extrapolated phase lag at zero frequency is zero for all the curves (Roger *et al.* 1987).

Up to now, the examples presented have been mostly analyzed in terms of 1-D models of heat diffusion. In many practical cases arising from sample geometry, or when one is interested in a *local* probing of thermal diffusivity, it is advisable to use a local excitation induced by a focussed laser beam and to interpret the data in terms of 3-D heat diffusion processes. This can be achieved with an IR camera, if the sample is a good IR emitter (which is not often the case for semiconductors), however, a very simple, precise and elegant method has been proposed by the Wayne State University group (Kuo, Lin, Reyes,

Favro, Thomas, Kim, Zhang, Inglehart, Fournier, Boccara, and Yacoubi, 1986; Kuo, Sendler, Favro and Thomas, 1986). These authors take advantage of the vectorial properties of the temperature gradient and record the transverse mirage signal as a function of the offset between the excitation and the probe beam. The distance between the two crossing points A and A' was found to be a linear function of $\sqrt{1/f}$ with a slope $\sqrt{D_{th}\pi}$ as discussed in detail in Section V of Chapter 4 in this volume.

We have described a few simple experimental schemes for measuring thermal diffusivity. More sophisticated instrumentation or analytical approaches have been used when the structure under study contains one very thin submicron layer or many layers. Coufal and Hefferle (1986) have demonstrated a new kind of calorimeter, by coating a pyroelectric membrane with fast thermal response with the film under study. Details of this technique are presented in Chapter 7. Rosencwaig (1983) has developed a photothermal displacement approach similar to the one by Olmstead, Amer, Kohn, Fournier and Boccara (1983) which, by a best-fit procedure, leads to the determination of the thermal and geometrical properties of a multilayered structure.

Finally, we point out that a careful analysis of the phase and amplitude of photothermal signals as a function of the wavelength can also lead to the determination of both optical and thermal properties of thin films and multilayered structures (Fournier and Boccara, 1987; Mandelis, Siu and Ho, 1984).

IV. Coupled Electronic and Thermal Transport Measurements

From the calculations developed in section II it appears that a careful analysis of a photothermal experiment can lead to a determination of both thermal and electronic diffusivity, minority carrier lifetimes and surface recombination velocity. Indeed we have shown how the photoacoustic or photothermal signals are affected by carrier diffusion. Nevertheless from an experimental point of view, as pointed out by Sablikov and Sandomiskii (1983), the difference is not large, at least at low frequency, between signals affected and those which are not affected by the diffusion process. Therefore, these signals can be hardly used for a precise determination of the electronic parameters.

For instance, Fig. (8) shows the PA signal obtained on a Si wafer subjected to two different surface preparations (Mikoshiba *et al.*, 1982). The overall shape of the amplitude and phase experiments are not drastically different from what is expected from a sample without carrier diffusion (which is represented here by the large surface recombination velocity limit).

As we shall see, it is more efficient to probe locally the thermal and electron-hole plasma waves induced at the sample surface by optical methods. This has been done in an "academic" experiment which has allowed us to obtain a clear vision of "perturbed thermal waves" accompanied by a good degree of precision in the determination of the electronic parameters (Skumanich, Fournier, Boccara, and Amer, 1985; Fournier *et al.*, 1986). Moreover, by using the modulation of reflectivity with temperature, Rosencwaig and coworkers have

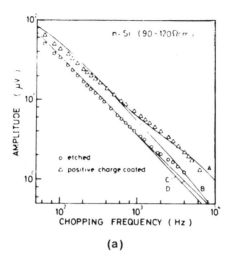

(a)

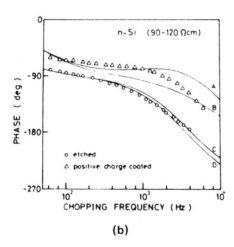

(b)

Fig. 8 Measured and calculated PA signal amplitude (a) and phase angle (b) as a function of chopping frequency. Circles, and triangles are experimental points of PA signals from etched and positive-charge coated surfaces, respectively. Curves A-D are calculated by the usual Jackson-Amer model using $\alpha = 6 \times 10^4$ cm^{-1} (A,B), and 1 cm^{-1} (C,D), respectively. Curves B and D incorporate a large surface recombination velocity. [From Mikoshiba *et al.*, (1982)].

been able to monitor a signal which was found to be highly sensitive to ion implantation (Smith, Rosencwaig, and Willenborg, 1985) and to surface contamination (Rosencwaig, 1987). These phenomena have opened the way to many exciting applications.

"MIRAGE" PROBING IN THE BULK OF SEMICONDUCTORS

The time dependent deflection of a narrow beam propagating through an inhomogeneous medium at a given depth x is given by:

$$\Theta(x,t) = \frac{l}{n}\frac{\partial n(x,t)}{\partial x} , \tag{8}$$

where Θ is the angular deflection, l is the interaction length ($l \gg x$, and also much larger than both the thermal μ and the carrier $\sqrt{\pi D/f}$ diffusion lengths), n is the local index of refraction, and $\partial n(x,t)/\partial x$ is the photo-induced gradient in the index of refraction. In the case of semiconductors, $\partial n(x,t)/\partial x$ has two contributions, a thermal and a free-carrier term, and is given by:

$$\frac{\partial n(x,t)}{\partial x} = \frac{\partial n}{\partial T}\frac{dT(x,t)}{dx} + \frac{\partial n}{\partial N}\frac{dN(x,t)}{dx} , \tag{9}$$

where $T(x,t)$ and $N(x,t)$ are the time-dependent temperature and minority-carrier density distribution respectively. From their values calculated in section II, one can obtain the thermal gradient:

$$\frac{dT}{dx}(x,t) = \mathrm{Re}\left[\left[\frac{\phi_o E_G/\lambda_{el}}{hv\tau Dk(s/D+\lambda_{el})[1/\tau D+i(\omega/D-\omega/D_{th})]}\right]\right.$$
$$\times e^{-x/\lambda_{el}}e^{i\omega t} - \left[\frac{\phi_o E_G/\lambda_{el}}{hv\tau Dk(s/D+1/\lambda_{el})[1/\tau D+i(\omega/D-\omega/D_{th})]}\right.$$
$$\left.\left. + \frac{\phi_o E_G s}{hvDk(s/D+1/\lambda_{el})} + \frac{(hv-E_G)}{hv}\frac{\phi_o}{k}\right]e^{-x/\lambda_{th}}e^{i\omega t}\right] , \tag{10a}$$

and the minority-carrier gradient:

$$\frac{dN}{dx}(x,t) = \mathrm{Re}\left[\frac{-\phi_o/\lambda_{el}}{hvD(s/D+1/\lambda_{el})}e^{-x/\lambda_{el}}e^{i\omega t}\right] \tag{10b}$$

The thermal term has two components, essentially one due to the thermal wave from the surface heating, which includes the immediate thermalization of carriers with energies greater than E_G to the bandgap energy, and the second due to the thermal energy released by nonradiative recombinations of the photoexcited carriers which have diffused into the semiconductor. The thermal term is only weakly dependent on the probe-beam wavelength. The free-carrier term, on the other hand, strongly depends on the probe-beam energy, strongly increasing at shorter wavelengths.

Eqs. (10a) and (10b) can be combined with Eq. (9) and simplified to give:

$$\frac{\partial n}{\partial x} = [C_1^{th}]e^{-x\sqrt{i\omega/D_{th}}} + [C_2^{th}]e^{-x\sqrt{(1+i\omega\tau)/D_{el}\tau}}$$

$$+ [C_3^{FC}]e^{-x\sqrt{(1+i\omega\tau)/D_{el}\tau}} , \tag{11}$$

$$= [C_1^{th}]e^{-x/\lambda_{th}} + [C_2^{th}]e^{-x/\lambda_{el}} + [C_3^{FC}]e^{-x/\lambda_{el}}$$

where C_1^{th} and C_2^{th} are coefficients of the thermal contribution and C_3^{FC} is the coefficient for the free-carrier contribution. There are several important physical implications of Eq. (11): The deflection has an exponential dependence on the distance away from the illuminated face, and also an exponential dependence on the square root of the modulation frequency of the pump beam illumination. This decay is characterized by the complex thermal and electronic diffusion lengths λ_{th} and λ_{el} respectively.

There are two contributions to the deflection, one from the thermal effects and one from the photoexcited carriers. However, as seen in Eqs. (10) and (11), the thermal term has two components - a "purely" thermal and an "electronic" component. The latter is due to the thermal energy released by minority-carrier recombination. It is important to understand the distinction between the electronic component in the thermal contribution and the free-carrier contribution although both exhibit the same exponential decay:

There are three general regimes: the first is associated with "purely" thermal behavior (term 1, Eq. (11)). The second regime is dominated by electronic behavior, i.e., associated with the minority carriers (term 2, which enters the thermal contribution, and term 3). These two regimes are most clearly seen at low and high modulation frequencies, respectively. At low frequencies, term 1 dominates as ω goes to zero. As $D_{el} > D_{th}$, the electronic behavior dominates at high frequencies through term 2 and/or term 3. The third domain is at intermediate frequencies, where there is a complex interaction which is highly sensitive to the carrier lifetime and surface recombination, as will be shown below.

Finally, in the "purely" thermal term 1, there is an excess energy part due to the rapid thermalization of energetic carriers to the band gap energy.

The predictions of the theory are presented graphically in Figs. 9-12. In these series of graphs the exponential dependence is emphasized by plotting the logarithm of the deflection signal amplitude as a function of the square root of the modulation frequency.

Fig. 9 shows the "interference" between the two components of the thermal gradient. Here even at relatively low frequencies the diffusion process affects the usual plane thermal wave behavior. Figs. 10, 11 and 12 indicate how the thermal and free carrier gradients vary with the position (x), the carrier lifetime (τ) and the surface recombination velocity (s).

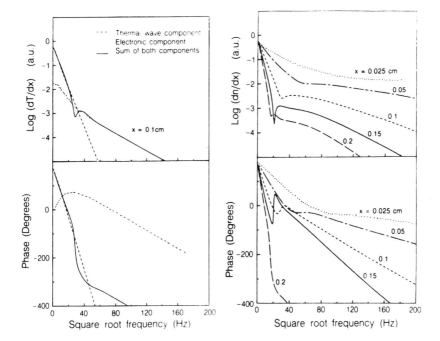

Fig. 9 Frequency dependence of the temperature gradient, Eq. (6), only, at a depth x = 0.1 cm, showing explicitly the "purely" thermal component and the electronic component [From Fournier *et al.*, (1986)].

Fig. 10 Frequency dependence of the sum of the temperature and free carrier contributions for various x [From Fournier *et al.*, (1986)].

Note that the limits associated with either small values of τ or large values of s correspond to the case where all the carriers recombine fast and close to the surface thus generating a pure thermal wave.

The experimental apparatus is shown in Fig. 13. This represents an extension of the "mirage" set-up (Boccara *et al.*, 1980). The excitation light source, i.e., pump beam, was the output from an Ar⁺-ion laser (5145 Å) uniformly illuminating the front face of the sample. Care was taken to prevent any light from falling on the other faces by use of an opaque mask. The probe beam used

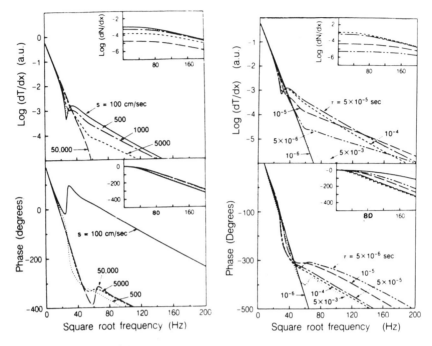

Fig. 11 Frequency dependence of the temperature gradient, and free carrier gradient (inset), for various s [From Fournier *et al.,* (1986)].

Fig. 12 Frequency dependence of the temperature gradient and free carrier gradient (inset), for various τ (at a depth of x = 0.1 cm) [From Fournier *et al.,* (1986)].

to detect the index of refraction gradient was the 1.15 or 3.39 μm line of a He-Ne laser (~1-3 mW) focused to ~ 30 μm beamwaist. The deflection was measured by a position sensitive detector, in conjunction with a conventional phase-sensitive detection scheme. In the case of the 1.15 μm line the position detector was a Si position sensor, whereas for the 3.39 μm line an effective position sensor resulted from the mirror configuration shown in the figure. This configuration is necessary to obtain rejection of the intensity modulation resulting from transmission changes induced both by gap shifts and free-carrier absorption which would obscure the deflection.

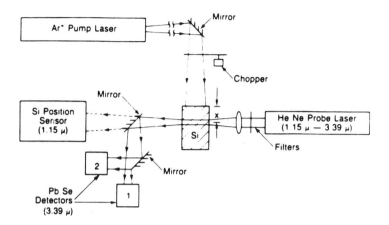

Fig. 13 The experimental configuration. Two detection schemes are shown: For the 1.15 μm wavelength a silicon position sensor is used while for the 3.39 μm line, a mirror separating the beam in two equal parts and two well matched PbSe detectors are used [From Fournier *et al.*, (1986)].

The solid examined is a p-type Si crystal. The Si sample had one face mechanically polished and treated with methanol to minimize surface recombination. The opposite face was unpolished and untreated. The power density on the Si sample varied from 30 to 800 mW/cm^2. The latter value corresponds to a minority-carrier density roughly one order of magnitude smaller than that of the majority carriers ($\rho \sim 6\Omega$–cm, $N_1 \sim 3 \times 10^{15}$ carriers/cm^3). Fig. 14 shows a comparison between experimental data and the theoretical model developed above. It is worth emphasizing that the parameters of the experiment are far from pure thermal wave behavior, and thus give easy access to τ, D_{el}, s and D_{th} determination.

We would further like to mention that parallel experiments have been performed in the time domain. Indeed, time resolved investigations of transport in semiconductors using the mirage effect have been studied by Pelzl, Fournier and Boccara (1987) using the apparatus of Fig. 15. The source is a frequency doubled Nd/YAG laser which produces free carriers in the polished or unpolished surface of a single silicon crystal by the absorption of a short (~ 15 ns) pulse. The time dependent diffusion and recombination of these carriers has been studied by means of the mirage effect as a function of the beam position. Deflection signals of opposite sign have been observed in the short and long time regime which are accounted for by the spatial variation of the refractive index due to changes of the carrier densities and temperature, respectively. Fig. 16 shows the results obtained when irradiating the polished and the unpolished surface of a Si single crystal. One can notice that the short time carrier contribution disappears for the unpolished surface because of the very large surface

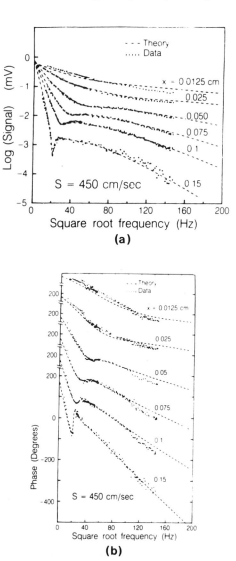

Fig. 14 Comparison of the experimental measurements (points) with the theory (dashed lines) for a well polished and methanol-treated surface and a 1.15 μm probe, showing excellent agreement for both the magnitude (a) and phase (b). The values used for calculating the theoretical curves are: $D_{el} = 30$ cm²/s, $D_{th} = 0.9$ cm²/s, $\tau = 4.5 \times 10^{-5}$ s and s = 450 cm/s [From Fournier *et al.* (1986)].

recombination velocity, thus only the thermal signal can be observed. From the time dependence of the deflection signal, values of the diffusion coefficients and of the recombination time have been evaluated.

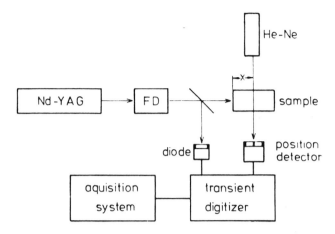

Fig. 15 Block-diagram of the experimental apparatus for the detection of the pulsed mirage signal [From Pelzl, Fournier and Boccara (1987)].

V. Future Trends and Conclusions

In this Chapter we have outlined the basis of transport measurements for semiconductor samples using light excitation. For bulk samples or for thin films deposited on a substrate we have shown how to get an *average* of the transport properties over the probed area. There is, however, a strong tendency to use highly localized optical or tip-like probes.

Fig. 16 Comparison of the time dependent deflection signals obtained at distance of 0.45 mm from a polished (p) and unpolished (np) surface of a Si single crystal. The upper and lower traces show the long and the short time behavior, respectively. [From Pelzl *et al.*, (1987)].

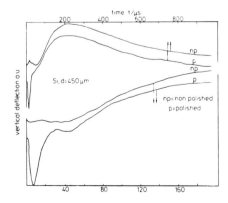

With these kinds of photothermal techniques it will soon be possible to get transport properties at a micron or submicron scale. Then it will be of interest to study the transition from *local* to *macroscopic* transport properties: (e.g., from a "grain" diffusivity to large scale diffusivity). This can be achieved by either using the recently introduced concept of diffusion in "fractal" space (Orbach, 1986; Mandelbrot, 1983) or by a more detailed analysis of the heat or carrier flux between grains. Most of the time when a heterogeneous sample exhibits a random distribution of links between "grains", diffusion processes are relevant to this type of analysis: in such systems the usual diffusion equations (e.g., Eq. 1 or Eq. 2) are no more valid, but the diffusivity D is now time (or frequency) dependent. Fig. 17 illustrates this point: a short (~ 10 ns) light pulse heats the surface of a sample consisting of an assembly of slightly compacted copper spheres whose surface temperature is measured as a function of time. At short time scale the diffusion takes place within a sphere, then the heat diffuses through the random links between grains and the slope (~ 0.22) is quite different from what we expect in a usual Euclidian space (0.5). At long times the space turns back to Euclidian (Stanley and Ostrowsky, 1985; Orbach, 1986).

We strongly believe that photothermal experiments will be of great help toward a better understanding of diffusion in complex structures and for characterization of such structures.

Fig. 17 Heat diffusion through an assembly of opaque spheres excited by a short laser pulse at time t = 0. For t < 10^{-4} s the diffusion takes place within a sphere, from a few 10^{-4} s to 10^{-1} s, the random distribution of links creates a fractal diffusion (slope ~ 0.22); at longer times the diffusion becomes Euclidian again (slope ~ 0.5).

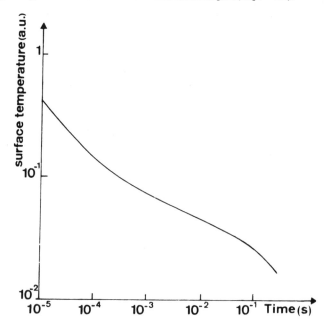

VI. References

Boccara, A.C., Fournier, D. and Badoz, J. (1980), Appl. Phys. Lett. **36**, 130.

Charpentier, P., Lepoutre, F. and Bertrand, L. (1982). J. Appl. Phys. **53**, 608.

Coufal, H. and Hefferle, P. (1986) Can. J. Phys. **64**, 1200.

Fournier, D. and Boccara, A.C. (1980). in *Scanned Image Microscopy*, (E.A. Ash, Ed.) Academic Press, p. 347.

Fournier, D., Boccara, A.C., Skumanich, A. and Amer, N.M. (1986), J. Appl Phys. **59**, 787.

Fournier, D. and Boccara, A.C. (1987). See the chapter on *"Photothermal Deflection Spectroscopy of Semiconductors"*.

Jackson, W. and Amer, N.M. (1980). J. Appl. Phys. **51**, 3343.

Jackson, W., Amer, N.M., Boccara, A.C. and Fournier, D. (1981). Appl. Opt. **20**, 1333.

Kanstad, S.O. and Nordal, P.E. (1986). Can. J. Phys. **64**, 1155.

Kordeki, R., Bein, B.K. and Pelzl, J. (1986). Can. J. Phys. **64**, 1204.

Kuo, P.K., Lin, M.J., Reyes, C.B., Favro, L.D., Thomas, R.L., Kim, D.S., Zhang, S.Y., Inglehart, L.J., Fournier, D., Boccara, A.C., and Yacoubi, N. (1986), Can. J. Phys. **64**, 1165.

Kuo, P.K., Sendler, E.D., Favro, L.D. and Thomas, R.L., (1986), Can. J. Phys. **64**, 1168.

Mandelbrot, B. (1983) *"The fractal geometry of nature"*, Freeman, New York.

Mandelis, A., Siu, E., and Ho, S. (1984). Appl. Phys. **A33**, 153.

Mikoshiba, N., Nakamura, H. and Tsubouchi, K. (1982). *Proc. IEEE Ultrasonics Symposium*, San Diego, California 1982, p. 580.

Miranda, L.C.M. (1982), Appl. Opt. **21**, 2923.

Murphy, J.C. and Aamodt, L.C. (1980), J. Appl. Phys. **47**, 402.

Olmstead, M.A., Amer, N.M., Kohn, S., Fournier, D. and Boccara, A.C. (1983), Appl. Phys. A **32**, 141.

Orbach, R. (1986) Science, **231**, 814.

Pankove, J.I. (1975), *"Optical processes in semiconductors"*. Dover publications, Inc. New York.

Pelzl, J., Fournier, D. and Boccara, A.C. (1987). CEE Report #4, STI - 006 J-C (CD) (to be published).

Pessoa, A., Ceasar, C.L., Patel, N.A., and Vargas, H. (1985). *Proc. 4th International Meeting on Photoacoustic, Thermal and Related Sciences*, Montreal, paper WA-9.1.

Roger, J.P., Lepoutre, F., Fournier, D. and Boccara, A.C. (1987). Thin Solid Films (submitted).

Rosencwaig, A. (1983), J. Phys. Paris Vol. C-6, 437.

Rosencwaig, A. (1987). See the Chapter on "*Thermal wave characterization and inspection of semiconductor materials and devices*".

Rosencwaig, A. (1980) "*Photoacoustics and Photoacoustic Spectroscopy*" in *Chemical Analysis* (P.J. Elving and J.D. Winefordner, eds.), Vol. 57, J. Wiley, New York, N.Y.

Rosencwaig, A. and Gersho, A. (1976), J. Appl. Phys. 47, 64.

Rosencwaig, A., Smith, W.L. and Willenborg, D.L. (1985), Appl. Phys. Lett. 46, 1013.

Rousset, G. and Lepoutre, F. (1982). Rev. Phys. Appl. 17, 201.

Sablikov, V.A. and Sandomirskii, V.B. (1983). Phys. Stat. Sol. B 120, 471.

Shank, C.V. (1984) *17th International Conference on the Physics of Semiconductors*, (Chadi and Harrison, Eds.) Springer-Verlag, p. 1545.

Smith, R.A. (1978). "*Semiconductors*". Cambridge University Press.

Smith, W.L., Rosencwaig, A. and Willenborg, D.L. (1985), Appl. Phys. Lett. 47, 584.

Skumanich, A., Fournier, D., Boccara, A.C. and Amer, N.M. (1985). Appl. Phys. Lett. 47, 402.

Stanley, H.E. and Ostrowsky, N. (1985). "*Growth and Form*", Nijhoff, Amsterdam.

Sze, S.M. (1969). "*Physics of semiconductor devices*". Wiley-InterScience.

THERMAL WAVE INVESTIGATION OF TRANSPORT PROPERTIES OF SEMICONDUCTORS II: ELECTRONIC AND NONRADIATIVE PHENOMENA

Helion Vargas

Instituto de Fisica
Universidade Estadual de Campinas
13100 Campinas, SP, BRAZIL

and

Luiz C.M. Miranda

Laboratorio Associado de Sensores e Materiais
Instituto de Pesquisas Espaciais
Caixa Postal 515
12201 - S.J. Campos, SP, BRAZIL

I. Introduction

A proper description of the photoacoustic (PA) signal in a solid should take into account the spatial and temporal dispersion of a distributed heat source resulting from the absorption of light. In the case of semiconductors this distributed heat source contains information on both transport and nonradiative properties of the sample. This renders the PA and the related photothermal spectroscopies unique tools for the investigation of the optical as well as the transport properties (e.g., carrier lifetime, diffusion length, surface recombination velocity) and nonradiative states of semiconductors. Physically, this dependence of the PA signal may be seen as follows: The absorption of light generates excess carrier distribution in the sample. These excess carriers diffuse through the sample and re-establish equilibrium by disposing of the energy in excess both by emitting radiation and by generation of heat. The PA technique responds only to the fraction of the energy that is converted into heat. The heat generation during this process of re-establishment of equilibrium is essentially due to the intraband transitions in the bulk, nonradiative bulk (band-to-band) transitions, and nonradiative surface recombinations generated within a diffusion length from the surface. A detailed description of these physical processes can be found in Chapter 15. Apart from this the PA effect in semiconductors is also sensitive to the presence of nonradiative states near the surface of the sample.

In this Chapter, we review some of the works describing the influence of transport and nonradiative processes on the PA signal of semiconductors. In Section II, we review the basic works on the subject, and in Section III we discuss the use of thermal wave detection for non-spectroscopic studies of transport phenomena in semiconductors.

II. Transport and Nonradiative Processes in PAS

The influence of the finite carrier lifetime, diffusion coefficient and nonradiative transitions in the PA signal of a semiconductor was first proposed and demonstrated by Ghizoni and co-workers (Bandeira *et al.*, 1982) and by Mikoshiba *et al.* (1982) in the case of piezoelectric detection. Following the work of Bandeira *et al.* (1982) several authors (Miranda, 1982; Sablikov and Sandomirskii, 1983a, 1983b; Vasilev and Sandomirskii, 1984a, 1984b) have extended their model to include the effect of surface band bending (Sablikov and Sandomirskii, 1983a, 1983b) as well as the influence of an alternating electric field (Vasilev and Sandomirskii, 1984b). In addition to the gas-microphone PA detection used in the above cited works, the effects of diffusion, nonradiative transitions and surface recombination of photoexcited carriers were also taken into account in the case of piezoelectric detection. This was accomplished by Mikoshiba *et al.* (1982) who adapted the Jackson-Amer model for piezoelectric detection (Jackson and Amer, 1980) to the case of semiconductors.

According to the Bandeira-Closs-Ghizoni (BCG) model, the PA signal from a semiconductor is described by the coupled set of thermal and carrier diffusion equations

$$\frac{\partial^2 T}{\partial x^2} = \frac{1}{\alpha_s} \frac{\partial T}{\partial t} - \frac{Q(x,t)}{k_s} \tag{1}$$

$$D \frac{\partial^2 n}{\partial x^2} = \frac{\partial n}{\partial t} + \frac{n}{\tau} - g(x,t) \tag{2}$$

where $T(x,t)$ is the temperature fluctuation in the sample of the thermal diffusivity α_s and thermal conductivity k_s, and $n(x,t)$ is the density of photoexcited carriers of diffusion coefficient D and recombination time τ. The carrier generation rate, $g(x,t)$, is given in terms of a sinusoidally chopped monochromatic light beam as

$$g(x,t) = \frac{\eta \beta}{h\nu} I_o e^{\beta x} e^{j\omega t} \tag{3}$$

where η is the nonradiative quantum efficiency, ν is the frequency of the incident light beam of intensity I_o, β is the optical absorption coefficient and ω is the modulation angular frequency. It is assumed that the sample of length l_s extends from $x=0$ to $x=-l_s$ and that the light beam is incident at the surface at $x=0$. The heat power density $Q(x,t)$ in Eq. (1) considered by BCG is a subset of the general PA signal source generation mechanisms discussed in Chapter 15. The BCG model considers the following processes: (i) instantaneous intraband nonradiative thermalization with energy greater than the gap energy E_g. This process, due to the electron-phonon collisions within the conduction band, occurs in a time scale of 10^{-12} sec and may be assumed instantaneous in the typical range of modulation frequencies of PAS. The heat power density is then

$$Q_T = \frac{\eta \beta (h\nu - E_g)}{h\nu} I_o e^{j\omega t} ; \tag{4}$$

(ii) nonradiative bulk recombination. This process is due to excess electron-hole pair recombination after diffusing through a distance $(D\tau)^{1/2}$, where τ is the band-to-band recombination time. The generated heat power density is

$$Q_{NR} = \frac{E_g}{\tau} n(x,t) , \tag{5}$$

(iii) nonradiative surface recombination. The heat power density due to the non-radiative carrier recombination at the surface is

$$Q_S = E_g \ [s_1 \delta(x) + s_2 \delta(x+l_s)] \ n(x,t) \tag{6}$$

where s_1 and s_2 are surface recombination velocities. Solving Eq. (2) subject to the boundary conditions

$$D \frac{\partial n}{\partial x} \bigg]_{x=0} = s_1 n(0)$$

$$D \frac{\partial n}{\partial x} \bigg]_{x=-l_s} = -s_2 \ n(-l_s)$$

and substituting the resulting $n(x,t)$ into Eqs. (5) and (6), the temperature distribution in the sample is then found by the solution of Eq. (1).

To demonstrate the influence of the carrier lifetime, diffusion length and surface recombination effects, as described above, Bandeira *et al.* (1982) have measured the PA spectrum of a CdS sample near the band gap region. In this case, because of the weak optical absorption, most of the photoexcited carriers are generated in the bulk and, consequently, the transport properties are expected to play a dominant role. When the optical absorption coefficient is large, heating is basically provided by the surface recombination of excess carriers which are generated very close to the illuminated surface, so that the transport properties are not expected to play an important role. However, at low absorption levels, the observation of the diffusive effects may be inhibited. To enhance the carrier diffusion effects at low absorption levels, Bandeira *et al.* (1982) have applied a D.C.-electric field transverse to the light path. Under these conditions, the photoexcited carriers are accelerated thereby Joule-heating the sample. The use of a D.C. electric field introduces an extra heat source, namely, $Q_J = e\mu E^2$, where μ is the sample mobility. This was also taken into account by these authors in the interpretation of their experimental results and is extensively discussed in Chapter 15. In Fig. 1, the experimental set-up of Bandeira *et al.* (1982) is shown schematically. In Fig. 2, we show the theoretical results for the PA signal, predicted by the BCG model, for several values of the applied electric field at a modulation frequency of 20 Hz. The values of the physical parameters used in the calculations were taken from McKelvey (1986) and Dutton (1958). We note that the PA signal exhibits a peak, in the long wavelength region, whose intensity increases with increasing electric field. This maximum of the PA signal is a consequence of carrier diffusion, occuring at a wavelength for which the absorption length β^{-1} is close to the carrier diffusion length L of the sample. This was proven by Bandeira *et al.* (1982) by expanding the expression for the PA signal in powers of β around $1/L$. The value of β which maximizes the PA signal was found to be

$$\beta \approx \frac{1}{L} [(1+(\omega\tau)^2]^{1/4} \tag{7}$$

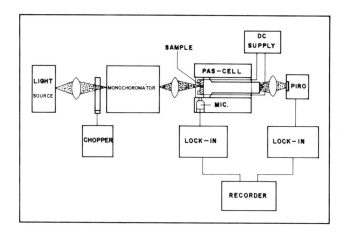

Fig. 1 Experimental set-up for the PA signal measurement [After Bandeira *et al.* (1982)]

Figure 3 shows the experimental results of Bandeira *et al.* (1982) at 20 Hz and for several values of the applied D.C. electric field. The good agreement between the experimental and theoretical results is quite apparent from Figs. 2 and 3.

Fig. 2 Theoretical curves for CdS at several applied electric fields for a frequency of 20 Hz. [After Bandeira *et al.* (1982)]

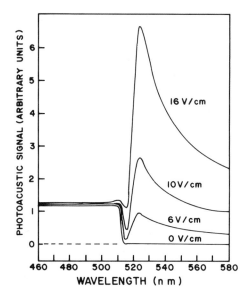

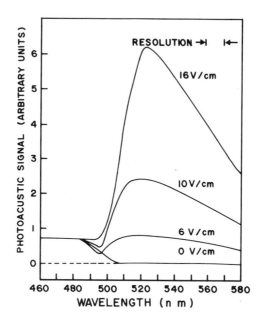

Fig. 3 Experimental results for CdS for several values of the applied electric field at a modulation frequency of 20 Hz. [After Bandeira *et al.* (1982)]

Using their data at 20 and 350 Hz, these authors were able to extract both the recombination time and the diffusion length of their sample. At the modulation frequency of 20 Hz, the peak below the gap occured at $\lambda = 523$ nm ($\beta \approx 30$ cm^{-1}). In this case $\omega\tau \ll 1$ and the carrier diffusion length was estimated from Eq. (7) to be $L \approx 0.03$ cm. At 350 Hz the peak occured at $\lambda = 521$ nm ($\beta \approx 50$ cm^{-1}) so that, using the above value of L, Eq. (7) yielded $\tau = 1$ ms for the recombination time. A similar experiment on CdS has recently been reported by Siu and Mandelis (1985). These authors have found a peak around $\lambda = 530$ nm, as in the experiment of Bandeira *et al.* (1982), as well as a secondary peak at 516 nm. A full discussion of these phenomena is presented in Chapter 15. In fact, several nonradiative states due to native defects near the band gap of CdS may be present. This has been shown by Wasa *et al.* (1980a) using piezoelectric detection, both for CdS and Si, as well as for GaAs and InP (Wasa *et al.* 1980b). In the case of GaAs, the influence of below-gap nonradiative states due to the surface damage, has also been pointed out by Eaves *et al.* (1981) as discussed in Chapter 16.

Apart from the gas-microphone and the piezoelectric detection, Amer and co-workers have recently proposed (Skumanich *et al.*, 1985) a spatially-resolved technique, based upon the Mirage effect, for investigating the transport properties of semiconductors. By employing a local probe beam they were able to investigate *in situ* information on the transport properties anywhere within, or at the surface of, the sample. They have applied their method to the case of a p-type Si sample with different surface treatments. The experimentally deduced values of D, τ, and *s* were in good agreement with the literature values.

III. Non-Spectroscopic Studies of Transport Properties Using Thermal Wave Detection

Thermal wave detection has also been used for investigating acoustic wave instabilities in semiconductors. The earliest experiment (Ghizoni *et al.*, 1978) reported on the use of a PA cell for studying acoustoelectric instability in Si. In this experiment, the periodic heating due to the electron drift in an electric field was studied. Under a D.C. electric field, the free carriers in a semiconductor, gaining energy from the field, ultimately establish a steady state population where the energy gained from the field is equal to energy lost to the lattice, by means of the electron-phonon interaction. Hence, by pulsing a D.C. voltage in a semiconductor, mounted in a PA cell as shown in Fig. 4, this periodic heating can be monitored by the microphone.

Fig. 5 shows the acoustic signal of a Si sample as a function of the pulse amplitude V_p for two different sample thicknesses and different values of the pulse duration τ_p. Fig. 6 shows the variation of the detected acoustic signal as a function of the pulse duration. These data indicate that at low values of the electric field, $E=V_p/d$, the acoustic signal S is simply given by the Joule heating, that is, proportional to $S \sim E^2$. However, at high fields, the signal increases exponentially, such that $S \sim \exp(E\tau_p)$. These results were explained (Ghizoni *et al.*, 1978) as a consequence of the onset of acoustoelectric instability in the sample. It is well known (Hutson *et al.*, 1961; Spector, 1966) that in the presence of drifting carriers, the phonon relaxation time, due to electron-phonon collisions, becomes negative (i.e., one has amplification) when the carrier drift velocity v_d exceeds the sound velocity v_s. Under such conditions, phonon-stimulated emission dominates over phonon relaxation and the acoustic signal grows exponentially: for fields greater than the threshold value given by $\mu E = v_s$, there is a phonon gain in the medium, as manifested by the exponential growth of the signal in Figs. 5 and 6.

Fig. 4 Acoustic cell used for investigating the transport properties in semiconductors. [After Ghizoni *et al.* (1978)]

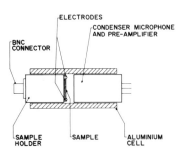

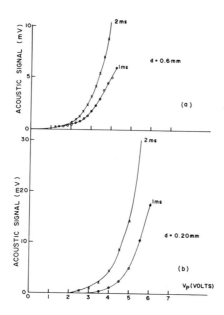

Fig. 5 Acoustic signal versus the applied peak voltage (V_p) for various pulse durations (τ_p). Sample thickness (a) d=0.6 mm, (b) d=0.2 mm. [After Ghizoni *et al.* (1978)]

Fig. 6 Acoustic signal versus pulse duration for different values of the amplitude V_p. [After Ghizoni *et al.* (1978)]

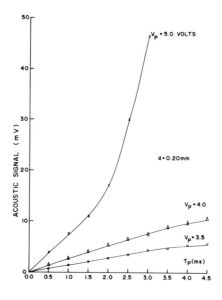

Several other mechanisms are also known to lead to amplification of acoustic waves in semiconductors and metals. In particular, the amplification of acoustic waves by a temperature gradient (the so-called thermoelectric amplification) has long been suggested by Bonch-Bruevich and Gulyaev (1963). Under the action of a temperature gradient a semiconductor or a metal exhibits an electric current such that the carrier drift velocity is given by (Ephstein, 1976):

$$v_d = \chi \tau k_B \, |\nabla T| /m, \tag{8}$$

where τ is the carrier relaxation time, ∇T is the temperature gradient in the sample, m is the carrier effective mass and χ is a factor which depends on the electron scattering mechanism and whether the carriers are non-degenerate or not. Eq. (8) shows that there is a thermoelectric field in the sample. This thermoelectric field will cause space charge drift; if the drift velocity is supersonic, the phonon-stimulated emission becomes greater than the phonon relaxation in complete analogy with the acoustoelectric instability. An experimental evidence of this thermoelectric amplification was recently reported by Rodrigues *et al.* (1986) for the case of a Ni sample, using piezoelectric detection. Fig. 7 shows the schematic arrangement of the experiment. The temperature gradient was established by means of 100 ns CO_2 laser pulses focused on one of the sample faces previously oxidized. The detection system consisted of a 28 μm thick PVF_2 piezoelectric film in intimate contact with the sample. The whole sample-detector system, as well as the focusing optics, were placed inside a vacuum chamber in order to prevent dielectric breakdown in air and to optimize the thermal and acoustic coupling of the CO_2 laser to the sample. Fig. 8 shows the piezoelectric signal as a function of the CO_2 laser pulse fluence J_p. This

Fig. 7 Schematic arrangement for studying acoustic wave instability due to a temperature gradient. [After Rodrigues *et al.* (1986)]

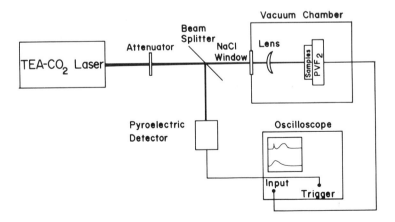

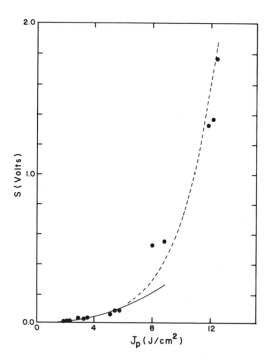

Fig. 8 Acoustic signal from the 28 μm thick PVF_2 piezoelectric film as a function of the CO_2 laser peak fluence for a 500 μm thick Ni sample. The CO_2 laser pulse duration is 100 ns. [After Rodrigues *et al.* (1986)]

data indicates that up to a pulse fluence of 5 J/cm^2 the signal behaves as J_p^2 and, therefore, as v_d^2; i.e., the signal is basically due to Joule heating. For pulse fluences greater than 5 J/cm^2 the detected signal grows exponentially with J_p (and, therefore, v_d) indicating that the thermoelectric amplification occurs at the threshold fluence of 5 J/cm^2. This value for the critical fluence agrees quite well with the one predicted by the theory.

IV. Conclusion

In this Chapter, the use of thermal wave detection for investigating the electronic transport and nonradiative processes in semiconductors was reviewed. Both spectroscopic as well as nonspectroscopic applications were discussed.

V. References

Bandeira I.N., Closs H. and Ghizoni C.C. (1982) J. Photoacoust. **1**, 275.

Bonch-Bruevich V.L. and Gulyaev Yu.V. (1963) Radiotek. Elektron. **8**, 1179.

Dutton D. (1958) Phys. Rev. **112**, 785.

Eaves L., Vargas H. and Williams P.J. (1981) Appl. Phys. Lett. **38**, 768.

Ephstein E.M. (1976) Sov. Phys.-Semicond. Engl. Transl. **9**, 1043.

Ghizoni C.C., Siqueira M.A.A., Vargas H. and Miranda L.C.M. (1978) J. Appl. Phys. **32**, 554.

Hutson A.R., McFee J.H. and White D.L. (1961) Phys. Rev. Lett. **7**, 237.

Jackson W. and Amer N.M. (1980) J. Appl. Phys. **51**, 3343.

McKelvey J.P. (1966) *Solid State and Semiconductor Physics*, Harper and Row, New York.

Mikoshiba N., Nakamura H. and Tsubouchi K. (1982) *Proc. IEEE Ultrasonics Symp.* San Diego, p. 580.

Miranda L.C.M. (1982) Appl. Opt. **21**, 2923.

Rodrigues N.A.S., Ghizoni C.C. and Miranda L.C.M. (1986) J. Appl. Phys. **60**, 1528.

Sablikov V.A. and Sandomirskii V.B. (1983a) Phys. Stat. Sol. (b) **120**, 471.

Sablikov V.A. and Sandomirskii V.B. (1983b) Sov. Phys.-Semicond. Engl. Transl. **17**, 50.

Siu E. and Mandelis A. (1985) *Proc. 4th International Meeting on Photoacoustic, Thermal and Related Sciences*, Montreal, paper WD-11.1.

Skumanich A., Fournier D., Boccara A.C. and Amer N.M. (1985) *Proc. 4th International Meeting on Photoacoustic, Thermal and Related Sciences*, Montreal, paper WD-2.1.

Spector H.N. (1966) in *Solid State Physics* (F. Seitz and F. Turnbull eds.) vol. 19, pp. 291, Academic Press, New York.

Vasilev A.N. and Sandomirskii V.B. (1984a) Sov. Phys.-Semicond. Engl. Transl. **18**, 1095.

Vasilev A.N. and Sandomirskii V.B. (1984b) Sov. Phys.-Semicond. Engl. Transl. **18**, 1221.

Wasa K., Tsubouchi K. and Mikoshiba N. (1980a) Jpn. J. Appl. Phys. **19**, L475.

Wasa K., Tsubouchi K. and Mikoshiba N. (1980b) Jpn. J. Appl. Phys. **19**, L653.

PHOTOTHERMAL DEFLECTION SPECTROSCOPY (PDS): A QUANTITATIVE CHARACTERIZATION TOOL OF OPTOELECTRONIC ENERGY CONVERSION PROCESSES AT SEMICONDUCTOR/ELECTROLYTE INTERFACES

Robert Wagner and Andreas Mandelis

Photoacoustic and Photothermal Sciences Laboratory
Department of Mechanical Engineering
University of Toronto
Toronto, M5S 1A4 CANADA

I. Introduction

In the past decade thermal wave phenomena have been shown to be capable of yielding quantitative and analytical information about the optical, thermal, and electrical properties of a system. Of major interest to us, several thermal wave (TW) techniques have been employed to characterize energy conversion in photoelectrochemical (PEC) cells and other related systems. Two representative pieces of work used TW methods to characterize specific optoelectronic energy conversion processes: Cahen (1978) developed a photoacoustic method to find the energy conversion efficiency of a photovoltaic cell; Fujishima, Maeda, Honda, Brilmyer, and Bard (1980b) used thermistor temperature measurements to find the internal quantum efficiency of a PEC cell. A review of the available literature would likely lead one to conclude that any of several TW methods can be used to monitor thermal processes in optoelectronic systems; therefore, a brief discussion in section VI will be devoted to the advantages and disadvantages of several TW probes applied to PEC systems. Although photothermal deflection spectroscopy (PDS) is the main technique covered in this Chapter because of its suitability to the study of PEC cells, several other TW methods such as gas-cell and piezoelectric photoacoustic spectroscopy (PAS) and thermistor photothermal spectroscopy (PTS) will also be discussed because most of the literature deals with these probes.

The major focus of this Chapter is PDS, a technique which provides a convenient, *in situ*, means of studying both the thermal loss mechanisms in a stable PEC cell, and the absorption spectrum of the PEC cell semiconductor electrode. Combining PDS with photocurrent and/or photovoltage measurements, one can account for two of the energy conversion channels in the PEC cell: the nonradiative, and carrier generation and transport processes; the other two conversion pathways in the cell, in the absence of photochemistry, are irreversible electrode corrosion and luminescence, and may or may not be significant in a specific PEC cell. Of main interest is the possibility of obtaining the internal quantum efficiency by PDS measurements, and hence the internal energy efficiency from current-voltage measurements for the cell under various loads.

The main topics to be covered in this Chapter include: A brief review of the PDS method and an examination of the relationship between PDS and conventional calorimetry (Section II); a literature review of thermal wave methods applied to electrochemical and optoelectronic energy conversion systems (Section III); the development of a model to quantify the PDS signal from a PEC cell operating under applied bias or under load, and the theory of PEC cell operation (Section IV); an analysis of some representative photothermal data for several n-CdS/electrolyte systems (Section V); a comparison of conventional and PDS methods for finding the quantum efficiency of a PEC cell and an examination of the applicability of various TW probes to PEC cell studies (Section VI); and, finally, a review of the Chapter's key points and an examination of future directions (Section VII).

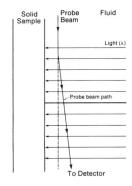

Fig. 1 Photothermal deflection effect for an illuminated solid/fluid interface.

II. The Nature of PDS and Its Relation to Conventional Calorimetry

Several authors have published works supplying the necessary theoretical framework and experimental verification leading to the establishment of PDS as an analytical technique (Jackson, Amer, Bocarra, and Fournier, 1981; Murphy and Aamodt, 1980; Mandelis, 1983). A detailed discussion can be found in Chapter 10. A brief description of the Mirage Effect (PDS) can be given as follows (Fig. 1). Consider a solid, in contact with a fluid medium, which is undergoing some heating process, with the heating source modulation assumed to be periodic. Regardless of the nature of the heating in the solid (usually optical), a periodically varying temperature gradient, and hence a refractive index gradient, will be induced in the fluid. A probe laser beam propagating through the fluid immediately adjacent to the solid will undergo a periodic deflection due to the refractive index gradient. This deflection can be detected by an appropriate optical sensing device, such as a position sensitive bi-cell photodetector. The magnitude of the beam deflection will be linearly proportional to the temperature gradient at the location of the probe beam (Mandelis, 1983). The phase of the PDS signal will provide information regarding the heat generation kinetics and/or the location of the heating "centroid" relative to the probe beam position. Although it has been assumed that the solid excitation is periodic, time-domain excitation (see Chapter 12) or other forms of frequency-multiplexing are also possible. With regard to the other TW techniques discussed in this chapter, PAS has been thoroughly dealt with in a book by Rosencwaig (1980), and thermistor PTS has been discussed by Brilmyer, Fujishima, Santhanam, and Bard (1977). Pertinent features of these techniques will be discussed later in the Chapter when it is deemed necessary.

Since the temperature gradient in a liquid surrounding a heated solid is proportional to the degree of heating of the solid, PDS can be considered a form of dynamic micro-calorimetry. In order to clarify this point, the nature of conventional calorimetry may be juxtaposed to that of modern thermal wave techniques. In conventional calorimetry (Busch, Shull, and Conley, 1978) one starts with a given system, at equilibrium, with the following properties: temperature, T_1, pressure P_1, volume, V_1, and constituents, n_1^i. The system is assumed to be thermally and chemically isolated from its surroundings so that no heat or mass transfer can take place into or out of the system. If a combination of optical, electrical, chemical, and/or nuclear processes is allowed to occur within the system, and the system is then allowed to come to a second equilibrium state, T_2, P_2, V_2, n_2^i, the energy change due to processes which took place within the system can be calculated by considering (T_1, P_1, V_1, n_1^i), (T_2, P_2, V_2, n_2^i), and the thermodynamics of the system. Although conventional calorimetry will provide unequivocal information regarding the amount of heat liberated/absorbed during the reactions occurring in the system, careful environmental control is required in order to obtain accurate results. No kinetic information, however, can be garnered from this approach since only equilibrium states are considered.

On the other hand, the PDS technique gives us a relative indication of the amount of heat generated in the solid during each modulation period. The PDS signal responds quickly to changes in the magnitude or spatial distribution of heat generation within the system, due to the close proximity of the probe beam to the active region of the system. The PDS probe is a non-intrusive, non-contact dynamic method, since a signal is only present when the system is not at thermal equilibrium (Mandelis, 1983). Thus, the PDS method allows one to obtain kinetic information (Mandelis and Royce, 1984). As a consequence of its fast response, the PDS signal allows fast data acquisition compared to conventional calorimetry. There is, however, a drawback to the PDS technique. The magnitude and phase of the PDS signal are complex functions of the heat generation temporal and spatial dependence, and depend on the location of the probe beam and instrumental transfer functions. In this respect, absolute information is difficult to extract and accurate calibration is only possible for a few simple situations.

One feature of PDS which has no parallel in conventional calorimetry is the possibility of performing thermal depth profiling. Like other thermal wave techniques, PDS is only sensitive to heat generation which occurs within approximately one thermal diffusion length from the heated surface. The thermal diffusion length, μ, is a function of several material parameters (Rosencwaig, 1980):

$$\mu(\omega) = (2k/\omega\rho C)^{\frac{1}{2}} \qquad (1)$$

where k is the thermal conductivity, ρ is the density, and C is the specific heat of the material; ω is the angular modulation frequency of the light intensity. Eq. (1) shows that by changing the modulation frequency of the excitation irradiance, one can obtain thermal information from various depths within the solid. Thus, at high frequencies the PDS signal will be primarily sensitive to surface heating processes, whereas at low frequencies bulk heating processes dominate, with the surface contributing a certain portion of the signal.

III. A Review of Thermal Wave Studies of Electrochemical and Optoelectronic Energy Conversion Systems

This section reviews the literature dealing with TW studies of electro-chemical, optoelectronic, and PEC systems. Frequency - and time-domain thermal wave and other micro-calorimetric techniques have been used to study electrochemical systems for several years. Representative experiments include: i) A measurement of the electrochemical Peltier heat of a specific redox reaction, with a given supporting electrolyte, by having a sensitive thermistor attached, or in close proximity to, the working electrode (Graves, 1972; Tamamushi, 1973, 1975; Ozeki, Watanabe, and Ikeda, 1979); ii) The *in situ* monitoring of the optical properties of electrodes during surface film chemical transformations. As a result of cyclic voltammetric measurements, the anodic and cathodic current peaks due to the formation or stripping of surface films have been correlated with changes in the reflectivity of the electrode monitored via PAS or PTS. For metal electrodes such changes have been measured by use of piezoelectric PAS (Malpas and Bard, 1980); gas-cell PAS (Masuda, Fujishima, and Honda, 1980b); and thermistor PTS (Fujishima, Masuda, Honda, and Bard, 1980c). With semiconductor electrodes such reflectivity measurements have been performed by gas-cell PAS (Masuda, Fujishima, and Honda, 1982; Masuda, Morishita, Fujishima, and Honda, 1981; Masuda, Fujishima, and Honda, 1980a); iii) The observation of changes in the PDS optical absorption spectrum of a semiconductor after electrochemical corrosion (Mendoza-Alvarez, Royce, Sanchez-Sinencio, Zelaya-Angel, Menezes, and Triboulet, 1983; Royce, Voss, and Bocarsly, 1983; Royce, Sanchez-Sinencio, Goldstein, Muratore, Williams, and Yim, 1982) or after electrochromic alteration of powdered electrochemical materials (Tamor and Hetrick, 1985); iv) The characterization of opaque PEC cell electrodes by obtaining PDS absorption spectra (Tamor and Hetrick, 1985; Roger, Fournier, Boccara, Noufi, and Cahen, 1985; Wagner, Wong, and Mandelis, 1986).

Several authors have combined optoelectronic with TW studies. Among representative experiments, those which have given quantum yield information are of particular interest for the purposes of this Chapter. Lahmann and Ludewig (1977) used PAS to determine the absolute fluorescence quantum yield of a Rhodamine 6G dye/water solution and Tam (1980) also employed PAS to find the photocarrier quantum yield of two dye films, chlorodiane blue and methyl squarylium. Furthermore, Chance and Shand (1980) studied the kinetics of a light-induced polymerization reaction of diacetylene by PAS. Cahen (1978); Cahen and Halle (1985); Faria, Ghizoni, Miranda, and Vargas (1986); and Flaisher and Cahen (1986) have been primarily concerned with finding the maximum energy conversion efficiency of a photovoltaic cell under load, employing PAS or photopyroelectric detection, as discussed in Chapter 7. Thielemann and Rheinländer (1985) further examined the mechanisms of heat generation and PAS signal variation with bias voltage in an illuminated photovoltaic cell.

With regard to PEC systems, one group of workers has developed a theoretical model which accounts for all of the significant heat generating sources in a PEC cell operated under monochromatic light excitation, as a function of applied bias and quantum efficiency (Fujishima, Maeda, and Honda, 1980a; Fujishima, Maeda, Honda, Brilmyer, and Bard, 1980b; Maeda, Fujishima, and Honda, 1982). These investigators further implemented a thermistor/electrode combination in order to monitor the heat generation in their PEC systems; the general theory of thermistor photothermal spectroscopy (PTS) was developed by Brilmyer and Bard (1980) and Brilmyer *et al.* (1977). Fujishima *et al.* (1980a,b) and Maeda *et al.* (1982) used their own PEC cell model and the PTS method to obtain the internal quantum efficiencies of several PEC electrode/electrolyte combinations. The term internal quantum efficiency refers to the fact that their model employed a self-normalization (a concept which will be discussed in greater detail in the next section), which did not take into consideration photons reflected at the electrode/electrolyte interface.

Most of the photothermal experiments performed by Fujishima *et al.* (1980a,b) and Maeda *et al.* (1982) involved the measurement of the temperature rise in their systems due to long-duration pulses of light, and employed slow response thermistors. Wagner *et al.* (1986) performed a similar experiment using PDS and obtained results consistent with those obtained using thermistors.

IV. Theoretical Calculation of PEC Energy Conversion Parameters Via PDS

In this section we derive a number of theoretical expressions for the PDS signal of a PEC cell operating under various conditions. Before we develop the

Fig. 2 The n-type semiconductor/electrolyte interface at equilibrium, where E is energy (vs. reference), V is potential (vs. reference), E_F is the semiconductor Fermi energy, E_R is the electrolyte redox level, CB is the conduction band, and VB is the valence band.

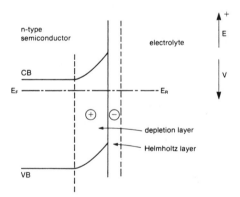

PEC cell model, it would be advantageous to look at the electronic and optical behavior of the n-type semiconductor/electrolyte interface. Consider the situation where a non-degenerate n-type semiconductor is brought into contact with an electrolyte having a redox potential energetically located within the forbidden gap of the semiconductor. Let the Fermi level of the isolated semiconductor be higher than the redox level of the electrolyte. Therefore, once equilibrium is established between the semiconductor and the electrolyte, a depletion layer will have formed at the semiconductor surface. The n-type semiconductor/electrolyte interface (Fig. 2) is similar in electrical and optical properties to the semiconductor/metal Schottky barrier diode. The electrical behavior of the n-type semiconductor/electrolyte interface under D.C. bias is as follows (Morrison, 1980): Under reverse bias the semiconductor is made positive relative to the counter electrode, the semiconductor bulk conduction band (CB) becomes lower in energy relative to the surface CB level and the surface barrier height is increased. The majority electron carriers are drawn away from the surface to the energetically favorable bulk CB states of lower energy. Electron flow out of the semiconductor into the electrolyte becomes more difficult and, in general, electrolyte states are not available to inject electrons into the CB. Further, the minority hole carrier density is too low to support appreciable current. Thus a small anodic current flows.

Under forward bias majority electron carriers are moved to the surface by a mechanism similar to that presented for reverse bias, and if energy levels are available in the electrolyte to accept them, high cathodic currents will be induced. When a large enough forward bias is applied, the energy levels in the semiconductor undertake the flatband configuration. If larger forward biases are applied, the semiconductor depletion layer will become an accumulation layer. For an n-type depletion (accumulation) layer a net positive (negative) charge is stored in the layer.

Like all diode-like semiconductor junctions, a semiconductor/ electrolyte junction is capable of creating a photocurrent and/or photovoltage when illuminated by superbandgap light. The photocurrent generated by all diode devices is a strong function of the degree of band bending in the space-charge region. The bias dependence of the photocurrent can be qualitatively understood as follows: For biases positive of the flatband potential, a depletion layer is present which induces efficient separation of photogenerated electron/hole pairs, electrons are pushed into the n-type bulk and holes move to the surface where they can accept electrons from an electrolyte reducing agent, hence the anodic current, Fig. 3a. For biases negative of the flatband potential an accumulation layer may form in the semiconductor, and photocarriers will again be separated, electrons flowing to the surface and holes to the bulk, Fig. 3b. Surface electrons can then be accepted by an electrolyte oxidizing agent leading to a cathodic current.

Large reverse biases are of interest to quantitative PDS measurements. At these biases, the photocurrent reaches a saturation level, indicating that increased band bending will not increase the efficiency of electron/hole separation. Thus, the quantum efficiency for current generation eventually becomes

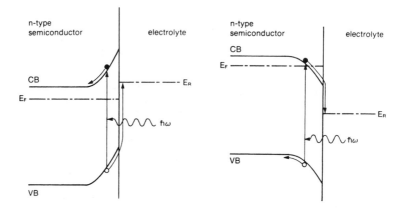

Fig. 3a The anodic current arising from the illumination of a depletion layer, under reverse bias, with superbandgap light ($\hbar\omega > E_g$). Symbols are as defined in Fig. 2.

Fig. 3b The cathodic current arising from the illumination of a forward-bias-induced accumulation layer ($\hbar\omega > E_g$). At the degenerate surface layer the Fermi level lies within the CB of the semiconductor. Symbols are as defined in Fig. 2

independent of reverse bias, a fact which helps in the determination of the quantum efficiency from PDS measurements. It is significant to note that for an experiment where modulated light illuminates a D.C. biased electrode, different combinations of anodic and cathodic A.C. and D.C. currents flow through the PEC cell. Using intensity modulated light and a frequency selective lock-in amplifier enables one to separate the A.C. photocurrent from the D.C. dark current.

For the development of a PEC cell photothermal laser beam deflection model, the following assumptions are used: a) The PEC electrode absorbs N photons per unit time at peak excitation, each with an energy $\hbar\omega$. Photons not absorbed by the electrode, due to transmission or reflection, are lost to the system and cannot be accounted for by PDS or photocurrent measurements; b) None of the optically generated electron/hole pairs recombine radiatively. Therefore, if Q is the quantum efficiency for current generation, QN carriers contribute to the current and $(1-Q)N$ carriers recombine nonradiatively; c) No corrosion takes place at the electrode and the assumed single chemical reaction in the cell is regenerative; d) All heat generating processes at the working electrode contribute to the PDS signal through the same proportionality constant, K, a complex function of geometry and several system parameters. The previous statement is equivalent to saying that all of the heating processes have the same temporal and spatial distributions; this assumption is generally valid for a PEC cell where most of the heating takes place within the very thin depletion layer of the working electrode, or at the electrode/electrolyte interface, and most of the

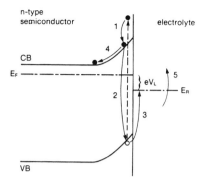

Fig. 4 Heat generating mechanisms which contribute to the PDS signal at the working electrode/electrolyte interface; refer to the text for an explanation of the five heat sources (PDS_j; $j=1,2,...,5$); refer to Fig. 2 for an explanation of the symbols. (After Maeda *et al.*, 1982).

de-excitation processes occur very soon after optical absorption.

The significant thermal processes at the working electrode/electrolyte interface are represented in the schematic diagram of Fig. 4. Quantitative expressions for the partial contributions of each of these processes to the PDS signal are now developed, following the notation of Maeda *et al.* (1982): i) Following optical absorption, nonradiative electron intraband de-excitation from higher states in the conduction band to states near the bandedge yields a signal component

$$PDS_1 = KN(\hbar\omega - E_g) \tag{2}$$

where E_g is the optical band gap of the semiconductor electrode, and K is the geometry-dependent proportionality constant. All optically generated CB electrons are assumed to undergo this fast transition. ii) Nonradiative interband de-excitation between the conduction and valence bands contributes a signal component

$$PDS_2 = KN(1-Q)E_g \tag{3}$$

iii) Electron injection from electrolyte species into the valence band (for n-type working electrode) gives

$$PDS_3 = KNQ(E_R - E_{VB}) \tag{4}$$

where E_R is the redox level of the electrolyte (vs. reference), and E_{VB} is the valence band (VB) energy level (vs. reference). iv) Carrier separation in the depletion layer, during which the carriers lose energy under the influence of the built-in field, plus the Peltier heat at the back Ohmic contact of the n-type electrode contribute

$$PDS_4 = KNQ(E_{CB} - E_R - E_L) \qquad (5)$$

or

$$PDS_4 = KNQe(V - V_{FB}) \qquad (6)$$

where E_{CB} is the CB energy level (vs. reference), $E_L \equiv V_L/e$, where V_L is the voltage drop across an external load attached to the cell, and e is the electronic charge, V is the applied bias (vs. reference), and V_{FB} is the flat band potential (vs. reference). Eq. (5), a relation applicable for the PEC cell under load, includes the Peltier heat for the back Ohmic contact; Eq. (6) does not include the Peltier heat, and is generally valid only for low-resistivity electrodes where the semiconductor Fermi level is close to the conduction band edge. Eq. (6) becomes valid for high-resistivity (more intrinsic) crystals by adding a term $KNQ(E_{CB}^{bulk} - E_F)$, where E_F is the semiconductor Fermi energy. v) The enthalpy change of the redox reaction (expressed as a change in entropy), the so-called electrochemical Peltier heat (EPH), gives

$$PDS_5 = KNQT\Delta S \qquad (7)$$

where T is the absolute temperature, and ΔS is the change in system entropy. The *EPH* is composed of three elements, the entropy of the electrode reaction, the entropy transported by the migration of ions and electrons, and the entropy due to electrochemical polarization (Tamamushi, 1973). Fujishima *et al.* (1980b) pointed out that the polarization entropy is already accounted for in Eq. (6) and the migration entropy would be negligible for a low resistivity system. Thus, the electrochemical Peltier heat is comprised mainly of the entropy of the electrode reaction. Also, the resistive heating of the electrode and electrolyte can be ignored for a low resistivity crystal and non-dilute electrolyte.

When the five partial PDS signals are summed up, the total PDS signal for a PEC cell under bias can be written as:

$$PDS(V) = KN[(\hbar\omega - E_g) + (1 - Q)E_g + Q(E_R - E_{VB}) + Qe(V - V_{FB}) + QT\Delta S] \qquad (8)$$

If the cell is at open circuit, the quantum efficiency, Q, is zero. Thus, the PDS signal is:

$$PDS_{OC} = KN\hbar\omega \qquad (9)$$

The open circuit PDS signal provides a convenient means for normalizing the PDS signal under bias, since under open circuit conditions all of the absorbed optical energy is converted to heat at the electrode/electrolyte interface. As mentioned previously, the concept of the internal quantum efficiency stems from the fact that the PDS signal under bias is normalized by the open-circuit PDS signal. Since neither signal is sensitive to the absolute number of photons incident upon the electrode, but only to absorbed photons, PDS measurements can only yield an internal quantum efficiency. Taking into account that K and N do not depend upon whether the system is at open circuit or under an applied bias, Eqs. (8) and (9) yield an expression for the normalized PDS signal under bias:

$$PDS(V)/PDS_{OC} = [(\hbar\omega - E_g) + (1-Q)E_g + Q(E_R - E_{VB})$$
$$+ Qe(V - V_{FB}) + QT\Delta S]/\hbar\omega \qquad (10)$$

Considering that the cell photocurrent and, hence, quantum efficiency are independent of bias for large reverse biases, we note that Eq. (10) becomes independent of Q for large reverse biases. Therefore, from the slope of that expression:

$$\partial[PDS(V)/PDS_{OC}]/\partial(V - V_{FB}) = eQ_{max}/\hbar\omega \qquad (11)$$

where Q_{max} is defined as the maximum internal quantum efficiency of the PEC cell under given experimental conditions. The value of e, the electronic charge,

Fig. 5 $PDS(V)/PDS_{OC}$ vs. $(V - V_{FB})$ behavior for an n-semiconductor/electrolyte interface illuminated by monochromatic superbandgap light; see Fig. 6 for relevant parameter values. (After Wagner *et al.*, 1986).

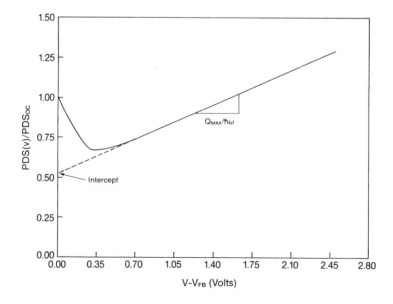

on the right side of Eq. (11) can be set to one if the unit for $\hbar\omega$ is chosen to be electron-volts. Eq. (11) suggests a method for measuring the internal quantum efficiency of the PEC cell under monochromatic excitation from the slope of the normalized PDS signal vs. the applied bias offset by the flatband potential. Fig. 5, a curve taken from Wagner *et al.* (1986), shows typical $PDS(V)/PDS_{OC}$ vs. $V-V_{FB}$ behavior; the slope at high positive bias should be equal to $Q_{max}/\hbar\omega$.

Once Q_{max} has been found, it is possible to obtain an approximate value for the energy change in the system due to the redox reaction, namely a value for $(E_R-E_{VB})+T\Delta S$, where (E_R-E_{VB}) is the change in energy of an electron moving from the redox level to the VB, and $T\Delta S$ is the EPH. The quantity $(E_R-E_{VB})+T\Delta S$ is experimentally determined from the intercept of the $PDS(V)/PDS_{OC}$ vs. $(V-V_{FB})$ curve, which is given by Eq. (12):

$$PDS_{intercept} = [\hbar\omega - Q_{max}E_g + Q_{max}(E_R-E_{VB}) + Q_{max}T\Delta S]/\hbar\omega \qquad (12)$$

Since the values of $\hbar\omega$, E_g, and Q_{max} are known, $(E_R-E_{VB})+T\Delta S$ can be derived.

In order to find the quantum efficiency for current generation as a function of applied bias, a general expression for Q will first be written, noting that Q is defined as the number of electrons per unit time passing through a plane section of the external PEC cell conductor, divided by the number of photons absorbed by the electrode per unit time:

$$Q = (N_A/F)I/N \qquad (13)$$

where N_A is Avogadro's number, F is one Faraday, I is the cell current in amperes, and N is the photon flux, in photons per second. Expressions for $Q(V)$, the quantum efficiency for the cell under bias, and Q_{max}, the maximum cell quantum efficiency under high reverse bias, can be written using Eq. (13). Noting that N_A, F, and N are independent of bias, $Q(V)$ can be written as a function of $I(V)$, I_{max}, and Q_{max}:

$$Q(V) = [I(V)/I_{max}]Q_{max} \qquad (14)$$

where I_{max} is the maximum photocurrent corresponding to Q_{max}. If Eq. (14) and the experimental determination of $(E_R-E_{VB})+T\Delta S$ are used in conjunction with the theoretical model for $PDS(V)/PDS_{OC}$, Eq. (10), it becomes possible to compare the experimental and theoretical $PDS(V)/PDS_{OC}$ vs. $(V-V_{FB})$ curves.

In addition to providing one with a value for the maximum internal quantum efficiency of a PEC cell, PDS measurements further allow one to determine the internal energy efficiency (the ratio of energy supplied to a cell load, to the energy of absorbed photons). The conventional methods of finding the quantum and energy efficiencies of a photovoltaic device are as follows: The quantum efficiency is found by measuring the number of photons incident upon the cell per unit time, and dividing this into the number of charge carriers passing through a load resistor per unit time. The quantum efficiency is a function of the load resistance, being greatest at short circuit and zero at open-circuit. Likewise, the energy efficiency is found by measuring the electrical output power to the cell load and dividing this by the incident optical power. The energy efficiency will depend upon the value of the load, and there will be some optimum load which results in a maximum energy efficiency (Fahrenbruch and Bube, 1983).

For the PEC systems under consideration in this Chapter, all cells were connected to potentiostats instead of isolated load resistors. The concept of a PEC cell energy efficiency has little meaning in this context, since under potentiostatic conditions the PEC cell is effectively at short circuit. Therefore, we will outline how the PDS-determined internal quantum efficiency could be used to determine the internal energy efficiency for a PEC cell. The basic relation used to obtain the internal energy efficiency (EE) is given by Eq. (15):

$$EE = E_{el}/E_{opt} = (E_{opt} - E_{th})/E_{opt} \qquad (15)$$

where E_{el} is the electrical energy dissipated in the cell load, E_{opt} is the optical energy incident upon the cell, and E_{th} is the thermal energy generated within the PEC cell, per modulation period. The thermal energy out of the cell, E_{th}, is made up of heat generated at the working, counter, and reference electrodes, and in the electrolyte. We have previously indicated that the heat generated in the low resistance, non-dilute electrolyte, due to ionic movement, is negligible. Since no current passes through the reference electrode, heat generation there can be ignored, too. On the other hand, the electrochemical Peltier heat at the counter electrode would be of the same magnitude but negative to that at the working electrode. Thus, the net thermal energy out of the PEC cell consists of the heat generation terms given by Eqs. (2)-(5) (minus the K factor):

$$E_{th} = N[(\hbar\omega - E_g) + (1-Q)E_g + Q(E_R - E_{VB}) + Q(E_{CB} - E_R - E_L)] \qquad (16)$$

The optical energy into the cell is given by:

$$E_{opt} = N\,\hbar\omega \qquad (17)$$

Thus, the internal energy efficiency is given by Eq. (18):

$$EE = Q_L\,E_L/\hbar\omega \qquad (18)$$

where E_L was defined in Eq. (5) and Q_L is the internal quantum efficiency under load. In order to find Q_L a relation similar to Eq. (14) can be used:

$$Q_L = (I_L/I_{max})Q_{max} \tag{19}$$

where I_L is the photocurrent through the load. In order for Eq. (19) to be valid one must find I_{max} and Q_{max} for the cell by PDS measurements, as outlined previously, using a potentiostat; the PEC cell must then be disconnected from the potentiostat and connected to a load, for power measurements. If the cell position relative to the monochromatic light source is not altered, the flux of photons into the cell will not be altered. By varying the load across the PEC cell and monitoring I_L and V_L, one can use Eqs. (18) and (19) to find the EE as a function of load, and hence, the maximum internal energy efficiency of the PEC cell.

Of additional interest is the measurement of the PDS signal for the PEC cell under various loads. Cahen (1978) found that the PAS signal from an operating silicon photovoltaic cell under load is minimum when the energy efficiency is greatest. He gave a relationship for the EE as a function of the PAS signal under load and at open circuit:

$$EE_{PAS} = [PAS_{OC} - PAS_L]/PAS_{OC} \tag{20}$$

where PAS_{OC} is the PAS signal at open circuit, and PAS_L is the PAS signal for the cell under load. Since the depletion layer in a silicon photovoltaic cell is normally quite close to its illuminated front surface, and the microphonic PAS signal, like the PDS signal, is sensitive to heat reaching the front surface of the photovoltaic cell, expressions for PAS_{OC} and PAS_L can be written which are similar to Eqs. (8) and (9) for the PEC cell:

$$PAS_L = K'N[(\hbar\omega - E_g) + (1 - Q_L)E_g + Q_L(E_g - E_L)] \tag{21}$$

$$PAS_{OC} = K'N \hbar\omega \tag{22}$$

K' is a geometry-related constant. As for the PEC cell, we have assumed that the fraction of heat reaching the front surface for each heating mechanism is the same. Combining Eqs. (20) to (22), we get:

$$EE_{PAS} = Q_L E_L/\hbar\omega \tag{23}$$

a PAS expression identical to Eq. (18) for the PEC cell, as expected. Noting the similarity between the information provided by the PAS and PDS techniques, it will be important to find a PDS relation for the EE of a PEC cell corresponding to Eq. (20). An expression for PDS_L can be found using Eqs. (2) to (5):

$$PDS_L = KN[\hbar\omega - Q_L (E_L - T\Delta S)] \tag{24}$$

Substituting PDS_L and PDS_{OC} for PAS_L and PAS_{OC} in Eq. (20), we obtain:

$$EE'_{PDS} = Q_L(E_L - T\Delta S)/\hbar\omega \qquad (25)$$

Eq. (25) is not the same as Eq. (18) because the derivation of Eq. (18) took into account the fact that the net EPH for the cell should be zero. Thus, Eqs. (20) and (25) would not be correct for PDS applications to a PEC cell. A modified form of Eq. (20) must be written for PDS measurements on a PEC cell. Noting that:

$$EE = EE'_{PDS} + Q_L\,T\Delta S/\hbar\omega \qquad (26)$$

we can write:

$$EE_{PDS} = (PDS_{OC} - PDS_L)/PDS_{OC} + Q_L\,T\Delta S/\hbar\omega \qquad (27)$$

where Eq. (27) is the PDS/PEC cell analog to Eq. (20) for a PAS/photovoltaic cell system.

Two more experimentally useful limiting expressions for the PDS signal from an operating PEC cell can be derived. Using Eqs. (9) and (24), a relation for PDS_L/PDS_{OC} is given by:

$$PDS_L/PDS_{OC} = [\hbar\omega - Q_L(E_L - T\Delta S)]/\hbar\omega \qquad (28)$$

At open circuit, Q_L is zero and Eq. (28) gives a value of one, as expected. At short circuit, $Q_L = Q_{SC}$ and $E_L = 0$; under this condition, PDS_{SC}/PDS_{OC} has the value:

$$PDS_{SC}/PDS_{OC} = [\hbar\omega + Q_{SC}T\Delta S]/\hbar\omega \qquad (29)$$

Both expressions (28) and (29) can be tested experimentally if the cell photocurrent is measured simultaneously with the PDS signal, and the maximum cell quantum efficiency and photocurrent are known.

V. Experimental Data Analysis from n-CdS PEC Cells

The following discussion will concentrate upon using the theoretical photothermal methods developed in Section IV to find the internal quantum efficiency of several PEC systems from published experimental data. The application of PDS to obtain the absorption spectra of semiconductors is covered in Chapter 10 of this book. Only one paper has been published, to the best of our knowledge, in which PDS was used to monitor heat generation in a PEC cell (Wagner *et al.*, 1986); therefore, some related data obtained by PTS will also be examined. Table I provides a description of the three n-CdS systems which were considered. n-CdS was chosen as the active semiconductor electrode for

Table I

System Property	PDS 1	PTS 2	 3
Crystal	single crystal n-CdS	single crystal n-CdS	single crystal n-Cds
Resistivity	20 Ω·cm	1-2 Ω·cm	Low
Electrolyte (aqueous)	1M Na_2S, 0.05M S, 1M $NaOH$	0.1M $K_4Fe(CN)_6$, 0.001M $K_3Fe(CN)_6$, 0.2M Na_2SO_4	0.1M Na_2SO_3, 0.2M Na_2SO_4
Redox reaction at photoanode	$2S^{2-}+2h^+\rightarrow$ S_2^{2-}	$Fe(CN)_6^{3+}+h^+$ $\rightarrow Fe(CN)_6^{4+}$	$2SO_3^{2-}+2h^+\rightarrow$ $S_2O_6^{2-}$
	a		b
E_{VB} $(NHE)^c$	1.38 V ^{f}	1.38 V ^{f}	1.38 V ^{f}
E_R^o $(NHE)^d$	-0.53 V ^{a}	0.36 V ^{f}	0.02 V ^{b}
E_R $(NHE)^e$	-0.53 V	0.24 V	0.02 V
Reference	Wagner *et al.* (1986)	Fujishima *et al.* (1980b)	Fujishima *et al.* (1980a)

[a] Gerischer (1977)

[b] Maeda *et al.* (1982)

[c] Valence band potential vs. neutral hydrogen electrode (NHE)

[d] Standard redox potential

[e] Redox potential, from Nernst equation (where possible)

[f] Morrison (1980)

this analysis for the following reasons: a) It has been studied extensively in both PEC and solid-state systems; b) There is a significant body of literature on n-CdS PEC cell/photothermal studies involving different electrolytes; and c) n-CdS thin film photovoltaic cells have been studied for several years and PDS studies may help in characterizing these systems.

In this section, both PTS and PDS signals will be termed PS (photothermal signal), in order to maintain consistency with the notation of both detection methods. The only photothermal data applicable to our analysis of n-CdS cells consists of $PS(V)/PS_{OC}$ vs. $(V-V_{FB})$ curves (refer to Table I for references). The examination of data will progress as follows: a) Plots of $PS(V)/PS_{OC}$ vs. $(V-V_{FB})$ and photocurrent vs. $(V-V_{FB})$ will be presented for the three systems identified in Table I; from these plots values for Q_{max} and $(E_R-E_{VB})+T\Delta S$, the system energy change due to the redox reaction, will be extracted; and b) Using

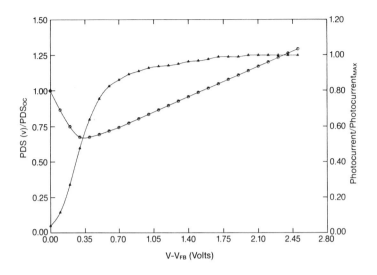

Fig. 6 PDS(o) and photocurrent (Δ) vs. voltage for the n-CdS, S_2^{2-}/S^{2-} system; 505 nm excitation. (After Wagner *et al.*, 1986).

Q_{max}, photocurrent data, and the experimentally derived values for $(E_R-E_{VB})+T\Delta S$, the $PS(V)/PS_{OC}$ vs. $(V-V_{FB})$ curves will be reconstructed from the theoretical model of section IV. No thermal efficiency analysis can be performed, since suitable PS data do not exist.

Fig. 7 PTS(o) and photocurrent (Δ) vs. voltage for the n-CdS, $Fe(CN)_6^{4+}/Fe(CN)_6^{3+}$ system; 490 nm excitation. (After Fujishima *et al.*, 1980b).

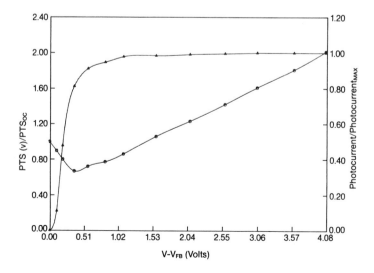

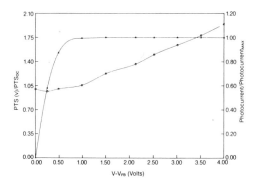

Fig. 8 PTS(o) and photocurrent (Δ) vs. voltage for the $n-CdS$, $S_2O_6^{2-}/SO_3^{2-}$ system; 400 nm excitation. (After Fujishima *et al.*, 1980a).

The PS and photocurrent plots for the three n-CdS crystals are shown in Figs. 6, 7, and 8. Note that all three samples show photocurrents which saturate at biases about 1 V positive of flatband, and have PS minima at about 0.3 V. In all three cells, care was taken to ensure that the systems were chemically stable and that little or no corrosion would occur during measurements. Using Eq. (11) the maximum internal quantum efficiencies of the three systems were obtained and are shown in Table II.

Table II

Determination of the maximum internal quantum efficiency of the three n-CdS PEC systems of Table I from TW photothermal measurements

System Property	PDS	PTS	
	1	2	3
$\hbar\omega^a$	2.455 eV	2.53 eV	3.10 eV
$SLOPE^b$	0.306	0.378	0.296
$Q_{max} = \hbar\omega \times SLOPE$	0.75	0.96	0.92

[a] Illuminating photon energy.
[b] Slope of the high positive bias segment of the $PS(V)/PS_{OC}$ vs $(V-V_{FB})$ curve.

The n-CdS electrodes have quantum efficiencies ranging from 0.75 to 0.96. The Q's for systems 2 and 3 of Table II are slightly less than one, although the value quoted for both systems, in the papers from which the data was taken, is one; the reasons for these discrepancies are the uncertainty in the slope measurements of the $PS(V)/PS_{OC}$ vs. $(V-V_{FB})$ curves, and the round-off errors associated with the exciting photon energies. The crystal having a quantum efficiency of 0.75 (crystal 1 of Table I) was illuminated with weakly absorbed light at 505 nm because higher energy light was strongly absorbed by the electrolyte, thus severely degrading the PDS signal. This weakly absorbed light would be expected to give a quantum efficiency below one because a significant number of carriers were generated in the bulk, away from the depletion layer; in fact, the absorption length (the inverse of the absorption coefficient) at 505 nm, 10^{-6} m (Dutton, 1958), is about the same as the depletion layer thickness, x_o, which can be calculated using the relation (Morrison, 1980):

$$x_o = (2\kappa\varepsilon_o V_s/eN_D)^{\frac{1}{2}} \tag{30}$$

where κ is the semiconductor dielectric constant, ε_o is the permittivity of free space, V_S is the surface barrier $(V-V_{FB})$, and N_D is the doping density. Using Eq. (30) along with parameter values taken from Wagner *et al.* (1986), the value of x_o is found to be about 10^{-6} m. In conclusion, it appears that low resistivity n-CdS crystals have a high internal quantum efficiency when used as PEC cell electrodes.

Using Eq. (12) the value of $(E_R-E_{VB})+T\Delta S$ was experimentally calculated for the three systems; the value of E_g used for these calculations, 2.4 eV, lies within the experimentally reported value range (Mandelis and Siu, 1986), and was chosen because this energy marked the onset of appreciable photocurrents in the photoaction spectra of Wagner *et al.* (1986). Then using tabulated values for E_R^o (the standard redox potential vs. the neutral hydrogen electrode (NHE)) and E_{VB} (the VB energy vs. NHE), the expected values of $(E_R-E_{VB})+T\Delta S$ were calculated and compared to the experimental values. E_R, the redox potential, can be calculated from E_R^o using the Nernst equation, as long as full information on the concentration of the redox species is available. In Table I, only crystal 2 was immersed in an electrolyte for which complete species concentration information was present, enabling E_R to be calculated; for crystals 1 and 3, E_R was assumed to equal E_R^o. In order to calculate $T\Delta S$ we must consider that $T\Delta S$ is defined as the change in enthalpy of the redox species following reaction at the working electrode. Since the PEC cells were not thermally insulated $T\Delta S$ will be assumed to be equal to the change in system free energy which is given by the simple relation (Busch *et al.*, 1978):

$$T\Delta S = nFE_R \tag{31}$$

where n is the number of electrons per redox reaction, F is Faraday's constant, E_R is the redox potential, in the oxidation sense for an anodic reaction, and $T\Delta S$ is the thermal energy added/removed from the system. Eq. (31) is really only valid for transitions between two equilibrium states, but should be applicable in cells exhibiting low photocurrents. Thus, the values of $(E_R-E_{VB})+T\Delta S$ found by PS measurements can be compared to those found from tabulated quantities, as shown in Table III. The comparison shows that there is good agreement between the PS experimental and tabulated results.

Eq. (10), the theoretical equation for $PS(V)/PS_{OC}$, when combined with experimentally determined values for $(E_R-E_{VB})+T\Delta S$ and $Q(V)$, yields results which are depicted in Figs. 9, 10, and 11. There is very good agreement between the model and experiment in all cases, especially for the data obtained by PTS (Figs. 10 and 11).

Table III

Comparison of the values of $(E_R-E_{VB})+T\Delta S$ by PS measurements and from tabulated values of E_R and E_{VB}.

System Property	PDS	PTS	
	1	2	3
PS intercept[a]	0.525	0.440	0.763
$(E_R-E_{VB})+T\Delta S$, from PS intercept, using Eq. (12)	0.845 eV	0.92 eV	1.60 eV
(E_R-E_{VB}), tabulated [b]	1.91 eV	1.14 eV	1.36 eV
$T\Delta S$, tabulated [c]	-1.06 eV	-0.24 eV	0.04 eV
$(E_R-E_{VB})+T\Delta S$, tabulated	0.85 eV	0.90 eV	1.40 eV

[a] Derived from $PS(V)/PS_{OC}$ vs. $(V-V_{FB})$ curves; see Fig. 5.
[b] Using values for E_R and E_{VB} from Table I.
[c] Using Eq. (27).

VI. Reliability of PDS as a Quantitative PEC Probe

In the previous section the heat generation model for a PEC cell, developed by Fujishima *et al.* (1980b,c) and Maeda *et al.* (1982), and the present PDS model, were shown to represent experimental PS data quite well. In this section the reliability of PS as a quantitative PEC probe will be discussed. Fujishima *et al.* (1980c) found that quantum efficiencies obtained by the PTS method correlated quite well with efficiencies calculated using an actinometric (potassium ferrioxalate) photon counting method, although in a few cases (GaAs, GaP, and MoS_2) the PTS method gave substantially higher values (see Table IV for a full summary of the electrodes studied). These authors suggested that the quantum efficiency discrepancy for MoS_2 was probably due to the high surface reflectivity of the very lustrous MoS_2. The differences in efficiency observed for both GaAs and GaP could also be partially explained by the high (>20%) reflectivity of the semiconductor/electrolyte interface due to refractive index mismatch (3.4 vs. 1.33). Another possible cause for quantum efficiency calculation error by the PS method is photoluminescence: if the electrode is

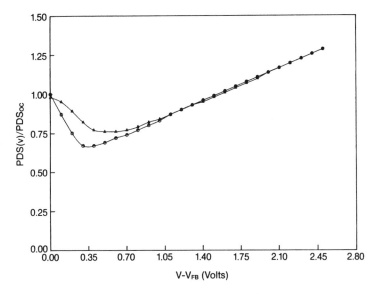

Fig. 9 Theoretical (Δ) and experimental (o) $PDS(V)/PDS_{OC}$ vs. $(V-V_{FB})$ curves for n-CdS, S_2^{2-}/S^{2-} system; 505 nm excitation. (Experimental curve after Wagner *et al.*, 1986).

Fig. 10 Theoretical (Δ) and experimental (o) $PTS(V)/PTS_{OC}$ vs. $(V-V_{FB})$ curves for n-CdS, $Fe(CN)_6^{4+}/Fe(CN)_6^{3+}$ system; 490 nm excitation. (Experimental curve after Fujishima *et al.*, 1980b).

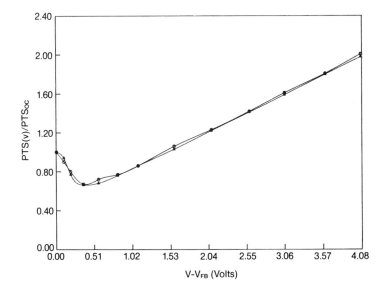

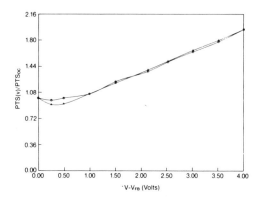

Fig. 11 Theoretical (Δ) and experimental (o) $PTS(V)/PTS_{OC}$ vs. $(V-V_{FB})$ curves n-CdS, $S_2O_6^{2-}/SO_3^{2-}$ system; 400 nm excitation. (Experimental curve after Fujishima *et al.*, 1980a).

appreciably luminescent, the radiative channel will be responsible for a substantial fraction of the de-excitation energy, and the above model will no longer be applicable without modification.

Table IV

Comparison of quantum efficiencies found by thermistor PTS and actinometric techniques for various electrode/electrolyte combinations (After Fujishima *et al.*, 1980a)

Electrode	$Q_{PTS}{}^c$	$Q_{ACT}{}^d$	Electrolyte
n-CdS s^a	1.00	1.00	Na_2SO_3
n-CdS p^b	0.67	0.67-1.00	Na_2SO_3
n-CdSe s	0.42	0.40-0.42	Na_2S
n-GaP s	0.85	0.62	Na_2S
n-GaAs s	0.80	0.47-0.49	Na_2S
n-TiO$_2$ s	0.70	0.70	H_2SO_4
n-ZnO p	0.85	0.76-0.78	Na_2SO_4
n-MoS$_2$ s	1.00	0.54	Na_2SO_4
p-GaP s	0.80	0.60-0.66	H_2SO_4
p-GaAs s	0.80-1.00	0.56-0.59	H_2SO_4

^{a} denotes single crystal

^{b} denotes polycrystal

^{c} quantum efficiency by PTS

^{d} efficiency by actinometry

In addition to providing one with reliable Q values, the PS-obtained values for $(E_R-E_{VB})+T\Delta S$ appear to be consistent with those found from tabulated values of E_R and E_{VB}, although a greater number of experiments with well characterized crystals and electrolytes will be necessary before it can be deduced whether PS-determined values are of higher or lesser reliability than those obtained from conventional calorimetry. The agreement between the experimental and theoretical values in Table III is quite good considering the uncertainty in the values for E_{VB} and E_R. Fujishima *et al.* (1980a) performed several other experiments to test the reliability of the PTS-obtained quantum efficiencies. They found that Q_{max} was independent of excitation light intensity, input pulse duration, and exciting wavelength, as long as the light was strongly absorbed.

The advantages of using the PS method to find Q for a PEC cell, rather than using conventional methods which require calibration of the light source, are several: 1. The PS method is simpler since it employs an internal calibration, namely, it uses the open-circuit PS signal as a normalizing value; 2. The PS method provides the internal quantum efficiency of the cell, which may be quite different from an efficiency found by photon counting techniques if the sample is highly reflective. Thus, the PS method *only* considers the efficiency of electronic processes resulting from photon absorption; 3. Other useful pieces of information can be obtained from PS experiments, such as the absorption spectrum of the electrode, and an approximate value for $(E_R-E_{VB})+T\Delta S$. The value of this sum can be compared to the theoretical value, providing a means for comparison with expected values; and 4. As mentioned in section II, TW techniques can be used to perform thermal depth profiling of the working electrode.

There is one potential disadvantage to PS techniques: the PS method may provide dubious Q values unless the following conditions are met: First, the thermal diffusion length must be long enough to ensure that the heat from all five of the heat generating mechanisms at the working electrode is detected by the probe. Since most of the heat at the working electrode is generated within the electrolyte or in the depletion layer very close to the probe, one can use higher modulation frequencies and still maintain good sensitivity. It must be emphasized that higher modulation frequencies are desirable in the case of PDS, because frequencies below *ca.* 15 Hz are susceptible to ambient noise. Second, for maximum sensitivity and interpretability, the various thermal waves from each of the five heating processes at the working electrode should arrive at the probe at about the same time, or the peak signal will not be as large as a coherent (scalar) addition of the waves would indicate, due to the vectorial nature of thermal wave superposition.

Considering that several different TW detection techniques may, in theory, be applicable to studying PEC cells, it is instructive to speculate on the relative advantages of the four techniques which have been applied to electrochemical systems: gas-cell PAS, piezoelectric PAS, thermistor PTS, and PDS. With regard to gas-cell PAS, Masuda *et al.* (1980a, 1981, 1982) developed a technique whereby the front-side of a thin semiconductor wafer was exposed to electrolyte and the back-side formed the wall of a photoacoustic cell; for low light modulation frequencies enough heat diffused through the thin semiconductor to give a measurable PAS signal. Although this cell proved to work well, it has several disadvantages: The detection probe was quite distant from where the heat generation occurred in the PEC cell. Also, a very thin working electrode was required in order to get an acceptable signal-to-noise ratio. In addition, thermal depth profiling would be difficult because the PAS signal would be strongly dampled for even moderate modulation frequencies. Finally, the apparatus was complicated enough so that sample preparation and changing would be quite difficult. Piezoelectric PAS shares the first three disadvantages with gas-cell PAS; in addition, preparing the semiconductor/piezoelectric combination would be complicated, especially considering that the piezoelectric may require electrical shielding and Ohmic contact must be made to the back of the semiconductor electrode. The piezoelectric method is also intrusive, since the transducer must be bonded to the sample; the bonding agent (grease, cement, etc.) may also chemically contaminate the electrode if proper care is not taken. Thermistor PTS requires that a thermistor be mounted close to the electrode surface where heat is being generated, but not be illuminated by the exciting radiation. This dual requirement may not be carried out without some compromise. Also, the thermistor's inherent limited frequency response may make the interpretation of data, obtained by varying the modulation frequency (depth profiling), more complicated. PDS, on the other hand, does not share any of the disadvantages of the other TW probes, as far as PEC cells are concerned. The probe beam can be brought very close to the semiconductor/electrolyte interface, exactly where the electrode is being illuminated, for maximum sensitivity and *in-situ* measurements. Furthermore, it is non-contact and does not alter, or restrict, the dimensions of the working electrode. The PDS system is, in addition, insensitive to scattered or reflected light. A certain percentage of scattered or reflected light, however, may be captured by the PTS thermistor probe, leading to spurious signals. No special PEC cell must be designed for PDS work, other than ensuring the presence of optical windows to allow the entry and exit of the probe beam, and also taking care to minimize the amount of reflected or scattered light which hits the probe beam position sensor. The working electrode may consist simply of a semiconductor, with a back Ohmic contact, glued to an insulating backing. Therefore, PDS emerges as a very suitable technique for studying PEC cells, in comparison with other thermal wave methods.

VII. Conclusions

The main points presented in this Chapter are as follows: PDS was presented as a form of dynamic micro-calorimetry with several features not common to conventional calorimetry, such as the possibility of performing kinetic studies and thermal depth profiling. Several TW methods were shown to be useful for studying electrochemical and energy conversion systems, with the main emphasis given to absorption spectra and efficiency analysis capabilities. In sections IV and V we presented a simple theoretical model for the heat-generating sources in a PEC cell, and showed that the model may be combined with a straightforward photothermal PEC experiment to find the internal quantum efficiency and the energy change in the system due to the redox reaction, including the EPH component for the cell under monochromatic excitation.

Further, the model was shown to predict the correct dependence of the PDS signal upon the applied bias for three documented n-CdS/electrolyte systems. A method for determining the maximum internal quantum efficiency was also developed, although no independent experimental data to corroborate this model is available up till the writing of this Chapter. A comparison was also made between the PDS method of determining cell efficiencies and conventional photon-counting techniques.

Although the application of PDS methods of PEC systems has been illustrated in sections IV and V, several aspects of the PDS/PEC cell relationship remain unstudied or not totally clear: The PDS model for heat generation at a PEC electrode integrates the total amount of heat created at the electrode, but does not take into account the spatial variation of the heating profile. A refined model would consider the different spatial dependences of the five major heat generation sources, and would utilize a general PDS model like that of Jackson *et al.* (1981). Also, the PDS method for determining the cell energy efficiency and the PDS dependence upon the cell load have not been experimentally confirmed. In addition, the difference between the Q calculated by PDS and by photon counting methods should be further studied; ideally one would expect the PDS quantum efficiency (Q_{PDS}) to be related to the photon counting quantum efficiency (Q_{COUNT}) through the relation:

$$Q_{COUNT} = (1-R)Q_{PDS} \tag{32}$$

where R is the overall reflectivity of the semiconductor/electrolyte interface. Eq. (32) is indicative of the fact that the PDS method is insensitive to reflected photons, unlike the photon counting methods. Finally, PDS holds great promise in performing thermal depth profiling of the PEC cell electrode. This would help one examine the spatial dependence of the heat generating mechanisms at the PEC electrode, by performing frequency scans at the different biases where the various heat generating mechanisms dominate. For example, near flatband conditions mechanism (ii) of section IV dominates, while for large positive biases mechanism (iv) dominates.

VIII. References

Brilmyer G.H., Fujishima A., Santhanam K.S.V., and Bard A.J. (1977). Anal. Chem. **49**, 2057.

Brilmyer G.H., and Bard A.J. (1980). Anal. Chem. **52**, 685.

Busch D.H., Shull H., and Conley R.T. (1978). "*Chemistry*", (Second Edition), Allyn and Bacon, Boston.

Cahen D. (1978), Appl. Phys. Lett. **33**, 810.

Cahen D., and Halle S.D. (1985), Appl. Phys. Lett **46**, 446.

Chance R.R., and Shand M.L. (1980), J. Chem. Phys. **72**, 948.

Dutton D. (1958), Phys. Rev. **112**, 785.

Fahrenbruch A.L., and Bube R.H. (1983). *"Fundamentals of Solar Cells"*, Academic Press, New York.

Faria, Jr. I.F., Ghizoni C.C., Miranda L.C.M., and Vargas H., (1986). J. Appl. Phys. **59**, 3294.

Flaisher H., and Cahen D. (1986), IEEE Trans. UFFC, **UFFC-33**, 622.

Fujishima A., Maeda Y., and Honda K. (1980a), Bull. Chem. Soc. Jpn. **53**, 2735.

Fujishima A., Maeda Y., Honda K., Brilmyer G.H., and Bard A.J. (1980b), J. Electrochem. Soc. **127**, 840.

Fujishima A., Masuda H., Honda K., and Bard A.J. (1980c), Anal. Chem. **52**, 682.

Gerischer H., (1977), J. Electroanal. Chem. **82**, 133.

Graves B.B., (1972), Anal. Chem. **44**, 993.

Jackson W.B., Amer N.M., Boccara A.C., and Fournier D. (1981), Appl. Opt. **20**, 1333.

Lahmann W., and Ludewig H.J. (1977), Chem. Phys. Lett. **45**, 177.

Maeda Y., Fujishima A., and Honda K. (1982), Bull. Chem. Soc. Jpn. **55**, 3373.

Malpas R.E., and Bard A.J. (1980), Anal. Chem. **52**, 109.

Mandelis A. (1983), J. Appl. Phys. **54**, 3404.

Mandelis A. And Royce B.S.H. (1984), Appl. Opt. **23**, 2892.

Mandelis A. and Siu E.K.M. (1986), Phys. Rev. **B34**, 7209.

Masuda H., Fujishima A., and Honda K. (1980a), Chem. Lett., 1153.

Masuda H., Fujishima A., and Honda K. (1980b), Bull. Chem. Soc. Jpn. **53**, 1542.

Masuda H., Fujishima A., and Honda K. (1982), Bull. Chem. Soc. Jpn. **55**, 672.

Masuda H., Morishita S., Fujishima A., and Honda K. (1981), J. Electroanal. Chem. **121**, 363.

Mendoza-Alvarez J.G., Royce B.S.H., Sanchez-Sinencio F., Zelaya-Angel O., Menezes C., and Triboulet R. (1983), Thin Solid Films **102**, 259.

Murphy J.C. and Aamodt L.C. (1980), J. Appl. Phys. **51**, 4580.

Morrison S.R. (1980), *"Electrochemistry at Semiconductor and Oxidized Metal Electrodes"*, Plenum Press, New York.

Ozeki T., Watanabe I., and Ikeda S. (1979), J. Electroanal. Chem. **96**, 117.

Roger J.P., Fournier D., Boccara A.C., Noufi R., and Cahen D. (1985), Thin Solids Films **128**, 11.

Rosencwaig A. (1980), *"Photoacoustics and Photoacoustic Spectroscopy"*, Wiley Interscience, New York.

Royce B.S.H., Sanchez-Sinencio F., Goldstein R., Muratore R., Williams R., and Yim W.M. (1982), J. Electrochem. Soc. **129**, 2393.

Royce B.S.H., Voss D., and Bocarsly A. (1983), J. de Physique **C6 44**, 325.

Tam A.C. (1980), Appl. Phys. Lett. **37**, 978.

Tamamushi R. (1973), Electroanal. Chem. and Interfac. Electrochem. **45**, 500.

Tamamushi R. (1975), J. Electroanal. Chem. **65**, 263.

Tamor M.A., and Hetrick R.E. (1985), Appl. Phys. Lett. **46**, 460.

Theilemann W., and Rheinländer B. (1985), Solid-State Electron. **28**, 1111.

Wagner R.E., Wong V.K.T., and Mandelis A. (1986), Analyst. **111**, 299.

V

Photothermal Wave Spectroscopies

of Semiconductors

PHOTOACOUSTIC MEASUREMENTS OF PHYSICAL PROCESSES IN CdS

Andreas Mandelis

Photoacoustic and Photothermal Sciences Laboratory
Department of Mechanical Engineering
University of Toronto
Toronto, M5S 1A4 CANADA

I. Introduction

The continuous need for nearly perfect CdS semiconducting material lattices for use in microelectronic and optoelectronic device technologies has brought to attention specific problems associated with the tendency of CdS to form various intrinsic defects and to deviate from stoichiometry, depending on the preparation method (Georgobiani, 1974). Pure and specially doped CdS single crystals or thin films have found increasing applications as photoconductive, photovoltaic and optoelectronic devices due to the relatively large bandgap of this material (Bube, 1960; Basov, Bogdankevich and Devyatkov, 1965; Reynolds and Czyzak, 1954). Recent applications with pure CdS as the active element in solid state excitonic lasers at low temperatures (Wünstel and Klingshirn, 1980; Song and Wang, 1984) have accentuated the demand for defect and impurity-free CdS semiconductors for high optical gain performance. Furthermore, reliable non-intrusive characterization of CdS powders has recently become a major thrust for investigation due to their applications in the phosphor, solar cell and photocell technologies which have been developed recently using screen printing of CdS (Nakayama, Matsumoto, Nakano, Ikegami, Uda and Yamashita, 1980). CdS is known to exhibit a complex native defect structure (Uchida, 1967) at wavelengths near the bandgap edge at ca. 2.4-2.5 eV (Klick, 1953; Davis, Cox and Nicholls, 1985). Over the past few decades a number of defect structure studies have been performed on CdS using photoconductivity (Uchida, 1967), luminescence (Lambe, 1955), capacitive photovoltage (Nesheva and Vateva, 1981), thermally stimulated currents (Bube, 1955a), scanning electron microscopy (Yang and Im, 1986), electrical and Hall conductivity (Morimoto and Kitagawa, 1985; Woodbury and Aven, 1974), and optical absorption coefficient modulation (Conway, 1970). These techniques yield very useful insights as to the origins and mechanisms associated with defect involvement in the electronic behavior of CdS. They are subject, however, to Fermi-level position statistics and thus they are only capable of detecting the radiative de-excitation channel (Amer and Jackson, 1984) or various energy conversion efficiencies (Cahen, 1978). The need for a complete picture of the de-excitation manifold of CdS requires additional knowledge of the nonradiative processes which are usually associated with energy loss mechanisms involving electronic defects in this and other II-VI direct gap semiconductors. To this end, in recent years photoacoustic spectroscopy has emerged as a technique sensitive to the nonradiative de-excitation channel (Mandelis and Royce, 1980; Rosencwaig, 1980). It has been readily and successfully applied to the characterization of electronic properties and defect structures of CdS, owing to its high sensitivity and ability to yield *direct* information about nonradiative de-excitation pathways operating in optically excited specimens of this material. This Chapter will examine the wealth of information which has been derived from the application of photoacoustic spectroscopy to CdS, which is unique in many ways and has helped enhance our understanding of the physical electronic behavior of this remarkable semiconductor by complementing and elucidating the partially known picture of the de-excitation manifold and defect state energetics of this material. Furthermore, the present Chapter should be regarded as a case study for the capabilities and potential of photoacoustic spectroscopy toward a better understanding of semiconductor physics.

In Section II of this Chapter the basic physical processes, which are responsible for photothermal wave signal source generation in general homogeneous semiconductors, are reviewed.

Section III part 1 presents in detail the state-of the-knowledge in spectroscopic and defect structure studies of CdS using microphone-gas coupled photoacoustic spectroscopy (MPAS). Emphasis is given to the nature and interpretation of the obtained experimental results, in view of the fact that spectral features have been found to be strongly dependent on the history of the sample. Combinations and comparisons of MPAS with other analytical spectroscopic methods are particularly emphasized, as they address the essential nature of the complementarity of photothermal wave processes to other de-excitation and energy conversion channels. Electronic transport properties in CdS can be studied using MPAS and the contribution of MPAS to mechanism elucidation is also discussed, in the light of existing theoretical models. This allows for configuration coordinate diagrams and theoretically simulated MPA spectra of extrinsic and nominally pure single crystalline CdS.

In Section III part 2 spectroscopic, defect center and nonlinear multiphoton absorption (mpa) photoacoustic phenomena in CdS are presented from the point of view of piezoelectric detection (TPAS). It is noted that spectroscopic features in CdS using the TPA method have been found to be quite different from MPA spectra of this semiconductor, depending on the detection geometry. This remarkable behavior has motivated the detailed discussion of the possible electronic and excitonic mechanisms generating the TPA spectra. As a result of the entire body of photoacoustic work in CdS up to the time of the writing of this Chapter, it appears that significant gains have been made with regard to our understanding of the electronic de-excitation processes of this material under selective probing of the extremely important, but previously difficult or impossible to measure, nonradiative channel. Our choice of CdS as a case-study II-VI semiconductor largely depends on our own interest in this material and reflects the excitement over its multifaceted spectroscopic response under different types of photoacoustic detection. Section IV, the conclusion, identifies the major gains in the state-of the-knowledge of the physics of CdS probed through photoacoustics. It indicates the type and extent of knowledge, that is typically expected to be gained when photoacoustic spectroscopy is applied to other defect-rich semiconductors, as beautifully exemplified further in Chapter 16 of this book.

II. Physics of Photoacoustic Signal Generation in Semiconductors

When a homogeneous semiconductor is optically excited with monochromatic photons of energy $\hbar\omega_o$ greater than, or on the order of, the fundamental energy gap E_g, several dynamic processes may occur, which result in the generation of thermal energy sources in the semiconductor:

1. At early times after the deposition of the optical energy, energetic electron-hole pairs are created possessing kinetic energies $\hbar\omega_o - E_g$. These carriers progressively suffer collisions with other carriers and lattice phonons, until thermal equilibrium is achieved. Some characteristic times involved in the intraband relaxation of carriers in silicon and germanium following the absorption of ultrashort pulses of coherent electromagnetic radiation are shown in Table I (van Driel, 1984). Assuming an average intraband relaxation time

Table I[a]
Important Carrier and Lattice Characteristic Times in Si and Ge

Phenomenon	Characteristic time (sec)
Carrier Coulomb thermalization	$<10^{-14}$
Carrier momentum relaxation	10^{-13}
Carrier-LO-phonon thermalization	10^{-12}
LO-phonon-acoustic-phonon interaction	10^{-12}-10^{-11}
Auger recombination ($N = 10^{20} \text{cm}^{-3}$)	10^{-10}
Impact ionization ($N = 10^{20} \text{cm}^{-3}$,	
$\quad T_e = 10^3 \text{K}$)	$>10^{-4}$
1-μm carrier diffusion time	10^{-10}
1-μm lattice heat diffusion time	10^{-8}

[a]   After van Driel (1984)

$\tau_{IB} \sim 10^{-12}$ sec (Bandeira, Closs and Ghizoni, 1982), effects due to intraband thermal diffusion may be neglected on the time scale of the conventional frequency domain photoacoustic response of the semiconductor and the heat source profile can be accurately described by the optical absorption distribution

$$\dot{H}_{IB}(\mathbf{r},t;\lambda) = \eta_G \, N_o(t)\exp[-\alpha(\lambda)\,|\,\mathbf{r}\,|\,]\,(\hbar\omega_o - E_g) \quad [J/cm^3 - s] \qquad (1)$$

where $\dot{H}_{IB}(\mathbf{r},t;\lambda)$ is the intraband heat release rate per unit volume of the optically excited semiconductor, η_G is the quantum efficiency for photogenerated carriers, $\alpha(\lambda)$ is the optical absorption coefficient at wavelength λ, and N_o is the photon deposition rate per unit volume.

2. Photogenerated electrons (holes) will diffuse during their lifetime τ in the conduction (valence) band through distances equal, on the average, to the diffusion length

$$L_c = (D_c \, \tau_c)^{\frac{1}{2}} \qquad (2)$$

c = n (electron) or p (hole).

Following diffusive spatial migration, excess electron-hole pairs generate a thermal wave source through nonradiative bulk interband recombinations (Quimby and Yen, 1980). The heat release rate per unit volume due to nonradiative recombination is given by (Bandeira *et al.*, 1982; Thielemann and Rheinländer, 1985)

$$\dot{H}_{BB}(\mathbf{r},t;\lambda) = \eta_G\,\eta_{NR}\,N_o(t)\exp[-\alpha(\lambda)\,|\,\mathbf{r}\,|\,]E_g \quad [J/cm^3-s] \qquad (3)$$

where η_{NR} is the nonradiative quantum yield. Eq. (3) is only relevant when $\hbar\omega_o \geq E_g$.

3. In the presence of an external electric field **E** applied across the semiconductor the instantaneous thermal power generated due to the Joule effect is **E · I**, where **I** is the current flowing in the external circuit due to the presence of photogenerated or dark carriers. The bulk heat release rate can then be written:

$$\dot{H}_J(\mathbf{r},t;\lambda) = e\,[n(\mathbf{r},t;\lambda)\mu_n + p(\mathbf{r},t;\lambda)\mu_p]E(t)^2 \quad [J/cm^3-s] \qquad (4)$$

where n (p) is the steady state electron (hole) population following optical absorption, generation and diffusion; μ_n (μ_p) is the electron (hole) mobility, and $E(t)^2 = |\,E(t)\,|^2$ for current flow in the direction of the field vector **E**.

4. Photoexcited carriers undergo front surface (FS) and rear surface (RS) nonradiative recombination resulting in the generation of surface heat release rates per unit area given by (Flaisher and Cahen, 1986; Bandeira *et al.* 1982).

$$\dot{H}_{FS}(\mathbf{O},t;\lambda) = \eta_{FS}[n(\mathbf{O},t;\lambda)-n_o]S_{FS}E_g \quad [J/cm^2-s] \qquad (5a)$$

$$\dot{H}_{RS}(\mathbf{L},t;\lambda) = \eta_{RS}[n(\mathbf{L},t;\lambda)-n_o]S_{RS}E_g \quad [J/cm^2-s] \qquad (5b)$$

where η_{FS} (η_{RS}) is the front (rear) surface nonradiative quantum yield, $n(\mathbf{r},t;\lambda)$ is the photogenerated electron density distribution, n_o is the equilibrium electron density and S_{FS} (S_{RS}) is the front (rear) surface recombination velocity. Eqs. (5) are valid for n-type semiconductors; a set of similar equations can be written for p-type materials with $n\rightarrow p$, as well as for intrinsic materials with $n\rightarrow n_i$ in Eqs. (5).

5. In certain cases an additional thermal source is introduced in a semiconductor due to optical absorption without free carrier generation, such as exciton formation in CdS (Wasa, Tsubouchi and Mikoshiba, 1980). Under those conditions the heat release rate following the absorption of photons of energy $\hbar\omega$ can be written (Thielemann and Rheinländer, 1985).

$$\dot{H}_{NF}(\mathbf{r},t;\lambda) = (1-\eta_G)N_o(t)\exp[-\alpha(\lambda)\,|\,\mathbf{r}\,|\,]\hbar\omega \quad [J/cm^3-s] \qquad (6)$$

In an actual experimental situation the thermal wave source generated due to modulated optical absorption in a semiconductor will be the result of all the above heating rates, Eqs. (1), (3)-(6):

$$\dot{H}_{Total}(\mathbf{r},t;\lambda) = \dot{H}_{IB} + \dot{H}_{BB} + \dot{H}_J + \dot{H}_{FS} + \dot{H}_{RS} + \dot{H}_{NF} \tag{7}$$

III. Spectroscopic and Defect Structure Photoacoustic Measurements in CdS

The vast majority of all reported photothermal wave measurements on CdS have been accomplished using either microphone gas-coupled photoacoustic spectroscopy (MPAS), or piezoelectric transducer-coupled photoacoustic spectroscopy (TPAS). Some photothermal deflection spectroscopic (PDS) measurements on CdS have also been reported recently, but they are primarily concerned with surface roughness effects (Hata, Hatsuda, Kawakami and Sato, 1985) and will not be considered any further in this Chapter which is concerned with photoacoustic studies of physical processes in CdS. Photothermal measurements effected through the use of thermistors and thermometers have largely been implemented in electrochemical environments and as such they are discussed elsewhere in this volume (See Chapter 14). The increasing amounts of photoacoustic experimental evidence to date indicate that the nature of the obtained results is technique specific with many large variations observed between the MPA and TPA spectra. In what follows we will, therefore, present the spectroscopic evidence by technique and discuss the differences among the observed and underlying physical phenomena, which are ultimately responsible for such variations.

1. MICROPHONE GAS-COUPLED PAS (MPAS)

A. SPECTROSCOPY.

A typical experimental arrangement using this type of thermal wave probe is shown in Fig. 1. The apparatus in Fig. 1 is further equipped with an additional external electrical circuit suitable for measuring photocurrent spectra as well. The earliest CdS spectra recorded photoacoustically (Rosencwaig, 1975) were obtained from samples in powdered form using MPAS. Those spectra were used as shown in Fig. 2 for the determination of the bandgap energy of the CdS powders, which was placed at 2.4 eV, in agreement with the value calculated from specular reflectance measurements. It has been noted, however, that there is an error committed on the determination of the bandgap width from photoacoustic spectra, due to the onset of photoacoustic saturation at the strong absorption side of the bandgap region of semiconductors (Rochon and Racey, 1984).

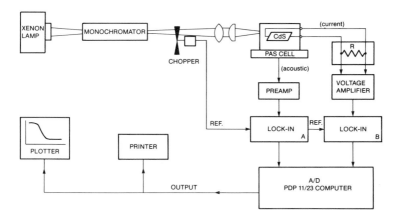

Fig. 1 Schematic diagram of semiconductor MPA experimental apparatus with photo-current spectra (PCS) measurement circuit attached [From Mandelis and Siu (1986)]

Fig. 2 MPA spectrum of the direct-band semiconductor CdS in powder form at 300°K. The bandgap energy derived from this spectrum is shown and compared to the value derived from specular reflectance measurements (in parentheses) [From Rosencwaig (1975)]

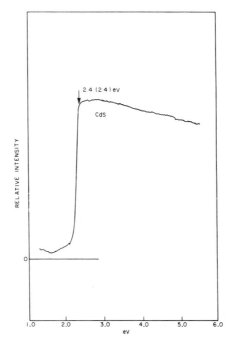

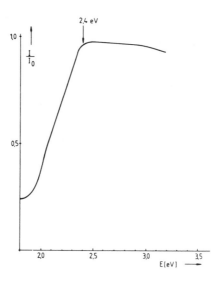

Fig. 3 TPA spectrum of a CdS single crystal with the position of E_g marked by an arrow [From Breuer (1980)].

The first published MPA spectrum of single crystals of CdS appears to be in the work of Lyamov, Madvaliev and Shikhlinskaya (1979). These authors reported agreement between the fundamental absorption edge of CdS calculated from the MPA spectrum and published data, but they did not give a numerical value for the bandgap energy of the single crystal. Breuer (1980) estimated photoacoustically the E_g of a CdS single crystal to be 2.4eV using a TPA method and based on the location of the "knee" of the spectrum, Fig. 3. CdS bandgap width values were obtained using various techniques previously (Klick, 1953; Gobrecht, 1953; Bube, 1955b; Cardona and Harbeke, 1965) and range between 2.39 eV and 2.53 eV at room temperature. Recently, Mandelis and Siu (1986) measured the MPA spectrum of well characterized (0001)-oriented pure n-CdS single crystals (Fig. 4) using an experimental apparatus similar to that of Fig. 1. Based on the reasoning of Rochon and Racey (1984), these authors concluded that the "knee" in the MPA absorption spectrum of CdS cannot be used to determine E_g and they argued that E_g of the single crystal in Fig. 4 may be somewhat larger than the 2.45 eV indicated by the "knee" of the MPA magnitude spectrum of Fig. 4a. Fig. 5 shows the CdS MPA transmission spectrum together with the MPA absorption spectrum from Fig. 4a. In the absence of polarization of the incident optical field, no exciton formation is apparent in the transmission spectrum in the 485 nm - 500 nm region (Dutton, 1958; Uchida, 1967). Uchida (1967) determined photoconductively the spectral positions of the A, B and C excitons at room temperature (501.5 nm, 498 nm, and 485.5 nm, respectively) due to the presence of three two-fold degenerate valence bands in CdS, the highest band having Γ_9 symmetry and the remaining two having Γ_7 symmetry (Thomas and Hopfield, 1959). The lowest energy excitons have been shown (Thomas, Hopfield and Power, 1960) to have ionization energies of *ca.* 0.028 eV. Based on the small ionization energies, Smith (1959) has shown that it was possible to assign exciton energy levels similar to shallow donor states very close to the conduction band. Exciton formation in CdS is thus expected to occur at photon energies slightly lower than the bandgap energy. Since the highest energy C exciton series have formation energy around 2.55 eV (Uchida, 1967), the value $E_g \approx 2.56$ eV was estimated as a more accurate representation of the bandgap energy than the 2.45 eV obtained from the MPA absorption spectrum "knee" of Fig. 4.

Takaue, Matsunaga and Hosokawa (1984) recorded MPA spectra of CdS single crystals using conventional (front side) and reverse-side detection. Typical absorption spectra from a $10\times10\times0.3$ mm^3 sample of unknown origin and orientation are shown in Fig. 6. The reverse side MPA signal was measured with the sample in contact with the beam entrance window inside the photoacoustic cell and with the microphone remotely located. The amount of thermal energy released in the back of the crystal following optical excitation and heat conduction throughout the sample thickness was thus probed. These authors found general agreement of the spectra obtained using front side detection with the Rosencwaig-Gersho (RG) model (Rosencwaig and Gersho, 1976; Eq. (9)) as represented by the solid lines in Fig. 6. A modified RG model was derived by the authors to account for the reverse side MPA signal. According to this model, the spatially averaged temperature θ in the gas is given by

$$\theta = \frac{2W}{1+g} \ \{(1+g)[(g-d)(d-r_2)\exp(-\sigma_1 l_1)+(g+d)(d+r_2)\exp(\sigma_1 l_1)]$$

$$+\tfrac{1}{2}(1+g)(1+r_2)[(1-d)(g-d)\exp(-\sigma_1 l_1)-(1+d)(g+d)\exp(\sigma_1 l_1)]$$

$$\times\exp(\sigma_2 l_2)-\tfrac{1}{2}(1-g)(1+r_2)[(1+d)(g-d)\exp(-\sigma_1 l_1)$$

$$-(1-d)(g+d)\exp(\sigma_1 l_1)]\exp(-\sigma_2 l_2)+(r_2-g)[(1+d)(g-d)\exp(-\sigma_1 l_1)$$

$$-(1-d)(g+d)\exp(\sigma_1 l_1)]\exp(-\sigma_2 l_2-\beta_2 l_2)\}\{(1-g)[(1+d)(g-d)$$

$$\times\exp(-\sigma_1 l_1)-(1-d)(g+d)\exp(\sigma_1 l_1)]\exp(-\sigma_2 l_2)-(1+g)$$

$$\times[(1-d)(g-d)\exp(-\sigma_1 l_1)-(1+d)(g+d)\exp(\sigma_1 l_1)]\exp(\sigma_2 l_2)\}^{-1}, \qquad (8)$$

where

$$g = \frac{k_g a_g}{k_2 a_2}, \ d = \frac{k_1 a_1}{k_2 a_2}, \ r_2 = \frac{\beta_2}{\sigma_2}, \ \text{and} \ \ W = \frac{\beta_2 I_o \eta_2}{2k_2(\beta_2^2-\sigma_2^2)} \qquad (9)$$

In Eqs. (8) and (9) notations similar to those found in the RG theory were used, with the subscripts 1 and 2 indicating window and CdS, respectively. The dashed lines in Fig. 6 are computer-generated plots of Eq. (8) for the amplitude

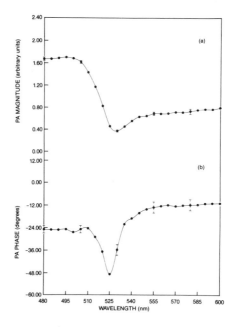

Fig. 4 MPA absorption spectrum of n-CdS Modulation frequency: 50 Hz; spectral resolution: 2 nm. a) Magnitude b) Phase [From Mandelis and Siu (1986)]

and phase variations of the MPA spectra near the absorption edge of CdS. The theoretical fits are poor in the long wavelength region, however, the general trends can be seen to be in agreement with the reverse side data of Fig. 6. The spectral amplitude maximum at *ca.* 532 nm in Fig. 6b is the result of the competition between i) increased thermal wave energy conduction to the back surface of the crystal with decreasing optical absorption coefficient (i.e. increasing

Fig. 5 Magnitudes of a) MPA absorption and b) MPA transmission spectrum of n–CdS. Modulation frequency: 50 Hz; spectral resolution; 2nm [From Mandelis and Siu (1986)]

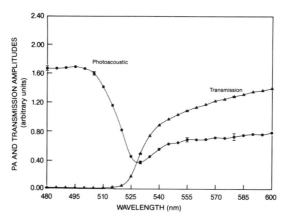

wavelength) due to motion of the sample heat centroid deeper into the bulk; and, simultaneously, ii) decreasing thermal wave amplitude with decreasing absorption. Takaue *et al.* (1984) pointed out that there exists a definite relationship between optical absorptance (βl_2) and thermal diffusion length $\mu_2(\omega)$ (Rosencwaig and Gersho, 1976) at the spectral position of the amplitude maximum in Fig. 6b. No explicit mathematical relationship was presented, however its existence is consistent with the appearance of the maximum at or above certain light intensity modulation frequency ω for which (βl_2) and $\mu_2(\omega)$ become comparable in magnitude. Takaue *et al.* (1984) claimed that the reverse side MPA spectra of CdS single crystals should be similar to TPA spectra of this semiconductor. In section III.2 below it will become apparent that this is the case for *only* specific types of CdS crystals, as the spectroscopic behavior of this material depends greatly on sample fabrication, treatment history and energetics of defect levels.

Fig. 6 MPA amplitudes and phases for CdS single crystal. -o-o-: conventional (front side) detection; $-\Delta-\Delta-$: reverse side detection. Dashed lines show fits to Eq. (8); solid lines show fits to the Rosencwaig-Gersho (1976) theory. Modulation frequency: 8 Hz (a); 80 Hz (b). [From Takaue *et al.* (1984)].

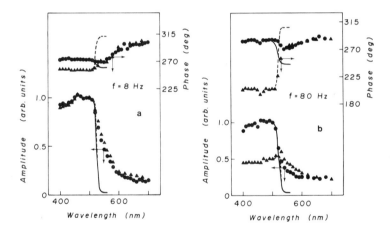

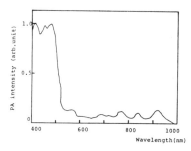
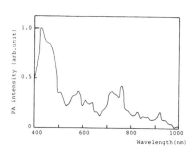

Fig. 7 MPA spectrum of CdS single crystal vapor grown by the modified Piper-Polich method. Modulation frequency: 40 Hz [From Morimoto *et al.* (1984)]

Fig. 8 MPA spectrum of Cu-doped CdS single crystal of the same origin as that of Fig. 7. [From Morimoto *et al.* (1984)]

The first MPA spectra of well characterized CdS single crystals were reported by Morimoto, Wada, Watanabe and Miyakawa (1984). Samples grown from the vapor phase from powders (Rare metallic, 99.999% pure) using the modified Piper-Polich method (Morimoto, Ito, Yoshioka and Miyakawa, 1982) were studied as shown in Fig. 7. The spectrum of Fig. 7 exhibits substantial structure in the subbandgap region. The peaks at *ca.* 840 and 920 nm seem to coincide with sharp spectral lines of the Xe lamp at these wavelengths. No further characterization of the subbandgap MPA features of pure CdS was attempted by those authors.

Considerable interest has lately been shown with respect to the system CdS:Cu due to the fact that copper doping forms deep nonradiative acceptor centers in CdS while it affects the luminescent properties of this material by the simultaneous formation of luminescent centers. This system has been used to demonstrate the ability of MPAS to detect nonradiative defect centers in semiconductors (Wasa, Yoshioka, Morimoto and Miyakawa, 1985; Morimoto *et al.*, 1984). Vapor phase grown, low resistivity n-CdS single crystals were doped with Cu upon evaporation of Cu_2S on the crystal surface and Cu diffusion into the samples for 1 hr in a Ar ambient of 1 atm at 430°C. The MPA spectrum of a typical Cu doped CdS crystal is shown in Fig. 8. Subbandgap contributions to the MPA signal of CdS:Cu are substantially higher than contributions to the pure crystal spectrum of Fig. 7. The peaks at *ca.* 840 and 920 nm are present in the Cu-doped crystal too, whereas the valley at *ca.* 500 nm anticorrelates with the peak in the excitation spectrum of the Cu center-induced broad red luminescence emission centered around 700 nm in this material (Morimoto, Ito and Miyakawa, 1982). Pronounced peaks at 580, 630 and 720 nm in Fig. 8 have been associated with defect-to-band electronic transitions from Cu impurity gap states at 2.14, 1.97 and 1.72 eV, respectively, below the conduction bandedge.

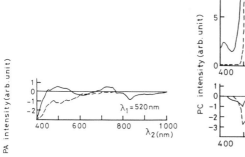

Fig. 9 Secondary MPA spectra of undoped CdS and CdS:Cu. Differences are recorded between spectra in the presence and in the absence of primary radiation λ_1 at 40 Hz [From Wada *et al.* (1985)].

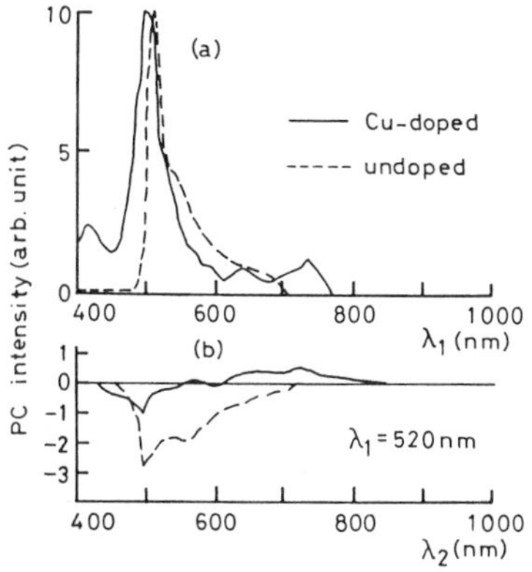

Fig. 10 Primary (a) and secondary (b) PC spectra at 1.5 V DC. Curves (b) obtained as in Fig. 9 [From Wada *et al.* (1985)]

Secondary MPA spectra (Wada *et al.* 1985) of the CdS:Cu system have further elucidated the contributions of defect states to the photoacoustic signal and have given valuable information regarding nonradiative defect physics in CdS. These authors used filter-monochromatized light from a Xe arc-lamp as the primary source at (λ_1) 520, 540 and 580 nm, as well as a He-Ne laser at 632.8 nm chopped at 40 Hz. Beam-splitted unmodulated (D.C.) light intensity from the same arc-lamp was further passed through a monochromator and used

Fig. 11 Primary (a) and secondary (b) PL spectra. Curves (b) obtained as in Fig. 9 [From Wada *et al.* (1985)]

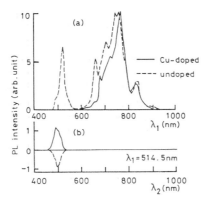

as the secondary source (λ_2: 400-1000 nm). An external circuit arrangement similar to that of Fig. 1 was also attached to the system to measure the induced photoconductivity (PC). A photoradiation meter was used to measure photoluminescent (PL) excitation and emission spectra. Low resistivity undoped CdS and doped CdS:Cu samples obtained through the previously mentioned Cu diffusion method were tested and their primary MPA spectra are shown in Figs. 7 and 8. Open-circuit secondary MPA spectra at λ_1 = 520 nm, Fig. 9, exhibit quenching in the region λ_2 < 750 nm (CdS) and λ_2 > 750 nm (CdS:Cu); the doped sample exhibits sensitization in the spectral ranges between 650 and 750 nm and around λ_g (= hc/E_g). PC spectra obtained with a 1.5 V D.C. electric field applied across the sample are shown in Fig. 10. PL spectra are shown in Fig. 11. The primary PC spectrum of the CdS:Cu sample shows sensitization in the λ_1 > 600 nm region; the secondary PC spectrum from the CdS sample exhibits quenching at λ_2 < 750 nm, while that of CdS:Cu exhibits sensitization between 600 and 800 nm. Important features of the primary PL spectra are green edge

Fig. 12 Secondary spectra of n-CdS with a 20 V D.C. applied electric field and 2 nm spectral resolution. Light modulation frequency: 20 Hz. a) MPA spectrum. No primary light source (–•–•–); with primary light source (–Δ–Δ–). b) PC spectrum. No primary light source (–•–•–); with primary light source (–Δ–Δ–). Experimental apparatus shown in Fig. 1 [From Mandelis and Siu (1986)].

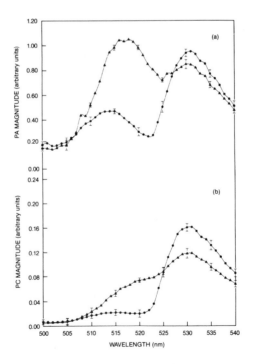

emission (CdS) and red emission (CdS:Cu). The secondary PL spectra exhibit quenching (CdS) and sensitization (CdS:Cu) around λ_g. In a study focussed around the bandgap region of CdS (500-540 nm), Mandelis and Siu (1986) used a broad-band light source (a table lamp) as the primary light source λ_1 to excite a high resistivity pure (0001)-oriented n-CdS crystal grown by the chemical vapor transport method (Nitsche, 1960). Crystals grown in this manner have been found (Boone and Cantwell, 1985) to contain large densities of native defects such as vacancies and interstitials of both elemental components and they exhibit a natural n-type character to an extent determined by the degree of self compensation and defect density. Fig. 12 shows MPA and PC quenching for $\lambda_2 > 525$ nm in agreement with undoped CdS trends in Fig. 9. The sensitization observed for $\lambda_2 < 525$ nm in both MPA and PC spectra is, however, contrary to CdS trends shown in Figs. 9 and 10 and may be related to the larger electric field applied across the sample of Fig. 12, as well as the defect morphology of the samples, as will be discussed in part (C) below.

Fig. 13 a) Magnitude spectra of MPA (–•–•–), and PC (–Δ–Δ–) responses with a transverse electric field of 10 V D.C. with the apparatus of Fig. 1. Light modulation frequency: 20 Hz; spectral resolution: 2 nm. b) phase spectra of MPA (–•–•–), and PC (–Δ–Δ–) responses [From Mandelis and Siu (1986)]

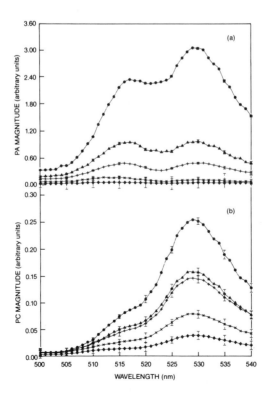

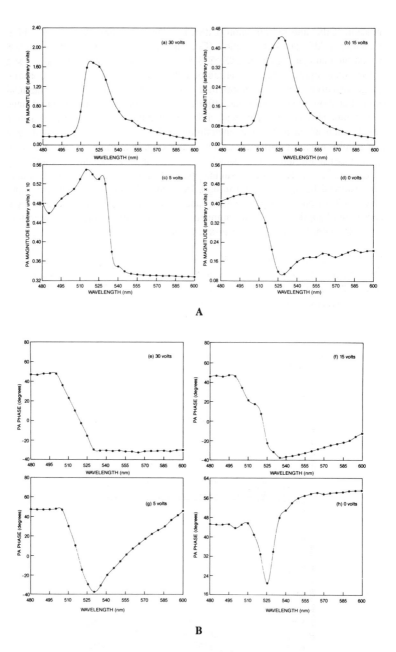

Fig. 14 MPA spectral responses of CdS under external D.C. transverse electric fields. Light modulation frequency: 20 Hz; spectral resolution: 2 nm A) MPA magnitudes; B) MPA phases [From Mandelis and Siu (1986)]

B. ELECTRONIC TRANSPORT PROPERTIES.

Interband and defect state induced nonradiative recombination and electronic transport phenomena in chemical vapor transport grown pure n-CdS crystals have been studied in a detailed set of combined MPA and PC experiments by Mandelis and Siu (1986) in conjunction with electronic transport properties in this material. These authors used the apparatus of Fig. 1 and established the non-equivalence between optical transmission and MPA spectra of high resistivity n-CdS, especially in the subbandgap region, Fig. 13, where lattice defects and wavelength-dependent nonradiative processes were expected to contribute a wealth of information to the MPA spectrum unparalleled by transmission spectroscopic information. Two resolved MPA peaks between 515 nm and 530 nm appeared under 2 nm spectral resolution, with either a D.C. (Figs. 13, 14) or A.C. (Fig. 15) electric field applied. The observed peaks were assigned to i) intrinsic band-to-band electronic transitions (For the high energy peak, Figs. 13, 15); and ii) an increase in the nonradiative quantum efficiency in the subbandgap region (for the low energy peak). These assignations were supported experimentally and theoretically by the modulation frequency responses of the peaks: The high energy peak exhibited a blue shift with increased modulation frequency, in agreement with results from the electron transport related theory of the MPA signal generation by Bandeira *et al.* (1982). These authors observed only one peak at *ca.* 523 nm with their uncharacterized CdS crystal, which they attributed to carrier diffusion within the crystal due to interband excitation under conditions maximizing the MPA signal. They found that the maximum in the MPA response occurred at a wavelength λ_{max}, such that the CdS absorption coefficient β was approximately equal to the inverse of the electron diffusion length L:

$$\beta(\lambda_{max}) \approx [1/L(\lambda_{max})][1+(\omega\tau_{BB})]^{1/4} \qquad (10)$$

where ω is the angular light modulation frequency and τ_{BB} is the electron lifetime in the conduction band. In the framework of that model the MPA response maximum is expected to shift to higher energies (higher β) with increasing modulation frequency. The low energy peak of Fig. 13 remained essentially independent of ω and was shown theoretically to be consistent with an increase in the nonradiative quantum efficiency at subbandgap wavelengths (Siu and Mandelis, 1986). Comparisons of the MPA with PC spectra, Figs. 13, 15, showed that the latter were unable to resolve the two peaks and thus MPAS was found to have the advantage of spectral resolution over PC spectroscopy, with modulation frequency f controlling peak separation: an $f^{-2.3\pm0.1}$ dependence for the 530 nm MPA peak; and an $f^{-1.8\pm0.1}$ dependence for the 517 nm MPA peak of Fig. 13, in agreement with the RG theory, with an additional f^{-1} dependence due to the Joule effect superposed on the expected thermal wave frequency response of the sample. Qualitative differences in the MPA and PC spectra of Fig. 13 at *ca.* 517 nm were sought in the nature of each type of response: The MPA technique probes a thickness in the material on the order of the thermal diffusion length

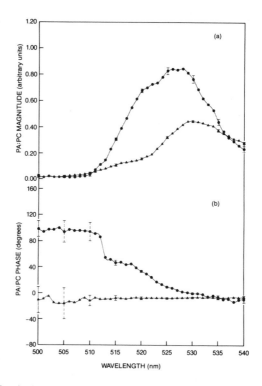

Fig. 15. a) Magnitude spectra of MPA (–●–●–), and PC (–Δ–Δ–) responses with an A.C. transverse electric field of 20 V peak-to-peak and unmodulated irradiation. Field modulation frequency: 20 Hz; spectral resolution: 2 nm, b) Phase spectra corresponding to (a) [From Mandelis and Siu (1986)].

$$\mu_s(f) = (\alpha_s/\pi f)^{\frac{1}{2}} \tag{11}$$

where α_s is the thermal diffusivity of the material. Using (Cesar, Vargas, Mendes Filha and Miranda, 1983) $\alpha_{CdS} = 0.15 cm^2/s$ and f = 100 Hz, we find $\mu_s(100 \text{ Hz}) \approx 22\mu m$. The PC technique is sensitive to carrier transport through distances on the order of the electron diffusion length (about $1\mu m$), which causes different depths of the sample to be probed by the two methods. The larger PA phase lag, Fig. 13 is also consistent with probing a larger depth than the PC phase spectrum. A similar argument has been advanced to account for differences in spectral features obtained through probing semiconductors with a combination of PC and photothermal deflection (PD) spectroscopies (Roger, Fournier, Boccara, Noufi, and Cahen, 1985).

Concerning the experimental mode of Fig. 15 (i.e. a D.C. optical field superposed on an A.C. electric field across the sample), there is a major difference from that of Figs. 13 and 14. In this mode *only* the thermal energy due to the Joule effect contributes to the photoacoustic and modulated PC signals. This is so because thermal energy from nonradiative carrier recombination is not temporally modulated and thus will not be detected by the MPA and PC probes. Therefore, a simpler interpretation of the experimental responses, consistent with the data of Part (B) is expected with this mode of experimentation. The MPA magnitude in Fig. 15a exhibits a maximum and an unresolved shoulder. A comparison with the MPA magnitude of Fig. 10a shows that the long wavelength maxima are located at 529-530 nm, i.e. in the same spectral location, while the peak position of the shoulder in Fig. 15a is located at $\lambda \geq 523$nm, a shift of $\approx$ 6nm from the secondary peak at 517 nm in Fig. 13a. The PC magnitudes, however, exhibit similar spectral features under both chopped and unmodulated illumination. Fig. 15b shows an increase in the lag of the MPA phase spectrum at subbandgap wavelengths, while the PC phase is spectrally flat. The presence of the MPA shoulder at *ca.* 523 nm was associated with the spectral shift of the peak at 517 nm in Fig. 10a attributed to the band-to-band electronic transition. According to Eq. (10) there will be a red spectral shift of the PA maximum due to interband excitation with decreasing optical modulation frequency. The spectrum of Fig. 15 can be thought of as the limiting case with $\omega \rightarrow 0$; in this limit the observed red shift of the 517 nm peak may be understood in terms of the model by Bandeira *et al.* (1982). The observed 3-4 nm red shifts of the high energy peak with decreasing light modulation frequency in the MPA spectra of CdS (Mandelis and Siu, 1986) render further credibility to the argument concerning the intraband excitation origin of the red-shifted shoulder in Fig. 15. The PC magnitude spectral similarity under both chopped and unmodulated optical excitation can be understood from the fact that the PC response is independent of the photoacoustic thermal diffusion length which is ultimately the reason for the frequency dependence of Eq. (10). Therefore, no frequency dependence of the PC signal maximum is expected and no red shift of the intrinsic should at $\omega_{AC} \rightarrow 0$, in agreement with Figs. 13a and 15a. Furthermore, Mandelis and Siu (1986) reported that the spectral features of the PC response in the presence of a D.C. electric field under unmodulated illumination were entirely similar to those of Figs. 13a and 15a.

Under unmodulated excitation and A.C. photocarrier orientation, the generation and recombination of free carriers are D.C. phenomena, which implies that the carrier lifetimes do not affect the MPA phase response (Mandelis, Teng and Royce, 1979). The increase in the PA phase lag observed in Fig. 15b is, thus, solely due to the finite diffusion time of the receding Joule effect-induced heat centroid in the crystal from the near surface to the bulk with increasing wavelength. The PC phase spectrum is independent of the photogenerated carrier density, which is temporally constant and varies with photon energy. It depends, however, on the carrier diffusion length (McKelvey, 1966)

$$L(\omega_{AC}) = \frac{L_n}{(1+i\omega_{AC}\tau)^{\frac{1}{2}}} \qquad (12)$$

and, therefore, in the limit $\omega_{AC} \rightarrow 0$ it is independent of the photocarrier lifetime τ. The PC phase is thus expected to be independent of the mechanism producing the photocurrent at $\omega_{AC} = 0$. It depends solely on the speed of photocarrier

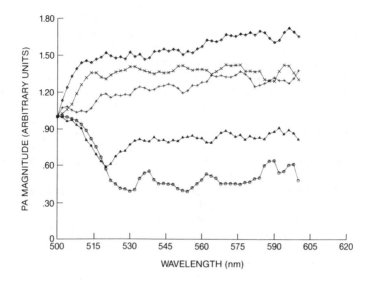

Fig. 16 Open-circuit MPA magnitude of pure n-CdS single crystal with temperature as a parameter; −o−o− T = 22°C; −Δ−Δ− T = -23°C; −+−+− T = -73°C; −×−×− T = -116°C; and −♦− T = -160°C [From Dioszeghy and Mandelis (1986)]

response to the A.C. electric field, which is essentially infinite at the modulation frequencies of our experiments. As a result of the above considerations, a spectrally flat PC phase response at zero lag would be expected, in agreement with Fig. 15b.

MPA and PC spectroscopic and electronic transport studies of high resistivity (1.3-1.9×10^5 Ohm-cm) pure n-CdS single crystals (Eagle-Pitcher Inc. Miami, Okla.) were performed isothermally with polarized light, and at various ambient temperatures down to *ca.* -160°C (Dioszeghy and Mandelis, 1986). The crystal growth details were described by Mandelis and Siu (1986), and the crystals were cut with the c-axis on the surface plane. The experimental apparatus was similar to Fig. 1, with the microphone separated from the photoacoustic cell by a thin teflon rod. The sample cell was housed in a thermally insulating container. Fig. 16 shows open-circuit unpolarized MPA spectra with temperature as a parameter. All curves have been normalized to the 500 nm value of the room temperature (T = 22°C) spectrum for direct comparison. Table II shows the spectral positions of the minima as a function of temperature. The spectral shift of the minima to higher photon energies was found to be higher than linear, especially at T < -80°C.

Table II[a]

MPA magnitude absorption minima dependence on temperature

Minimum spectral position		Temperature
(nm)	→ (eV)	(°C)
526	2.36	22
520	2.38	-23
514	2.41	-73
504	2.46	-116
490	2.53	-160

[a] After Dioszeghy and Mandelis (1986)

Fig. 17 Magnitudes of a) MPA and b) PC spectral responses with a transverse electric field of 20 V D.C.; $-o-o-$ random, polarization; $-+-+-$ $E_\perp C$; $-\Delta-\Delta-$ $E//C$. Light modulation frequency: 23 Hz; Spectral resolution: 2 nm; T = 22°C [From Dioszeghy and Mandelis (1986)]

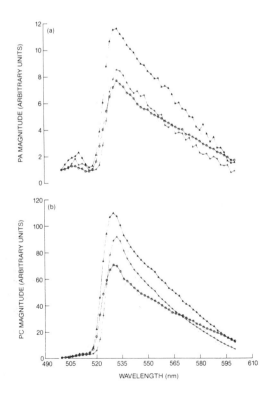

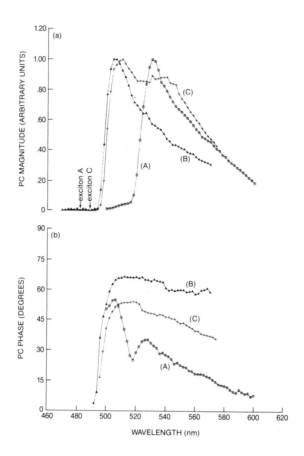

Fig. 18 Magnitudes (a) and phases (b) of MPA and PC spectra with a transverse electric field of 20V D.C.; curves (A): T = 22°C; curves (B): T = -160°C; curves (C): T = -160°C after fracture. Light modulation frequency: 23 Hz; Spectral resolution: 2 nm [From Dioszeghy and Mandelis (1986)]

The gradual nature of the spectral inversion observed in Fig. 16 with decreasing temperature was consistent with a monotonic increase in subbandgap photosensitivity at lower temperatures, in agreement with experimental observations of the low temperature sensitization of the photoconductivity (Bube and Barton, 1959; Bube 1955c) due to the defect structure in CdS. The open-circuit MPA spectra of Fig. 16 were thus shown to be a very sensitive tool for monitoring the degree of sensitization at low temperatures where the defect structure of CdS dominates the visible MPA spectrum.

MPA and PC spectra at room temperature in the presence of a transverse D.C. electric field were also recorded with random polarization, with **E** ∥ C and with $E_\perp C$ by the same authors (Fig. 17). The observed wavelength difference $\Delta\lambda \cong 3$ nm between the rising edges of the $E//c$ and $E_\perp C$ spectra was found to be in agreement with polarization studies of the photoconductivity of pure ("undoped") CdS single crystals at 77°K (Yoshizawa, 1976; Park and Reynolds, 1963) and at 293°K (Park and Reynolds, 1963). The high energy secondary peak exhibited by the MPA spectra at *ca.* 508 nm - 510 nm and the unresolved PC shoulder in the same energy range were interpreted in terms of electronic mechanisms similar to those of Fig. 13.

The temperature dependence of the MPA and PC spectra in the presence of 20V D.C. is shown in Fig. 18. The low temperature MPA spectrum (B) in Fig. 18a is blue-shifted by ca. 26 nm with respect to the room temperature spectrum (A). The acoustic noise associated with the low temperature spectrum did not allow a complete characterization of the secondary high energy peak at the temperature. The position of the main MPA peak is at 506 nm. Spectrum (C) was taken after deliberate mechanical damage was introduced in the crystal through fracture, followed by substantial MPA signal enhancement. In comparison with spectrum (B) the greatest enhancement was observed in the sub-bandgap region, $\lambda \geq 510$ nm. This MPA magnitude increase was accompanied by a phase lag increase at $\lambda \geq 520$ nm, as shown in Fig. 18b, curve (C). The T = -160°C PC spectrum of the intact crystal shows a very low signal at photon energies above 2.51 eV with two minima at ca. 482 nm and 488 nm. Tentative assignations of these minima were made in terms of exciton A and C absorption lines, respectively; Gross and Novikov (1959) established a correlation between the exciton emission spectrum and the PC spectrum. These authors showed that all pure CdS crystals can be divided into two groups: in crystals of the first group exciton absorption lines at 487 nm, 484 nm and 479 nm at 77°K were associated with PC maxima, and in crystals of the second group these same wavelengths were associated with PC minima. At 77°K Uchida (1967) placed excitons A, B and C at 487.5 nm, 484.5 nm and 472.5 nm, respectively. At -160°C and using the temperature coefficient of all three exciton photoconductivities established by Uchida (=3×10^{-4} eV/°K), PC minima due to A and C exciton absorption and thermal release to the conduction band are expected at 489 nm and 482 nm, respectively, for crystals of the second group, consistently with the PC spectrum of Fig. 18, curve (B). No PC minima could be observed in the low temperature spectrum of the fractured crystal in Fig. 18, curve (C). Subbandgap wavelength sensitization, however, of the PC spectrum of the damaged crystal was observed with spectral features similar to those of the PA spectrum in Fig. 18 curve (C).

Dioszeghy and Mandelis (1986) further applied transverse A.C. electric fields across CdS crystals irradiated with continuous monochromatic light, and concluded that this technique was capable of deconvoluting the Joule effect contributions to the MPA signal from other nonradiative mechanisms in agreement with results by Mandelis and Siu (1986). Therefore, it can be inferred that this mode of experimentation has a strong potential for directly studying solid state scattering mechanisms which are responsible for producing thermal effects and setting limitations on the magnitude of the PC response and the crystalline quantum efficiency.

C. NONRADIATIVE PHYSICS OF THE PHOTOACOUSTIC EFFECT IN CdS SINGLE CRYSTALS

Information concerning nonradiative processes in CdS has been largely obtained through the photoconductive behavior of single crystals coupled with the MPA response of secondary spectra, Figs. 9, 10 and 12. In Fig. 12 (Mandelis and Siu, 1986) MPA and PC sensitization was observed for $\lambda \leq 526$ nm and a response magnitude reversal for $\lambda > 526$ nm. The behavior of Fig. 12 has been discussed in terms of a hypothesis concerning the nature of bandgap-active defects in CdS, which has been advanced by Rose (1955). According to this hypothesis, two generic classes of defects, acting as recombination centers, are predominant throughout the bandgap. Class I defects have a small capture cross-section for holes and a large capture cross-section for electrons, once they have captured a hole:

$$[V_s^-]^+ + h^+ \rightarrow [V_s]^{++} \tag{13}$$

$$[V_s^-]^o + h^+ \rightarrow [V_s^-]^+$$

V_s represents an anion (here assumed sulphur) vacancy, with the sign inside the bracket representing the number of trapped electrons, and the sign outside the bracket indicating the effective charge of the defect with respect to the rest of the crystal. Class II defects have a large capture cross-section for holes and are filled with electrons in the dark. These defects lie close to the valence band edge and they capture holes according to:

$$[V_{Cd}]^{--} + h^+ \rightarrow [V_{Cd}^+]^- \tag{14}$$

$$[V_{Cd}^+]^- + h^+ \rightarrow [V_{Cd}^{2+}]^o$$

where V_{Cd} represents a cation (here assumed cadmium) vacancy in the pure crystal. In the proper spectral range ($\lambda \leq 526$ nm in Fig. 12), holes formed by the primary excitation will concentrate in the class II centers via Eq. (14), while electrons, which occupied these centers in the dark, will be transferred via the conduction and valence bands to the energetically higher lying class I centers, due to the large electronic capture cross-section of these centers. These defects will thus be eliminated as efficient recombination centers due to their occupancy by the transferred electrons, with a resulting increase in the lifetime of free electrons and, therefore, an increase in the photoconductivity of the crystal. A simultaneous increase in the MPA signal will ensue as the amount of nonradiative recombination will be proportional to the total (increased) number density of free electrons. This defect structure was also used to explain the quenching observed in Fig. 12 for $\lambda > 526$ nm. This phenomenon will occur only when class II defects are acting as recombination centers, i.e. when electrons are excited from the valence band to class II centers in the proper spectral range ($\lambda > 526$ nm), thus freeing holes which will eventually be transferred to class I defects and turn these defects into efficient traps for electronic capture. Bube

(1955c) has given the requirements imposed on electron and hole Fermi level positions for such excitations to occur, and has identified the spectral positions of three class II energy levels in the CdS bandgap, associated with infrared quenching. Both the PC and MPA secondary signals are expected to decrease with quenching with respect to the dark (primary zero) level, as the number density of free electrons will decrease in the presence of the class I centers of Eq.

Fig. 19 a) Franck-Condon configuration coordinate diagram of possible sensitization mechanism in high resistivity pure n-CdS. Process A: primary optical excitation of electron from class II defect state to excited state in the conduction band. Process B: intraband de-excitation and cross-over to class I defect state. Process C: secondary excitation from low lying acceptor level(s) (AL) to excited state in the conduction band. Process D: intraband de-excitation to the bottom of the conduction band with no cross-over to occupied class I states, resulting in free electron density increase (sensitization); ν_p, ν_s are primary and secondary optical excitation frequencies, respectively. $\nu_s \rightarrow \lambda_s \leq 526$ nm; (●) electron; (o) hole b) Franck-Condon configuration diagram of possible quenching mechanism in high resistivity pure n-CdS. Process A, A': primary optical excitation of electron from the valence band or low lying acceptor level(s) (AL) to class II defect state. Process B, B': hole transfer from AL or VB to class I defect state. Process C: secondary excitation from low lying AL to excited state in the conduction band. Process D: intraband de-excitation and cross-over to unoccupied class I defect state resulting in free electron density decrease, through trapping (quenching); $\nu_s \rightarrow \nu_s > 526$ nm. [From Mandelis and Siu (1986)].

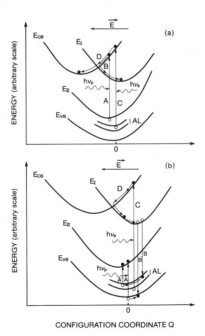

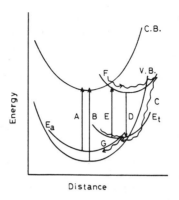

Fig. 20 Configuration coordinate diagram for the CdS:Cu system. The parameters used are as follows: $A = \sigma_a N_a f_a I$, $B = \sigma_b I$, $C = c_3 N_a(1-f_a)n$, $D = c_1 N_t(1-f_t)n$, $E = \sigma_t N_t f_t I$, $F = c_4 n$, $G = c_2 N_t f_t N_a(1-f_a)$, where N_t: trap density; N_a: acceptor density; f_i: occupation probability, σ_i: photoionization cross section, and c_i: capture rate of i-th level; n: electron density [From Wada *et al.* (1985)]

(13). Both sets of data, Fig. 12 (Mandelis and Siu, 1986), and Figs. 9,10 (Wasa *et al.*, 1985), indicate that quenching in pure n-CdS crystals can also occur under primary irradiation in the red spectral region ($\lambda < 600$ nm), followed in the high resistivity sample of Mandelis and Siu (1986) by a sensitized spectral region with a threshold wavelength at *ca.* $\lambda_t \approx 526$ nm. Fig. 19 is a configuration coordinate diagram indicating a possible transition mechanism leading to a qualitative explanation of the secondary MPA and PC sensitized and quenched spectral responses of Fig. 12, based on Rose's hypothesis. The effects of the applied electric field in Fig. 19 are primarily i) an enhancement of the velocity distribution of free carriers, with a mean velocity increase and the resulting lowering of the capture rate at the shallow class I defect centers (Bonch-Bruevich and Landsberg, 1968); and ii) the Poole-Frenkel effect, due to which the ionization energy along the field direction **E** is reduced (Stoneham, 1975). Electric field effects for deep levels such as class II defect centers are expected to involve multiphonon mechanisms (Koral, 1977). Wada *et al.* (1985) also presented a configuration coordinate diagram to explain the secondary MPA and PL behavior of the CdS:Cu system of Figs. 9-11. These authors assumed two localized levels, one acceptor impurity at E_a and one trap level at E_t, Fig. 20. From the de-excitation rate equations, the MPA signal can be written as

$$S_a \propto K I(1+\alpha I) \tag{15}$$

where K is the sensitivity, I is the intensity of the incident radiation and α is the non-linear parameter. Under the small signal approximation

$$\alpha = -\frac{\sigma_a}{c_3 n_o}\left[1-\frac{N_t}{n_o}\left[\frac{\sigma_t}{\sigma_a}+\frac{c_2 N_a}{c_3 n_o+c_2 N_t}\right]\right] \tag{16}$$

where $\sigma_i(c_i)$ is the photoionization cross section (capture rate) of level (i); $N_t(N_a)$ is the trap (acceptor) density; and n_o is the equilibrium concentration of conduction band electrons. For undoped CdS single crystals, $N_t \approx 0$ and $\alpha < 0$ leading to quenching. In samples with $N_t \gg 0$, α can be positive, leading to sensitization. Transition E from E_t (Fig. 20) tends to reduce the rate of the nonradiative de-excitation process G from E_t to E_a. This mechanism tends to increase the number density of conduction band electrons (sensitization) and acceptor level hole density, resulting in the enhancement of the MPA signal in the 700-800 nm region of Fig. 9. Similarly, enhancement in the A transition would result in an increase of the nonradiative de-excitation rate C, and thus spectral MPA sensitization would ensue in the bandgap region λ_g, in agreement with Fig. 9. Wada *et al.* (1985) further showed that the electronic energy manifold of Fig. 20 was consistent with mechanisms explaining their PL data from CdS:Cu and estimated the capture cross section σ_a at 580 nm to be *ca.* 7.2×10^{-17} cm^2 from Eq. (16) and from the experimental value of α, in reasonable agreement with the results obtained from photocapacitance spectroscopy (Grimmeiss, Kullendorff and Broser, 1981).

D. CdS POWDER AND THIN FILM MPAS.

Lyamov *et al.* (1979) showed that MPA spectra of CdS powders exhibited amplitudes *ca.* four times larger than those of CdS single crystals. Morimoto *et al.* (1984) obtained similar results and showed that this difference could be semi-quantitatively accounted for by light scattering effects according to a formula derived by Helander (1983), concerning the ratio Γ of the MPA signal from light scattering samples to that from non-scattering crystals:

$$\Gamma = (\alpha + \alpha_o) \cdot \left[\frac{3(1+\Delta)}{(1-3\alpha/\beta)[1+\Delta(3\alpha/\beta)^{\frac{1}{2}}\{\beta_t(3\alpha/\beta)^{\frac{1}{2}}+a_o\}]} + \frac{1}{\beta_t+a_o}\left[1 - \frac{3}{1-3\alpha/\beta}\right]\right] \quad (17)$$

$$\Delta = \frac{2}{3}(1+2R)$$

$$\beta = (\pi d^2/4)K v$$

where d is the diameter of the (assumed) spherical scatterers and v is the number of scatterers per unit volume. K was defined as the ratio of the scattering cross-section to the geometrical cross-section. It is known to have values between 0 and 5 (Helander, 1983). a_o is the thermal diffusion coefficient of the sample (Rosencwaig and Gersho, 1976). β_t is the effective absorption coefficient:

$$\beta_t = \alpha + \beta \quad (18)$$

where α is the absorption coefficient and β is the scattering coefficient. Eq. (17) predicts that the MPA signal intensity of CdS powder is about $[3(1+\Delta)-2]$ times larger than that of a single crystal for $\lambda > \lambda_g$. Taking the reflectivity $R \leq 0.2$, Γ

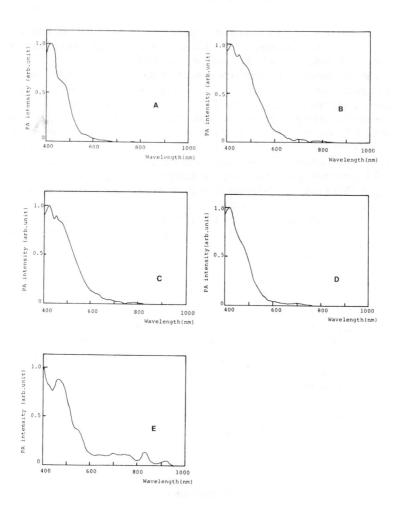

Fig. 21 MPA spectra of CdS powders. (a)-(e) correspond to sample preparation according to Table III [From Morimoto *et al.* (1984)]

gives a factor of *ca.* three in signal enhancement for powders, in good agreement with the experimental results. Table III and Fig. 21 show MPA spectra of various CdS powders. The spectra of samples No. 1 and 4 are similar, since the difference in E_g between α-CdS and β-CdS is only 0.03 eV, which is probably too small to be detected. Samples No. 2 and 3 exhibit large MPA signals in the

Table III[a]

Powdered samples used for MPAS measurement.

Sample No.	Form	Crystal structure	Remarks	Fig. No.
1		β	chemically deposited	4
2		almost α	sintered at 780°C	5
3	powder	α	polycrystalline remainder	6
4		α	single crystal	7
5		α	Cu-doped single crystal	8

[a] After Morimoto *et al.* (1984)

500-600 nm region, which, according to Morimoto *et al.* (1984) are due to electronic transitions from deep level Cd vacancies to the bottom of the conduction band. This conjecture was verified using X-ray analysis of the samples. The CdS:Cu powder No. 5 shows subbandgap characteristics similar to, but smaller in magnitude than, those of the single crystal, Fig. 8. These features were attributed to the same electronic mechanisms as those of the single crystal. Yoshida *et al.* (1985) further examined the MPA and PL responses of 10 nm CdS particles. They used the MPA signal delay produced by gaseous desorption, coupled with excitation from an Ar-ion laser, and the PL spectra to study the coupling of optical energy to the CdS matrix, and to probe localized electron states in the microparticles.

Granular thin CdS films with grain sizes on the order of the effective thickness of the films (several microns) were prepared by vacuum deposition on glass substrates, and the coefficient A of dissipative light losses in the form of thermal energy was determined using He-Ne laser light or Xe lamp excitation and MPA detection (Zarembo, Merkurova and Shikhlinskaya, 1985). These authors discussed the advantages of MPAS as a direct method of measuring A in comparison with spectrophotometry, including its ability to i) give higher accuracy in investigating very thin films; and ii) sensitively probe structural imperfections of the CdS films through the MPA signal enhancement from such films, compared to substrate and polished CdS single crystal signals.

2. TRANSDUCER-DETECTION PHOTOACOUSTICS (TPAS)

A. SPECTROSCOPY.

An apparent complication arising in the application of TPA techniques to CdS crystals lies in the seemingly great discrepancies between spectra obtained from "class A" experiments, by using these methods of photoacoustic detection

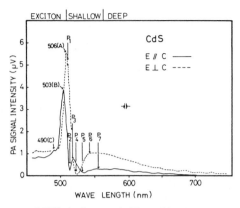

Fig. 22 TPA spectra of CdS single crystal. The solid and dashed lines show the spectrum observed with the light electric field vector $E//c$ and $E\perp c$, respectively. P_j denote the states observed in the PC measurements by Uchida (1967). [From Wasa *et al.* (1980)]

with the transducer attached to the back of a specimen which is illuminated from the front (Wasa *et al.*, 1980; Hata, Sato, Nagai and Hada, 1984; Hata, Sato and Kurebayashi, 1983); and spectra obtained from "class B" experiments using front-side illumination and MPA detection (Bandeira *et al.*, 1982; Mandelis and Siu, 1986). In class B are also included spectra obtained piezoelectrically with the sample illuminated from the back (Hata *et al.*, 1983; Hata *et al.* 1984). In Section III.1 the presented experimental evidence was shown to support the view that MPA spectra of CdS correspond primarily to the fundamental absorption characteristics at λ_g and superbandgap wavelengths, as well as to nonradiative energy release due to the electronic defect structure of this material at subbandgap wavelengths (Mandelis and Siu, 1986; Bandeira *et al.*, 1982). Hata *et al.* (1983) performed a series of TPA experiments with CdS single crystals to identify the origin(s) of the abovementioned discrepancies between class A and class B experiments. These workers concluded that class A experiments yield photoacoustic spectra related to the exciton levels and nonradiative electronic states in CdS, while their interpretation of class B experiments is essentially in agreement with the view described above. Fig. 22 shows the TPA spectrum of a single crystal of CdS with its c-axis parallel to the surface (Eagle-Pitcher Inc. Miami, Okla. Grade A), using back-side detection with an attached ZnO thin film transducer, and front-side illumination (Wasa *et al.* 1980; Mikoshiba, Wasa and Tsubouchi, 1980; Mikoshiba, 1982). Considerable physical insight toward the interpretation of the spectra of Fig. 22 under polarized light was obtained from comparisons with the CdS photoconductive response, which was extensively studied by Uchida (1967). That author reported PC spectra of several pure CdS crystals from Eagle-Pitcher, Inc., heat-treated in various partial pressures of sulphur vapor. A particular crystal was heat-treated in p_s=20 mm Hg for one hour at 850°C and exhibited a dramatic increase in resistivity, such that $\rho \geq 10^6$ Ohm-cm. The PC spectral response of that crystal under an applied voltage of 6V D.C. showed (Uchida, 1967: Fig. 2) a broad band with a maximum located at 532 nm. The position of that peak, identified by Uchida as P_5, was

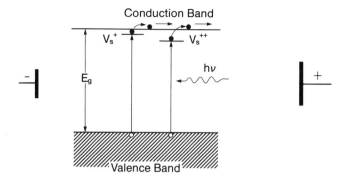

Fig. 23 Energy diagram of the proposed (Uchida, 1967) transition mechanism for photo-excited electrons to native defect sulphur vacancy donor states: V_s^{++}: doubly ionized defect responsible for spectral peak P_5 (532 nm); V_s^+: singly ionized defect responsible for peak P_3 (517 nm) observed in some sulphur vapor treated CdS crystals. [From Mandelis and Siu (1986)]

associated with the photo-excitation of a valence band electron to a doubly ionized donor defect state identified as a sulphur vacancy V_s^{++}, followed by thermal release from this level to the conduction band as shown in Fig. 23. Wasa *et al.* (1980) showed the luminescent nature of the spectral peak P_5 through its absence (a "trough") from the CdS TPA spectrum at 534 nm and after comparison with the PL spectrum of that material, in which P_5 appeared as a peak at 531 nm. These authors used the same comparison between PL and TPA spectra to establish the nonradiative nature of Uchida's PC peak P_3 (517 nm). The TPA spectral position of P_3 was thus found to be 518 nm. The nonradiative nature of P_3 was later confirmed photoacoustically by Hata *et al.* (1984), who placed its

Table IV[a]

The values of wavelength corresponding to the light absorption due to seven states in CdS measured by various methods. (units:nm)

	Photoconductivity[b]	Photoluminescence[c]	PAS	Nature
P_1	507.5	...	509	nonradiative
P_2	513.5	514	513	radiative
P_3	517	...	518	nonradiative
P_4	522.5	524	523	radiative
P_5	532	531	534	radiative
P_6	542	...	544	nonradiative
P_7	555	...	554	radiative

[a] After Wasa *et al.* (1980)

[b] After Uchida (1967)

[c] Obtained by the same sample as that used in PAS measurement

spectral position at 519 nm. Table IV shows the assignations of peaks P_1-P_7 by Wasa *et al.* (1980) arrived at by comparisons among TPA, PC and PL methods. Uchida (1967) also reported that excitonic peaks appear at room temperature in the PC spectrum of CdS crystals with very high sulphur content, with the lowest energy exciton A located at 501.5 nm. Normally, excitons are not expected to exist in pure CdS crystals at room temperature (Park and Reynolds , 1963). In Fig. 22, Wasa *et al.* confirmed Uchida's report, as they observed the A excitonic peak disappear in the E//c field configuration, in agreement with the quantum mechanical selection rule governing the interband transition from the A-valence band to the conduction band (Hopfield, 1960). Thus, they placed exciton A at 506 nm. Table V shows the wavelength assignations of excitons A, B and C by Wasa *et al.* (1980).

Table V[a]

The values of wavelength corresponding to A, B, C-exciton absorption in CdS measured by various methods. (units:nm).

	Light absorption[b]	Electroreflectance[c]	Photoconductivity[d]	TPAS
A-exciton	501.5	505.7	501.5	506
B-exciton	498.5	502.8	498	503
C-exciton	486.1	491.0	485.5	490

[a] After Wasa *et al.* (1980)

[b] Gutsche E. and Voigt I. (1967). "II-VI Semiconducting Compounds" (D.G. Thomas, ed.) p. 337. Benjamin, New York

[c] Cardona M., Shaklee K.L. and Pollak F.H. (1967). Phys. Rev. **154**, 696

[d] Uchida I. (1967)

Hata *et al.* (1983) performed similar experiments to that by Wasa *et al.* (1980) using Grade A CdS crystals from Eagle-Pitcher, Inc. with the c-axis parallel to the surface plane. The TPA detector was a transparent $LiNbO_3$ wafer with aluminum electrodes. The crystals were irradiated from either the front (A) surface, or the back (B) surface through the transducer material, Fig. 24a. Fig. 24b shows the obtained CdS spectrum for irradiation in the geometry (A): Two peaks were identified, $P_I = 517$ nm and $P_{II} = 542$ nm, with a valley V_I between them at 524 nm. These authors identified their peak P_I with the P_3 peak of Wasa *et al.* (1980) (Table IV); V_I was identified with P_4 and P_{II} with P_6. The wavelength correspondence between these assignations is in agreement to within 1 nm. No identification of the excitonic peaks A, B and C could be made in the data of Fig. 24b, due to the sharp emission lines of the Xe-lamp spectrum superposed on the CdS spectrum at *ca.* 500 nm, which severely degraded resolution at the wavelength region. Furthermore, no comment was made as to the

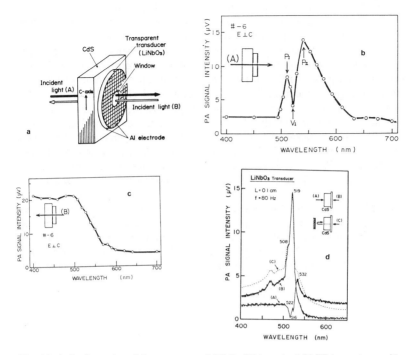

Fig. 24 a) Configuration of the transparent LiNbO₃ TPA method; b) TPA spectrum with incident beam perpendicular to the CdS surface: E⊥c, geometry (A), f=10 Hz; c) TPA spectrum with incident beam through the transparent window of the transducer: E⊥c, geometry (B), f=10 Hz. [From Hata *et al.* (1983)]; d) TPA spectra for CdS crystal with incident beam direction along the c-axis: Geometries (A) and (B) as above; (C) mirror reflection method [From Hata *et al.* (1984)]

absence of peaks P_1 and P_7 from the data of Fig. 24b. A similar spectral structure from CdS single crystals in the shape of platelets of thickness 100 μ*m*, grown by sublimation in a closed quartz vessel, was observed independently by Goede, Heimbrodt and Sittel (1986), also in the geometry (A). These workers obtained TPA signals from pure CdS using an aluminized PZT transducer and reported two peaks at *ca.* 537 nm and 508 nm with a valley in between at *ca.* 525 nm. Hata *et al.* (1983) further obtained spectra after irradiation in the geometry (B), shown in Fig. 24c. Their results are consistent with fundamental absorption spectra, similar to those obtained using MPAS. It is interesting to note that the TPA measurements in the geometry (A) have *not* proven to be equivalent to back-surface detected MPA spectra (Takaue *et al.*, 1984), in that the latter do not carry nonradiative peak information. Hata *et al.* (1984) have shown both theoretically (using the RG model) and experimentally, that in the TPA method peaks due to nonradiative de-excitation processes are superposed on the ordinary absorption spectra of CdS in the geometry (B) and render their interpretation complicated. These peaks (e.g. a sharp feature at 519 nm) cannot be predicted by the RG approach, however, their presence or absence from spectra in the geometry (B) must depend largely on the electronic defect history of the material, as both situations have been reported by these workers (compare Figs. 24c and 24d, curve B). Data obtained in the geometry (A) tended to support the RG model: spectra thus recorded exhibited a narrow valley at 516 nm and a broad peak at 532 nm, Fig. 24d. Both features were also present in a theoretical curve from the RG calculation. Hata *et al.* (1984) demonstrated that the 532 nm peak position depends on the sample thickness and, therefore, it cannot be due to any nonradiative de-excitation process, such as the peaks in Table IV. This spectral feature was, therefore, associated with purely optical and thermal processes in the CdS bulk. The preliminary TPA study in the geometry (A) by Breuer (1980), Fig. 3, supports the absorption-like spectral interpretation. On the other hand, TPA spectra of three CdS thin single crystalline platelets of

Fig. 25 A) Normalized TPA spectra of CdS:Se crystals with the same thickness $d \approx 100\mu m$. Se content (mol. %): 0 (#1), 10 (#2), 25 (#3) and 55 (#4). B) Normalized TPA spectra of CdS:Te crystals with $d \approx 1mm$. Te content (mol. %): 0 (#1), 6.0×10^{-3} (#2), 1.8×10^{-1} (#3), 1.2 (#4), 4.0 (#5) and 20 (#6). LA (HA) ≡ Low (High) Absorption region. In (B) the LA and HA regions are separate; f = 12.3 Hz [From Goede *et al.* (1986)]

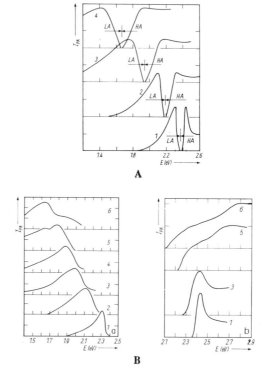

unknown characteristics, obtained also in the geometry (A), have shown sub-bandgap features different from each other which have been attributed to different values of the bulk-to-surface recombination rate ratio for these platelets (Konstantinov, Hinkov, Burov and Borissov, 1979). It thus definitely appears that the lack of a general comprehensive theoretical basis on which one can interpret TPA measurements of CdS in both geometries (A) and (B), is a major obstacle to acquiring a unified qualitative appreciation of this method of photoacoustic experimentation. Specialized TPA theories, in addition to the work by Hata *et al.* (1984), with direct applications to, and fair agreement with, CdS experimental results have been advanced by Konstantinov and Khinkov (1980) and Mikoshiba (1982).

B. IMPURITY CENTER STUDIES IN CdS

Goede *et al.* (1980) performed detailed studies of the systems CdS:Se and CdS:Te. Spectra obtained from CdS platelets, as described in section (A) above, with 0% impurities were similar in structure to those reported by Hata *et al.* (1983). Impurity spectra are shown in Fig. 25. The peaks of the high absorption region (HA) spectra were attributed to excitonic transitions corresponding to free exciton energies in the case of the isovalent impurity Se. The red shift of these peaks with increasing Se concentration, Fig. 25A, was consistent with similar conclusions arrived at through purely optical measurements: by observing the dependence of the TPA peak intensity on the polarization of the exciting radiation, Goede *et al.* (1986) were able to resolve A and B type excitons, as suggested by Wasa *et al.* (1980). The low absorption region (LA) spectrum of Fig. 25A exhibits absorption maxima with sharp decreases to zero signal intensity at the high photon energy side. Similar results were obtained qualitatively from the photoluminescent CdS:Cu system's TPA response by Morimoto *et al.* (1982) and the spectral maxima were tentatively assigned to the wavelength dependence of the nonradiative quantum efficiency, $\eta_{NR} (\lambda)$, due to the presence of large concentrations of native defects. The red shift and spectral broadening of the Te-doped CdS crystals (LA spectra), Fig. 25Ba, was found to be consistent with the spectral dependences of the overlapping absorption bands of various Te_m clusters having exciton binding energies which increased with increased Te concentrations, i.e. increased m (Goede, Heimbrodt and Müller, 1981). These authors pointed out that their Te_m cluster absorption model further predicted successfully that the exciton peak of the HA spectra in Fig. 25Bb would only be broadened with increased m, but its spectral position would remain essentially fixed. No explanation, however, for the origin of the short wavelength shoulder in curves #5 and 6, Fig. 25Bb, could be given. The most apparent quantitative value of the TPA impurity spectra in Fig. 25 is their potential to be used for quantitative estimates of impurity concentrations upon suitable calibration, especially for impurity densities too low to be detected by conventional absorption measurements. Morimoto *et al.* (1982) showed that such a calibration may be possible for the CdS:Cu system by using SIMS profiles alongside TPA results, provided that excessive background signals are not a problem with SIMS data. These authors argued that the spectroscopic differences between their TPA spectra of pure CdS and CdS:Cu (cf. Fig. 26) and

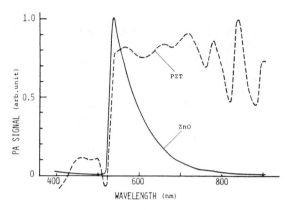

Fig. 26 Quadrature (A cos θ) TPA spectra of undoped CdS detected by a ZnO (solid curve) and a PZT (dashed curve) piezoelectric transducer. Unpolarized light; f = 40 Hz [From Morimoto *et al.* (1982)]

MPA spectra of the same materials (cf. Figs. 7,8) are partially due to differences in the relative weights of the acoustical and thermal wave branching ratios in each transducer. The example of Fig. 26, showing TPA spectral dependence on the nature and geometry of the piezoelectric transducer used, further clarifies this argument and certainly points to MPA measurements as the preferred experimental detection scheme, in the absence of comprehensive quantitative TPA detection models. Miyakawa (1982) has shown that the thermal wave branching ratio (TPA detection) is proportional to $\eta_{NR}(\lambda)/\alpha(\lambda)$ in the superbandgap region ($\lambda<\lambda_g$), while it depends on $\eta_{NR}(\lambda)\alpha(\lambda)$ in the subbandgap region ($\lambda>\lambda_g$). That same ratio in the MPA scheme is proportional to $\eta_{NR}(\lambda)\alpha(\lambda)$ for $\lambda>\lambda_g$, and it depends only on $\eta_{NR}(\lambda)$ for $\lambda<\lambda_g$ (photoacoustic saturation). These MPA dependences are attractive in that they render the task of deconvoluting $\eta_{NR}(\lambda)$ from $\alpha(\lambda)$ easier than the respective TPA dependences, especially if $\alpha(\lambda)$ is known from other techniques.

C. MULTIPHOTON ABSORPTION (mpa) IN CdS.

Interest in the photoacoustic detection of multiphoton absorption (mpa) processes in crystalline CdS has been recently stimulated by the importance of nonlinear absorption as a limiting factor in the ability of this semiconductor to remain transparent in optical window applications. Furthermore, fundamental aspects of the solid state physics of CdS can be explored via mpa detection (Nathan, Guenther and Mitra, 1985), such as new information on band structure constants. Laser induced damage mechanisms have been the subject of considerable controversy, with mpa being a prime candidate for the initiation of electronic avalanches (Bloembergen, 1974). Thus, the magnitudes of second, third and higher order nonlinear absorption processes in CdS are of importance and a few such processes have been investigated photoacoustically. Kawai, Morimoto, Yoshioka and Miyakawa (1982) used a PZT ceramic transducer with a resonance frequency of 9.6 MHz, and low resistivity (0.4-4 Ohm-cm) CdS single crystals. A Q-switched ruby laser generated pulses of 80 MW peak power

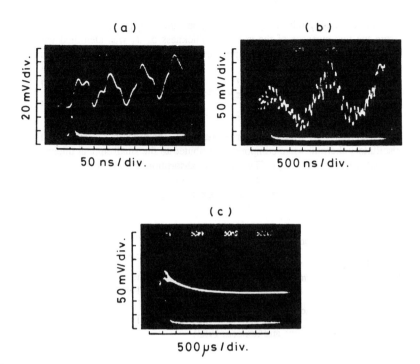

Fig. 27 Pulsed TPA waveforms for $E//c$ incident ruby laser radiation: (a) 50 ns/div. A_3 component; (b) 500 ns/div. A_2,A_3 components; (c) 500 μs/div. A_1,A_2 components [From Kawai *et al.* (1982)]

with 15 ns duration. Waveforms of the CdS time-domain TPA response for $E//c$ polarization in the geometry (A) (Fig. 24) are shown in Fig. 27. Three time scales were identified with distinct components: oscillatory A_3 with time constant $T_3 \approx 100ns$, Fig. 27a; A_2 with $T_2 = 2\mu s$, Fig. 27b; and a non-oscillatory time decay A_1 with a constant $T_1 \approx 1ms$, Fig. 27c. A theoretical heat conduction/surface displacement model was developed and used to identify the three components as i) an oscillatory component due to the thermally induced forced oscillation of the crystal with period

$$t_1 = \frac{2\pi}{\alpha_1 c} \tag{18}$$

where c is the longitudinal sound velocity in CdS and α_1 is the root of the equation

$$\tan(\alpha_1 a) = \alpha_1 k(h_1 + h_2)/(\alpha_1^2 k^2 - h_1 h_2); \tag{19}$$

k is the thermal conductivity, h_1 and h_2 are the thermal emission coefficients from the two surfaces of the crystal of thickness α. t_1 was identified with the process A_3; ii) an oscillatory component due to the free oscillation of the sample with period

$$\tau_1 = \frac{4a}{c} \approx 2.5\mu s \tag{20}$$

which agrees well with the observed period T_2 of process A_2; and iii) a non-oscillatory thermal wave component, attributed to the small optical thickness of the sample (process A_1). Two-photon absorption effects were invoked to explain the observed nonlinear signal amplitude dependence of the input optical pulse intensity. For two-photon absorption at a depth z in CdS:

$$\frac{dI(z)}{dz} = -\alpha_o I(z) + \alpha_2 I^2(z) - \sigma_{ex} N_{ex} I(z) \tag{21}$$

where α_o is the linear absorption coefficient, α_2 is the two-photon absorption coefficient, σ_{ex} is the mean linear absorption cross-section for two-photon excited electrons and holes, and N_{ex} is the number of two-photon excited carriers. Neglecting the excited state absorption ($\sigma_{ex}=0$), Eq. (21) gives (Van Stryland, Woodall, Williams and Soileau, 1983)

$$I(a,r,t) = \frac{I(O,r,t)(1-R)^2 e^{-\alpha_o a}}{[1+(\alpha_2/\alpha_o)I(O,r,t)(1-R)(1-e^{-\alpha_o a})]} \tag{22}$$

Fig. 28 Calculated intensity dependence of thermal and acoustic components in CdS with linear absorption coefficient α_o as a parameter, $\alpha_2 = 5\times10^{-8}$ cm/W and $(h_1 a/k) = 0.5$. Solid (dashed) lines denote acoustic (thermal) components [From Kawai *et al.* (1982)]

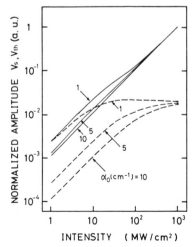

where R is the surface reflectivity and $I(O,r,t)$ is the irradiance of a gaussian laser beam of radius r, incident on the front surface of the sample. A similar calculation was performed by Kawai *et al.* (1982) and resulted in Fig. 28. This figure was found to be in excellent qualitative agreement with experimental results and yielded values α_o of several cm^{-1} at 694 nm upon assuming $\alpha_2 = 0.5 \times 10^{-7}$ cm/W (Lotem and de Araujo, 1977). In a careful combined transmission/TPA absorption experiment using single pulses from a mode-locked, Gaussian mode Nd:YAG laser at 1.06 μm and 30-200 ps (FWHM) duration, Van Stryland *et al.* (1983) found that the distinct advantage of TPAS over transmission was the higher sensitivity of the former, which renders it extremely useful for the determination of small linear absorptivities, down to 10^{-6} cm^{-1}.

For observation of nonlinear absorptivities, however, the total linear absorption appears as a background which limits the TPA applicability to low loss samples with nonlinear absorption cross sections comparable to the linear absorption. These authors showed that the energy absorbed, E_{abs}, by a sample exhibiting multiphoton absorption is directly proportional to the TPA signal and given by

$$\frac{E_{abs}}{E_{inc}} = (1-R)L\left[\alpha_o + \frac{\alpha_n I_o^{n-1}}{n\sqrt{n}}(1-R)^{n-1}\right] \qquad (23)$$

where I_o is the incident irradiance and E_{inc} is the incident energy on the sample. Eq. (23) shows the increased sensitivity of the TPA technique in that the term proportional to α_n is comparable to α_o, which can be very small, rather than being compared to unity as is the case with the transmission technique. The TPA experiments on CdS yielded information related to three-photon absorption parameters, namely $\alpha_3 = 0.06$ cm^3/GW^2 and $\sigma_{ex} = 1.5 \times 10^{-17}$ cm^2. This information was similar to that obtained through transmission, however the use of TPA in conjunction with lower irradiance repetition rate short laser pulses was claimed to offer the best possibilities for non-destructive direct mpa studies in CdS and other semiconductors susceptible to avalanche breakdown.

IV. Concluding Remarks

Even though photoacoustic spectroscopy has been applied to CdS for over ten years, it is within the last five years that its contributions to the advancement of our understanding of the physics, in general, and the electronic behavior, in particular, of this semiconductor have started to unravel.

As shown in this Chapter, several widely different types of photoacoustic spectra can be observed with CdS, depending on the detection mode: Conventional microphone gas-coupled detection (MPAS) and piezoelectric transducer detection (TPAS) with optical excitation from the back surface generate spectra similar to the optical absorption spectrum of CdS at superbandgap energies, as well as information on exciton structures; TPAS with front optical excitation has been shown to generate sharp spectral features due to nonradiative and impurity center transitions at photon energies above and below E_g; MPAS at subbandgap wavelengths can yield information on the spectral dependence of the nonradiative quantum efficiency and, when coupled with photocurrent spectroscopy, can be used as a sensitive tool to study electronic transport properties and nonradiative scattering mechanisms in the CdS lattice which give rise to the Joule effect; MPAS with back surface detection gives a spectral maximum which is sensitive to thermal parameters of CdS. Pulsed laser TPA studies of CdS are still at their infancy, however, they promise to deliver a wealth of information on nonlinear multiphoton absorption processes and on the dynamics of the de-excitation manifold of this semiconductor through *direct* monitoring of the extremely important, and frequently rate limiting, nonradiative channel.

A few seemingly conflicting TPA spectra, which appeared in the literature along with conflicting interpretations from different laboratories, are indicative of the need to develop a general TPA theory of defect rich semiconductors incorporating such features as deep defect centers, exciton levels and realistic band structures. Such a theoretical approach has been developed for MPA detection, which at this time appears to be the optimal experimental method from the point of view of interpretability.

Finally, the widely different, and at times conflicting, photoacoustic results which have been obtained from CdS crystals with different growth and fabrication histories point to the importance of detailed identification of sample history with every published report. Unfortunately, many otherwise very worthwhile studies have emerged over the past ten years with complete or partial lack of such crucial information. This omission, in turn, diminishes to a large extent the value of the overall contribution and renders the task of those who will endeavor to construct a coherent photoacoustic picture of the essential physics of CdS difficult at best. Nevertheless, given the large gains in our knowledge of physical processes in CdS as a result of the photoacoustic probes to-date, the future presence of the photoacoustic technique in the field of CdS and other compound semiconductors appears well established as a unique tool for the advancement of basic material physics and in important technological applications such as device substrate characterization and processing control.

V. References

Amer, N.M. and Jackson, W.B. (1984). *"Optical Properties of Defect States in a-Si:H in Semiconductors and Semimetals"* (J.I. Pankove, ed) Vol. 21 B, 83. Academic Press, Orlando, Fla.

Bandeira, I.N., Closs, H. and Ghizoni, C.C. (1982). J. Photoacoust. 1, 275.

Basov, N.G., Bogdankevich, O.V. and Devyatkov, A.G. (1965). Sov. Phys.-JETP 20, 1067.

Bloembergen, N. (1974). IEEE J. Quant. Electron. QE-10, 375.

Bonch-Bruevich, V.L. and Landsberg, E.G. (1968). Phys. Stat. Sol. **29**, 9.

Boone, J.L. and Cantwell, G. (1985). J. Appl. Phys. **57**, 1171.

Breuer, H.D. (1980). Naturwissenschaften **67**, 91.

Bube, R.H. (1955a). J. Chem. Phys. **23**, 18.

Bube, R.H. (1955b). Phys. Rev. **98**, 431.

Bube, R.H. (1955c). Phys., Rev. **99**, 1105.

Bube, R.H. and Barton, L.A. (1959). RCA Rev. **111**, 564.

Bube, R.H. (1960). *Photoconductivity of Solids* Wiley, New York, p. 391.

Cahen, D. (1978). Appl. Phys. Lett. **33**, 810.

Cardona, M. and Harbeke, G. (1965). Phys. Rev. **A137**, 1467.

Cesar, C.L., Vargas, H., Mendes Filha, J., and Miranda, L.C.M. (1983). Appl. Phys. Lett. **43**, 556.

Conway, E.J. (1970). J. Appl. Phys. **41**, 1689.

Davies, J.J., Cox, R.T. and Nicholls, J.E. (1985). *Proc. 13th Intl. Conf. Defects in Semiconductors* (L.C. Kimerling and J.M. Parsey Jr., eds) p. 1237. Mettalurgical Soc. AIME Publ., Warrendale, Penn.

Dioszeghy, T. and Mandelis, A. (1986). J. Phys. Chem. Solids, **47**, 1115.

Dutton, D. (1958). Phys. Rev. **112**, 785.

Flaisher, H. and Cahen, D. (1986). IEEE Trans. UFFC **UFFC-33**, 622.

Georgobiani, A.N. (1974). Sov. Phys. - Usp. **17**, 424.

Gobrecht, H. (1953). Z. Physik **136**, 224.

Goede, O., Heimbrodt, W. and Müller, R. (1981). Phys. Stat. Sol. (b)**105**, 543.

Goede, O., Heimbrodt, W. and Sittel, F. (1986). Phys. Stat. Sol. (a)**93**, 227.

Grimmeis, H.G., Kullendorff, N. and Broser, R. (1981). J. Appl. Phys. **52**, 3405.

Gross, E.F. and Novikov, B.V. (1959). Fiz. Tverdogo Tela **1**, 357.

Hata, T., Sato, Y. and Kurebayashi, M. (1983). Jpn. J. Appl. Phys. **22**, Supplement 22-3, 205.

Hata, T., Sato, Y., Nagai, Y. and Hada, T. (1984). Jpn. J. Appl. Phys. **23**, Supplement 23-1, 75.

Hata, T., Hatsuda, T., Kawakami, M. and Sato, Y. (1985). Jpn. J. Appl. Phys. **24**, Supplement 24-1, 204.

Helander Per, O.F. (1983). Ph.D. Thesis, Linköping Univ., Sweden, ch. 4.

Hopfield, J.J. (1960). J. Phys. Chem. Solids **15**, 97.

Kawai, M., Morimoto, J., Yoshioka, T. and Miyakawa, T. (1982). Jpn. J. Appl. Phys. **21**, Supplement 21-3, 104.

Klick, C.C. (1953). Phys. Rev. **89**, 274.

Konstantinov, L.L., Hinkov, V.P., Burov, J.I. and Borrisov, M.I. (1979). *Physics of Semiconductors 1978* (B.L.H. Wilson, ed.) p. 219. Inst. Physics Publ.,

London, England.

Konstantinov, L.L. and Khinkov, V.L. (1980). Annu. Univ. Sofia Fac. Phys. **72**, 51.

Koral, E.N. (1977). Sov. Phys. - Engl. Transl. **19**, 1327.

Lambe, J. (1955). Phys. Rev. **98**, 985.

Lotem, H. and de Araujo, C.B. (1977). Phys. Rev. **B16**, 1711.

Lyamov, V.E., Madvaliev, U. and Shikhlinskaya, R.E. (1979). Sov. Phys. Acoust. - Engl. Transl. **25**, 241.

Mandelis, A., Teng, Y.C., and Royce B.S.H. (1979). J. Appl. Phys. **50**, 7138.

Mandelis, A. and Royce, B.S.H. (1980). J. Appl. Phys. **51**, 610.

Mandelis, A. and Siu, E.K.M. (1986). Phys. Rev. **B34**, 7209.

McKelvey, J.P. (1966). *"Solid State and Semiconductor Physics"* Harper & Row, New York, sec. 13.6.

Mikoshiba, N., Wasa, K. and Tsubouchi, K. (1980). *Proc. IEEE Ultrasonics Symp.* (B.R. McAvoy, ed.) p. 658 IEEE Publ., New York.

Mikoshiba, N. (1982). *"Investigation of Nonradiative Processes and Defects in Semiconductors by Photoacoustic and Current-Injection-Induced Acoustic Spectroscopy"* in *Semiconductor Technologies* (J. Nishizawa, ed.) Vol. 1982, 13. North-Holland, Amsterdam, Netherlands.

Miyakawa, T. (1982). Mem. Natl. Def. Acad. Jpn. **22**, 83.

Morimoto, J., Ito, T., Yoshioka, T. and Miyakawa, T. (1982). J. Cryst. Growth **57**, 362.

Morimoto, J., Ito, T and Miyakawa, T. (1982). Jpn. J. Appl. Phys. **21**, Supplement 21-3, 101.

Morimoto, J., Wada, H., Watanabe, A. and Miyakawa, T. (1984). Mem. Natl. Def. Acad. Jpn. **24**, 211.

Morimoto, K. and Kitagawa, M. (1985). J. Phys. Soc. Jpn **54**, 4271.

Nakayama, N., Matsumoto, H., Nakano, A., Ikegami, S., Uda, H. and Yamashita, T. (1980). Jpn. J. Appl. Phys. **19**, 703.

Nathan, V., Guenther, A.H. and Mitra, S.S. (1985). J. Opt. Soc. Am. **B2**, 294.

Nesheva, D. and Vateva, E. (1981). C.R. Acad. Bulg. Sci. **34**, 767.

Nitsche, R. (1960). J. Phys. Chem. Solids **17**, 163.

Park, Y.S. and Reynolds, D.C. (1963). Phys. Rev. **132**, 2450.

Quimby, R.S. and Yen, W.M. (1982). J. Appl. Phys. **51**, 4985.

Reynolds, D.C. and Czyzak, S.J. (1954). Phys. Rev. **96**, 1705.

Rochon, P. and Racey, T.J. (1984). J. Photoacoust. **1**, 475.

Roger, J.P., Fournier, D., Boccara, A.C., Noufi, R., and Cahen, D. (1985). Thin Solid Films **128**, 11.

Rose, A. (1955). Phys. Rev. **97**, 322.

Rosencwaig, A. (1975). Anal. Chem. **47**, 592A.

Rosencwaig, A. and Gersho, A. (1976). J. Appl. Phys. **47**, 64.

Rosencwaig, A. (1980). *"Photoacoustics and Photoacoustics Spectroscopy"* in *Chemical Analysis* (P.J. Elving and J.D. Winefordner, eds). Vol. **57**. J. Wiley, New York, N.Y.

Siu, E.K.M. and Mandelis, A. (1986). Phys. Rev. **B34**, 7222.

Smith, R.A. (1959). *"Semiconductors"* Cambridge Univ. Press, Cambridge, Grea Britain, ch. 3.

Song, J.J. and Wang, W.C. (1984). J. Appl. Phys. **55**, 660.

Stoneham, A.M. (1975). *"Theory of Defects in Solids"* Clarendon Press, Oxford, ch. 12.

Takaue, R., Matsunaga, M. and Hosokawa, K. 1984). J. Appl. Phys. **56**, 1543.

Thielemann, W. and Rheinländer, B. (1985). Solid-State Electron. **28**, 1111.

Thomas, D.G. and Hopfield, J.J. (1959). Phys. Rev. **116**, 573.

Thomas, D.G., Hopfield, J.J. and Power, M. (1960). Phys. Rev. **119**, 570.

Uchida, I. (1967). J. Phys. Soc. Jpn. **22**, 770.

van Driel, H.M. (1984). *"Physics of Pulsed Laser Processing of Semiconductors"* in *"Semiconductors Probed by Ultrafast Laser Spectroscopy* (R.R. Alfano, ed.) Vol. **II**, 61. Academic Press, Orlando, Fla.

Van Stryland, E.W., Woodall, M.A., Williams, W.E. and Soileau, M.J. (1983). *Proc. Symp. Laser Induced Damage in Optical Materials: 1981* (H.E. Bennett, A.H. Guenther, D. Milam and B.E. Newnam, eds.) p. 589 NBS Publ. (NBS-SP-638) Washington, DC.

Wada, H., Yoshioka, H., Morimoto, J. and Miyakawa, T. (1985). Jpn. J. Appl. Phys. **24**, Supplement 24-1, 217.

Wasa, K., Tsubouchi, K. and Mikoshiba, N. (1980). Jpn. J. Appl. Phys. **19**, L475.

Woodbury, H.H. and Aven, M. (1974). Phys. Rev. B., **9**, 5195.

Wünstel, K. and Klingshirn, C. (1980). Opt. Commun. **32**, 269.

Yang, H.G. and Im H.B. (1986). J. Electrochem. Soc. **133**, 479.

Yoshida, T., Ogawa, T. and Arai, T. (1985). J. Vac. Soc. Jpn. **28**, 563.

Yoshizawa, M. (1976). Jpn. J. Appl. Phys. **15**, 2143.

Zarembo, L.K., Merkurova, S.P. and Shikhlinskaya, R.E. (1985). Opt. Spektrosc. (USSR) - Engl. Transl. **58**, 505.

16

PHOTOACOUSTIC SPECTRA OF SEMICONDUCTORS I: ROOM TEMPERATURE SPECTRA

Tetsuo Ikari, Shigeru Shigetomi and Yutaka Koga

Department of Physics
Kurume University
1635 Mii, Kurume, Fukuoka, JAPAN

I. Introduction

Recent developments in the photoacoustic spectroscopy (PAS) of semiconductors are the result of not purely physical interest but also of industrial requirements. The photoacoustic effect was first discovered by A.G. Bell and others in the 1880's and generated considerable interest at that time. Sunlight was focused onto a sample contained in a cell that was connected to a listening tube. When the sunlight was repeatedly blocked and unblocked, sound could be heard through the listening tube at the sunlight chopping frequency. However, the phenomenon was regarded as a curiosity of no great functional or scientific value and was quickly forgotten.

The PA spectroscopy of solid materials has been revived in the seventies and has been used to investigate physical and chemical phenomena in a number of fields including biology and medicine (Rosencwaig, 1977). This may be due to rapid developments in the experimental instrumentation, such as laser sources or highly sensitive detectors of sound or elastic waves.

The basic concept of PA detection is simple: Light absorbed by a sample will excite a fraction of particles in the ground state into higher energy states. These excited states will subsequently relax through a combination of radiative and nonradiative pathways. The nonradiative component will ultimately generate heat in the localized region of the excitation light beam and generate a pressure or an elastic wave which subsequently propagates away from the source. The pressure or the elastic wave is then detected by a suitable sensor such as a microphone for gaseous samples. In the present case for semiconductors, the discussion is mainly focused on the relaxation of optically excited electrons from occupied states. One of the principal advantages of PA spectroscopy over other conventional optical measurements is that it enables one to obtain spectra similar to the optical absorption spectra on any type of solid or semisolid material, be it crystalline, powder, amorphous, gel, etc. And the photoacoustic detection is unique in that it is a direct monitor of the nonradiative relaxation channel and, hence, complements the absorption and fluorescence spectroscopic techniques.

In section II, theoretical models of the PA spectroscopy are reviewed for both the microphonic and piezoelectric detection techniques. Since full expressions are somewhat difficult to interpret because of their complicated nature, important special cases are discussed to gain a physical insight. The explanation of recently accumulated PA spectra is mainly based on the theory given there.

Experimental photoacoustic arrangements for the microphonic and piezoelectric detection techniques are discussed in section III.

In section IV, we show typical PA spectra at room temperature for elementary and compound semiconductors including heterostructure devices which are interesting due to recent industrial applications. We also show how the observed experimental results may be interpreted by theoretical predictions discussed in section II, or their modifications. Finally, the usefulness of PA spectroscopy in investigating band structures or impurity states of semiconductors will be demonstrated.

II. Theoretical Models

1. MICROPHONIC DETECTION TECHNIQUE

The basic concepts of the theory of microphone gas-coupled PA spectroscopy are given by Rosencwaig and Gersho (1976), hereafter referred to as the RG theory, and by Rosencwaig (1977). In the following, we briefly review the RG theory which is crucial to understanding the experimental results of PA spectra. Consider a simple cylindrical cell as shown in Fig. 1. The total length of the cell is assumed to be small compared to the wavelength of the acoustic signal, which implies that the microphone, not shown in the figure, will detect the spatially averaged pressure produced in the gas cell. The sample is considered to be in the form of a disk having a diameter D and a thickness L. The sample is mounted so that its front surface is exposed to the gas (air) within the cell and its back surface is supported by a poor thermal conductor. A further assumption is made that the gas and backing material are not light absorbing. The following parameters are defined here: k_i, the thermal conductivity; ρ_i, the density; C_i, the specific heat; β_i $(=k_i/\rho_i C_i)$, the thermal diffusivity, $a_i = (\omega/2\beta_i)^{1/2}$, the thermal diffusion coefficient; and $\mu_i = 1/a_i$, the thermal diffusion length of material (i). The subscripts i = s, g and b of these parameters which will appear in the following stand for solid (sample), gas and backing material, respectively, and ω is the angular modulation frequency of the incident light.

When a sinusoidally modulated monochromatic light of wavelength λ is incident on the solid sample surface, the temperature in the solid sample as a function of position and time is calculated by solving the thermal diffusion equation. The temperature and flux continuity conditions at the sample surface (Z=0 and -L) were used for boundary conditions. The complex amplitude of the periodic temperature at the solid-gas interface (Z=0) can thus be calculated explicitly. Subsequently, the periodic temperature variation in the gas cell can be calculated. Rosencwaig and Gersho have hypothesized that the primary

Fig. 1 Cross-sectional view of a simple cylindrical PA cell. The positions of the solid sample, backing material and gas column are shown. [From Rosencwaig (1977)]

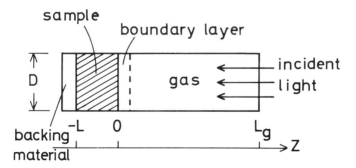

source of the acoustic signal in the PA cell arises from the periodic heat transfer from the solid sample surface to the surrounding gas as the solid is cyclically heated by the chopped light. Furthermore they have assumed that the thermal expansion and contraction, as well as any thermally induced mechanical vibration, of the solid are generally too small in magnitude to affect the observed acoustic signal. At a distance of $2\,\pi\,\mu_g$ from the sample surface, the periodic temperature variation in the gas is effectively fully damped out. Thus, a thermal boundary layer can be defined from Z=0 to $2\,\pi\,\mu_g$ and the spatially averaged temperature of the gas within the boundary layer can be determined. Because of its periodic heating and the long acoustic wavelength, the boundary layer of the gas expands and contracts periodically and can thus be thought of as an acoustic piston acting on the rest of the gas column, producing an acoustic pressure signal that travels instantaneously through the entire gas column. This acoustic pressure signal is then detected by the microphone as the PA signal.

Although the full expression for the PA signal is complicated and somewhat difficult to interpret, physical insight may be gained by examining a few special cases. These cases were grouped according to optical opacity of the solid, as determined by the relation of the optical absorption length $\mu_\alpha = 1/\alpha$ to the thickness L of the solid sample, where α is the optical absorption coefficient of the sample.

Six cases of experimental importance discussed in the RG theory are as follows:

Case I: Optically transparent solids $(\mu_\alpha > L)$. In these cases, the light is absorbed throughout the length of the sample.

Case Ia: Thermally thin solids $(\mu_s \gg L;\, \mu_s > \mu_\alpha)$ Here, the PA signal Q is

$$Q \sim [(1-j)\,\alpha L/2a_g](\mu_b/k_b). \tag{1}$$

where, $j^2 = -1$. The acoustic signal is thus a linear function of the absorptance (αL) and is proportional to ω^{-1}. The thermal properties of the backing material also come into play in the expression.

Case Ib: Thermally thin solids $(\mu_s > L;\, \mu_s < \mu_\alpha)$:

$$Q \sim [(1-j)\,\alpha L/2a_g)](\mu_b/k_b). \tag{2}$$

The signal is again proportional to (αL) and ω^{-1}.

Case Ic: Thermally thick solid $(\mu_s < L;\, \mu_s \ll \mu_\alpha)$:

$$Q \sim -j(\alpha\mu_s/2a_g)(\mu_s/k_s). \tag{3}$$

The signal is now proportional to $(\alpha\mu_s)$ rather than (αL). That is, only the light absorbed within the first thermal diffusion length μ_s contributes to the signal, in spite of the fact that light is being absorbed throughout the solid. The frequency dependence is $\omega^{-3/2}$.

Case II: Optically opaque solids ($\mu_\alpha \ll L$). In these cases, most of the light is being absorbed within a distance small compared with L.

Case IIa: Thermally thin solids ($\mu_s \gg L$; $\mu_s \gg \mu_\alpha$):

$$Q \sim [(1-j)/2a_g](\mu_b/k_b). \tag{4}$$

The acoustic signal is independent of α (photoacoustic saturation), depends on the thermal properties of the backing material, and varies as ω^{-1}.

Case IIb: Thermally thick solids ($\mu_s < L$; $\mu_s > \mu_\alpha$):

$$Q \sim [(1-j)/2a_g](\mu_s/k_s). \tag{5}$$

The signal is independent of α and varies as ω^{-1}. The thermal parameters which affect Q are now those of the solid.

Case IIc: Thermally thick solids ($\mu_s \ll L$; $\mu_s < \mu_\alpha$):

$$Q \sim [(-j\alpha\mu_s)/2a_g](\mu_s/k_s). \tag{6}$$

This is an interesting and important case. Optically we are dealing with a very opaque solid. However, as long as $\mu_s < 1/\alpha$, this solid is not photoacoustically opaque. In spite of the optical opacity of the solid, the PA signal is proportional to $\alpha\mu_s$ and varies as $\omega^{-3/2}$. In practice, however, extremely high light modulation frequencies are required to attain this limit experimentally.

These formulas developed for the special cases may give numerical results for the amplitude and phase of the PA signal. However, an important factor has been omitted from the RG theory: the reflectance of the sample surface. The simplest way to take into account this effect is to multiply the intensity of the incident light by a factor of $[1-R(\lambda)]$, where $R(\lambda)$ is the reflectivity at wavelength of λ. The PA signal Q then becomes proportional to $[1-R(\lambda)]$. An interesting case can be obtained for optically opaque solids where most of the light is absorbed within a thin layer beneath the sample surface (Cases IIa and IIc). In this limit, the PA signal is independent of α and the only wavelength dependent term is $[1-R(\lambda)]$. In the case of optically transparent solids, an internal multiple reflection in the sample should be also considered (Fujii *et al.*, 1981; Mandelis *et al.*, 1984).

A theory of the PA effect which includes the contribution of mechanical vibrations of the sample, which had been ignored in the RG theory, has been proposed by McDonald and Wetsel (1978). These authors showed that experimental results could be accurately reproduced by an extension of Rosencwaig and Gersho's piston model. The piston-like motion of the gas boundary layer adjoining the sample is superimposed on the mechanical vibration of the sample surface to give a composite piston displacement which produces the pressure signal in the gas. The RG theory has also been extended to include effects of a coating layer with different thermal and optical properties (Mandelis, Teng and Royce, 1979; Fernelius, 1980).

The internal multiple reflection effect becomes important to the study of PA spectra of samples which form a thin film. Yamashita *et al.* (1981) have extended the RG theory to the case of thin films and derived a general expression for the PA signal. The main factors affecting the relationship between the PA signal $Q(\alpha)$ and the optical absorption coefficient α are the thickness of the sample L, the penetration depth of the incident light α^{-1} and the sample thermal diffusion length $\mu_s = [2k_s/\omega\rho_s C_s]^{\frac{1}{2}}$. In the strong absorption region where α is high, the PA signal is saturated irregardless of whether the sample is bulk ($L \gg \mu_s$) or thin film ($L \ll \mu_s$) since both L and μ_s are much larger than α^{-1}, i.e. $Q = Q_s$, where Q_s is the saturation value of $Q(\lambda)$. On the other hand in the weak absorption region with low values of α, Q is nearly proportional to α and is reasonably approximated by the following equations:

$$Q \sim L\alpha; \quad \alpha^{-1}, \mu_s \gg L, \tag{7}$$

for a thin film, and

$$Q \sim \alpha\mu_s/(1+j); \quad L, \alpha^{-1} \gg \mu_s, \tag{8}$$

for a bulk sample, respectively. In Eqs. (7) and (8) the proportionality constants are determined by the reflectivities for the normally incident light beam from the sample - gas and from the sample - backing material interfaces. It should be noted that Eq. (7) does not involve μ_s, which is directly associated with the sample thermal parameters. This feature means that one can directly determine α from the normalized PA signal Q/Q_s without any detailed information on the thermal properties of the thin sample and is one of the great advantages of PA measurements of thin samples.

Rear side PA spectroscopic detection has been employed to determine the low absorption coefficient of thick semiconductor samples (Fathallah and Zouaghi, 1985). These authors derived theoretically the PA signal in the gas cell behind the unilluminated surface of thick samples and showed a simple analytical procedure for determining the optical absorption spectra. By using their technique, the low value of absorption coefficient of 15 cm^{-1} could be measured for a thick sample of GaP.

2. PIEZOELECTRIC DETECTION TECHNIQUE

When an intensity modulated light beam is absorbed by a medium, a part or all of the excitation energy is converted to thermal energy. In the microphonic PA spectroscopy, the generated heat is coupled to the optically nonabsorbing gas and the ensuing time dependent pressure fluctuation is detected by a microphone. An alternative PA technique has been developed utilizing a piezoelectric transducer which is attached to the sample. The absorption induced heat causes the samples to develop thermal stresses and strains which are subsequently converted to a measurable voltage by the transducer.

There are several advantages to the piezoelectric PA technique over the microphonic one: i) Sensitivity similar to that attained with the microphone is possible; ii) Piezoelectric transducers have a wide frequency response range from a few Hz to MHz and this fact enables one to obtain the PA signal in either the A.C. or the impulse response mode. The details of the pulsed PA spectroscopy of condensed matter were extensively reviewed by Patel and Tam (1981); iii) The transducer can be operated over a wide range of temperatures and pressures; iv) Since the sample-transducer configuration is compact, it is useful in space limited experiments. The latter two advantages allow us to mount a transducer inside a low temperature optical cryostat and thus obtain low temperature PA spectra, which will be discussed in Chapter 17.

The fundamental theory of piezoelectric PA spectroscopy was developed for the condensed mater samples by Jackson and Amer (1980). Hereafter we refer to this theory as the Jackson and Amer model. These authors used PZT (Lead Zirconate Titanate) as a transducer. We first discuss sources of surface strain as shown in Fig. 2. The transducer may be attached to either side of the sample. The signal is generated when a light beam is incident on an absorbing solid with thickness L. The temperature of the illuminated volume increases, leading to the expansion of that region as well as to the outflow of heat. The expansion of the central region causes displacement of the sample surface by two separate mechanisms. First, the enlargement of the central region causes the general expansion of both surfaces of the sample (Z=O and Z=L). Second, in the case of strongly absorbing samples, the heat in the illuminated region decays spatially through the thickness of the sample [Fig. 2(b)]. Consequently, the front position of the sample expands more than the rear, resulting in a bending of the sample. Such bending compresses the rear surface of the solid (Z=L) which oppose the general expansion. The bending also causes expansion of the front surface (Z=O), thus adding to the general expansion. These displacements of the sample surfaces are then sensed by a sufficiently thin transducer attached directly to the sample at Z=O or Z=L. The voltages in the Z-direction are detected between the two surfaces of the transducer by appropriate electrodes. An annular transducer may be attached to the sample surface for this purpose at Z=O toward the light beam.

Treating the sample as an elastic layer and neglecting the transducer's effect on the sample, the three dimensional uncoupled quasistatic thermoelastic equations were solved using a Green's function technique for the stress (Jackson and Amer, 1980). An expression for the dependence of the PA signal on absorption, modulation frequency, thermal properties and the mechanical properties was thus derived. The approach of the calculation is as follows. First, the temperature distribution in the sample was calculated by assuming that the incident light intensity decays exponentially in the sample. The strain developed by the rise of temperature was then calculated by solving the strain equation. Using the

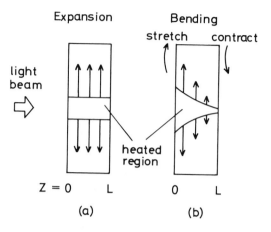

Fig. 2 Sources of surface strain for average expansion (a) and bending (b). The transducer may be attached to either side of the sample. [From Jackson and Amer (1980)].

appropriate stress-strain equation, a particular solution for the stress in the sample was found. Having solved for the stress induced by the absorption of light, the corresponding stress sensed by the transducer could also be determined. The strain at the sample-transducer interface was then converted to the voltage between the electrodes of the transducer using PZT equations of state.

Since the general expression of the PA signal is complicated, some special cases were discussed in the Jackson and Amer model. For simplicity, they neglected the surface thermal conduction from the sample to the surroundings. The light beam was illuminated on the sample surface at Z=O in all cases.

Case I: For a thermally and optically thick sample with the transducer located at Z=L,

$$Q \sim \frac{-2MPa_t}{j\omega L(\rho C)_s} \tag{9}$$

where a_t is the linear expansion coefficient of the solid, P is the incident power corrected for the sample reflectivity $R(\lambda)$ and M is a constant calculated by means of piezoelectric and dielectric constants.

Case II: For a thermally thin but optically thick sample with the transducer at Z=L,

$$Q \sim \frac{MPa_t}{j\omega L(\rho C)_s} \tag{10}$$

Case III: For thermally thick but optically thin samples,

$$Q \sim \frac{MPa_t}{j\omega L(\rho C)_s} [1-\exp(-\alpha L)\pm(6/L\alpha)[(1-L\alpha/2)$$

$$-\exp(-\alpha L)(1+L\alpha/2)]], \tag{11}$$

where α is the optical absorption coefficient. The negative sign applies in the case when the transducer is away from the laser beam (Z=L) and the positive sign is applicable when the transducer is toward the laser beam (Z=O).

The general form of the signal from the above cases can be written as

$$Q \sim M\, a_t\, \frac{1}{\rho C}\, \frac{P}{\omega} \tag{12}$$

P/ω is the energy deposited per cycle, $1/\rho C$ converts the energy to a temperature and a_t transforms the temperature rise to a strain. Jackson and Amer predicted some interesting results for the PA signal from their calculations as follows:

1) The signal amplitude is proportional to the reflection corrected incident power.

2) The signal amplitude is related to the material parameters through the quantity $a_t(1-R)/(\rho C)$.

3) The signal amplitude has a $1/L$ dependence for opaque samples. Hence, thin samples tend to yield higher signals.

4) The signal has approximately a $1/\omega$ dependence.

5) The phase for metal data undergoes a 180° shift as the thermal length becomes smaller than the sample thickness (cases I and II in the preceding paragraph).

6) For small values of α, the PA signal is directly proportional to α. For high values of α, the position of the PZT with respect to the direction of the incoming beam (Z=O or Z=L) yields significantly different results. When the transducer is away from the beam (Z=L), as α increases the signal should decrease, eventually passing through zero and changing sign at higher values of α. In this geometry the compression by bending exceeds the expansion due to heating. On the other hand, when the transducer is attached to the sample surface toward the beam (Z=O), the signal shows little saturation until α reaches high values. Here the bending and the expansion terms add, thus increasing the observed signal.

The Jackson and Amer model for piezoelectric PA spectroscopy was later extended to include intraband transitions, effects of diffusion, bulk lifetimes and surface recombination of the excess minority carriers (Mikoshiba *et al.*, 1983). The minority carriers generated by the intrinsic light illumination whose energy was higher than the band gap energy relax to the band edge by releasing their excess energies to the lattice system. These carriers are then recombined by a nonradiative transition process in the bulk or on the surface. The heat generated by these nonradiative transitions was taken into account in the calculations. Mikoshiba *et al.* (1983) found that the modulation frequency dependence of the PA signal was very sensitive to the presence of surface states. This feature allows one to estimate the bulk lifetime and the surface recombination velocity by measuring the PA signal as a function of the modulation frequency.

Since the transducer configuration itself affects the results significantly, piezoelectric detection of the PA signal sometimes offers too complicated a spectrum to understand quantitatively. The peaks and the dips are sometimes inverted in the PA spectra (Hata *et al.*, 1984). An alternative spectroscopic technique using a pyroelectric thin film has recently been developed to detect the temperature change in solids (Mandelis and Zver, 1985; and Chapter 7). The pyroelectric detection has distinct advantages over piezoelectric PA spectroscopy. These include the simplicity of calibration, insensitivity to acoustic noise and mechanical resonance, a high potential for signal to noise ratio improvement and a flat response in the range 10^{-1} to 10^{-7} K/s. This technique may be useful for samples which exhibit complicated PA spectra.

III. Experimental Configurations

A photoacoustic spectrometer is composed of three main parts; a source of incoming radiation, the experimental chamber and a data acquisition system. A typical PA spectrometer is shown in Fig. 3 as a block diagram. The optical radiation from ultraviolet to far-infrared is provided through a suitable monochrometer by a conventional light source such as an arc lamp, an incandescent lamp or a glow bar. A laser is alternatively used to obtain intense and coherent radiation or pulsed excitation for the PA measurements. The PA signal is detected by a microphone or a piezoelectric transducer in the PA cell. Then the signal is amplified by a preamplifier in the cell and filtered by a lock-in amplifier to obtain a high signal to noise ratio. Autophase lock-in amplifiers are preferable, as they enable one to measure the amplitude and the phase of the PA signal at the same time.

A small portion of the incident beam is split off and detected by a pyroelectric detector to normalize the PA signal by the intensity of the incident light. For the microphonic technique, the pyroelectric detector can be replaced by a PA cell which contains a reference material as a sample. Carbon black is a good reference material because its PA spectrum is almost the same as the power spectrum in the visible region (Rosencwaig, 1977). A more conventional method, however, is usually employed: The PA signals of the sample and the carbon black are separately measured and the normalized PA signal is obtained by dividing the signal of the sample by that of the carbon black.

Photoacoustic cells for the microphone PA technique generally incorporate a suitable microphone with its preamplifier. Both conventional condenser microphones with external biasing and electret microphones with internal self-biasing provided from a charged electret foil are appropriate PA detectors.

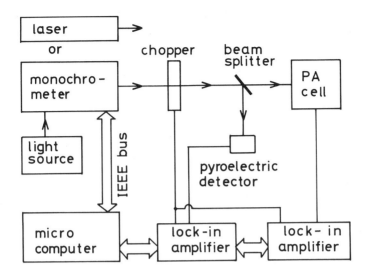

Fig. 3 A block diagram of the PA spectrometer. The PA signal is detected by a microphone or a piezoelectric transducer in the PA cell.

Some criteria governing the actual design of the PA cell were given by
Rosencwaig (1977) in detail. They include the following:

(1) The cell should be designed with good acoustic seals and with walls
of sufficient thickness to form good acoustic barriers. The cell should also be
isolated from room vibrations.

(2) The extraneous PA signal arising from the interaction of the light
beam with the walls, windows and microphone in the cell should be minimized.
A large thermal mass should be avoided, as it results in a small temperature rise
at the surface and thus a small PA signal. One should also minimize the amount
of scattered light that can reach the microphone diaphragm.

(3) Since the signal in the PA cell varies inversely with the gas volume in
the cell, the gas volume should be minimized. However, one must take care not
to minimize this volume to the point that the acoustic signal produced at the
sample suffers appreciable dissipation to the cell walls before reaching the
microphone. A typical design of the PA cell is drawn in Fig. 4(a). This
configuration of the PA cell is generally used and is also commercially available.

As far as the piezoelectric PA technique is concerned, the transducer is
coupled directly to the sample. Thus we obtain a good acoustic impedance
matching and a high acoustic transmission efficiency. This is in sharp contrast
to the transmission across an interface between the solid and the gas in the
microphonic technique. The criteria (1) and (2) discussed above for the microphonic technique are also applicable to the present case. One example for the
arrangement of a sample and a transducer is shown in Fig. 4(b). This apparatus

was developed to observe the absorption spectra of rare-earth oxide powders by pulsed PA measurements (Patel and Tam, 1981). A L-shaped piece of fused quartz is cut from a plate of 0.15 cm thickness. A suspension of the sample powder with a transparent viscous carrier liquid (ethylene glycol) is put on one end of the piece and a cover plate of quartz is tightly clamped onto the sample. When the sample powder is irradiated, the induced elastic wave can be transmitted through the L-shaped piece of the PZT transducer attached to the other end of the piece where the transducer is spring-loaded with a thin layer of silicon grease at the PZT-quartz interface to ensure good acoustic contacts. The advantage of an L-shaped piece is that scattered light cannot directly reach the transducer due to the bend and thus it cannot cause large spurious signals at the detector surface. Further reduction in light scattering effects can be obtained by locating the transducer away from the bend, or by using a substrate with multiple bends. More detailed discussions for the design of the PA cell were reviewed elsewhere (Rosencwaig, 1977; Patel and Tam, 1981; and Chapter 8).

IV. Room Temperature Photoacoustic Spectra

1. THE ELEMENTARY SEMICONDUCTORS Si and Ge

The photoacoustic spectra of Si and Ge have been measured by the microphonic technique in the wavelength region shorter than the fundamental absorption edge (Tokumoto *et al.*, 1981). In this region, the PA signal is regarded as independent of the absorption coefficient (saturation region). The PA spectra are shown in Figs. 5(a) and 6(a). In the spectra for Si, an abrupt change of the PA signal was observed at 1100 nm (1.12 eV) and this corresponds to the fundamental absorption edge. The observed energy gap of 1.12 eV is in good agreement with the published value despite an uncertainty in determining the band gap energy from the PA spectra. This uncertainty mainly arises from the fact that silicon is an indirect band gap semiconductor. For Ge, no such abrupt change was observed because the absorption edge was out of the experimental range.

In the saturation region, all the incident light is absorbed within a thin layer beneath the sample surface. The only wavelength dependent term of the PA signal is the $[1-R(\lambda)]$ term. Calculated spectra of $[1-R(\lambda)]$ as a function of wavelength for Si and Ge are also shown in Figs. 5(b) and 6(b) (Tokumoto *et al.*, 1981). All the dips observed in the PA spectra are well reproduced in the $[1-R(\lambda)]$ spectra. This shows that the dips in the PA spectra in the saturation region can be ascribed to optical reflection effects inherent to the electronic band structure.

Next, we briefly consider the peaks observed in the reflectivity spectra which cause the dips in $[1-R(\lambda)]$ spectra above the fundamental absorption edge. The reflectivity amplitude at normal incidence is related to the index of refraction $N(\lambda)$ by the Fresnel formulas,

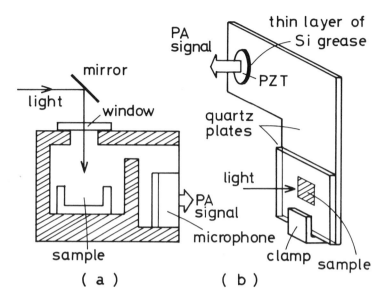

Fig. 4 Typical designs of a PA cell for the microphonic (a) and piezoelectric detection technique (b). [From Patel and Tam (1981)]

$$R(\lambda) = \frac{N(\lambda)-1}{N(\lambda)+1} = \frac{n(\lambda)-jk(\lambda)-1}{n(\lambda)-jk(\lambda)+1},\qquad(13)$$

and the use of the dispersion relation allows a determination of $n(\lambda)$ and $k(\lambda)$ (Ziman, 1972). Since the absorption coefficient $\alpha(\lambda)$ is given by the relation

$$\alpha(\lambda)=4\pi k(\lambda)/\lambda,\qquad(14)$$

the absorption coefficient is closely related to the reflectivity. The visible and UV reflectance spectra from polished, cleaved or etched surfaces of bulk semiconductors show a series of peaks which correspond to direct transitions at higher energy regions. The intensity and sharpness of these transitions result form the high density of initial and final states involved. Thus the peaks are quite intense for transitions where the two levels have the same energy difference over a large range of position in momentum space, where the valence and the conduction subbands have parallel branches. Therefore, interband transition related, band structure information for many semiconductors can be gained from the structures of the reflectivity spectra, i.e. the PA spectra in the saturation region.

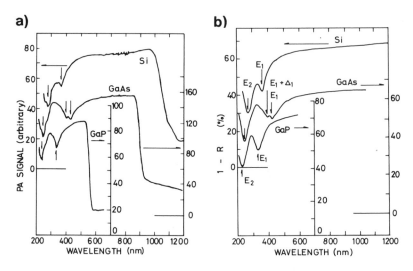

Fig. 5 The microphonic PA spectra of etched Si, GaAs and GaP samples (a), and [1-R(λ)] spectra (b) as a function of wavelength. The basic lines of these spectra are different and are drawn in the figure. [From Tokumoto *et al.* (1981)]

Fig. 6 The microphone PA spectra of Ge and InSb samples (a), and [1-R(λ)] spectra (b) as a function of wavelength. [From Tokumoto *et al.* (1981)]

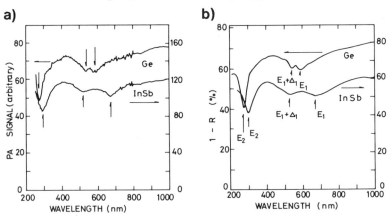

2. III-V COMPOUNDS

a. Microphonic photoacoustic detection

Microphonic photoacoustic measurements in the saturation region where the PA signal is independent of the absorption coefficient were carried out for III-V semiconducting compounds of GaAs, GaP and InSb (Tokumoto *et al.*, 1981). The samples were lapped by alumina powder and etched by proper etchants. The PA spectra and the calculated $[1-R(\lambda)]$ spectra are shown in Figs. 5 and 6. Abrupt changes in the PA spectra of GaAs and GaP are shown at 870 and 540 nm, respectively. These correspond to the fundamental absorption edge and lead to a calculation of the direct band gap of 1.43 eV of GaAs and indirect band gap of 2.30 eV for GaP which are in agreement with those reported in the usual textbooks (Pankove, 1971). For powdered GaP samples, Rosencwaig (1977) also observed a "knee" in the PA spectra at 2.80 eV which he attributed to the forbidden direct band transition.

In the PA saturation region, the dips correspond well to the minima in the $[1-R(\lambda)]$ spectra in a fashion similar to the case for Si and Ge. Each spectrum exhibits two or three minima which correspond to E_1 and E_2 or E_1, $E_1+\Delta$ and E_2 transitions. Comparing the PA spectra with the $[1-R(\lambda)]$ spectra, it can be seen that not only the structure but also the relative amplitudes are the same. This means that a PA study in the saturation region above the fundamental absorption edge is very useful to the investigation of interband transitions, i.e., the band structure especially for samples whose reflectivity spectra are not easy to obtain.

Next, we discuss extrinsic properties due to an impurity or a defect. The extrinsic absorption spectra arising from deep levels associated with the presence of Cr in GaAs were first reported by Eaves *et al.* (1981). These authors observed a broad absorption band at 1 μm in the PA spectra. Since a similar absorption band structure was also observed in the optical absorption measurements, they concluded that this band was due to the optical transition from the ground state of a deep level associated with the Cr impurities. More detailed PA measurements of the chromium level in GaAs were carried out by Tokumoto and Ishiguro (1983). The investigation of the Cr in GaAs has received considerable attention because the Cr doped GaAs has been used as a semi-insulating substrate material for high frequency devices. The Cr state in the Ga substitutional site in the GaAs crystal has three charge states. Since the neutral Cr atom has an electron configuration of $(3d)^5$ $(6s)^1$, the Cr^{3+} center with $(3d)^3$ configuration forms a neutral acceptor when the Cr content significantly exceeds the content of the shallow donors. The Cr^{2+} center is most probably a singly ionized acceptor and the Cr^{1+} center is a doubly ionized center. The latter Cr^{1+} center can exist only under strong light illumination.

The PA spectra of Cr-doped GaAs with different concentrations in the wavelength region from 800 to 1600 nm are shown in Fig. 7 (Tokumoto and Ishiguro, 1983). The measurements were carried out with etched samples at room temperature by a microphonic technique. The abrupt change of the signal amplitude at 870 nm was attributed to the direct fundamental absorption edge.

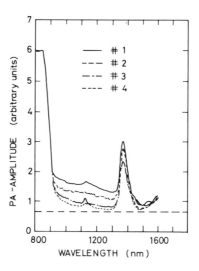

Fig. 7 The PA spectra of Cr-doped GaAs with different Cr concentrations of 2.0×10^{17} (#1), 7.0×10^{16} (#2), 5.7×10^{15} (#3) and 0 cm^{-3} (#4). The dashed line represents the noise level. The PA peaks at 1120 and 1380 nm are due to water vapor in the sample cell. [From Tokumoto and Ishiguro (1983)]

For the wavelength region longer than the absorption edge, a small and broad band was observed. The intensity of the band increases with the increase of the doping Cr impurity concentration. The apparent peaks at 1120 and 1380 nm were due to the PA signal of water vapor in the sample cell (Herzberg, 1950).

For the absorption band related to the Cr^{3+} center, the internal transition was thought to appear near the wavelength region of 2 μm. Since there are usually many donor impurities such as oxygen in the actual GaAs crystal, the Cr^{3+} level can be compensated by these donors. Since one electron can move from the donor to Cr^{3+} level by the compensation process, the Cr^{2+} level with the electron configuration of $(3d)^4$ appears. Due to the crystal field of T_d symmetry of the Ga site, the lowest energy state splits into two. One is the triply degenerate ground state of $^5T_2$ and the other is the doubly degenerate excited state of ^5E symmetry. Tokumoto and Ishiguro (1983) argued that the internal transition between these two states causes the absorption band at 1200 nm. Since the excited ^5E state merges in the conduction band, a broad band is observed in the PA spectra. A more intensive discussion concerning the effect of the crystal field on the Cr level in GaAs is given by Deveaud *et al.* (1984).

MacFarlane and Hess (1980) have investigated the damaged layer on GaAs substrates formed by ion implantation of Si atoms by using the microphonic PA technique. When the GaAs substrate was implanted, the amplitude of the PA signal at the 1.06 μm wavelength of a Nd:YAG laser became substantially larger than that from single crystalline material. Dose levels as low as

10^{12} cm^{-2} for the 150 keV Si$^+$ ions could be detected by the PA technique, which dose level is usually difficult to obtain by the Rutherford backscattering technique (RBS). Therefore, these authors demonstrated that through the use of this PA technique for the study of recrystallization of the implanted layer, measurements could be accomplished much more rapidly, sensitively and conveniently than by many other techniques such as electron and x-ray diffraction and RBS.

Similar PA measurements for a Si$^+$ ion implanted GaAs substrate were carried out to study a recrystallization process by Morita and Sato (1983). The surfaces of the semiconducting and seminsulating GaAs substrate were investigated and Fig. 8 shows their results. The surface I is as-implanted, and III and IV are laser annealed under 1- and 100-bar Ar atmospheres, respectively. The amplitude of the PA signal in the long wavelength region increased by the ion implantation with 10^{15} cm^{-2} Si$^+$ ions at 100 ~ 180 keV. The signal then decreased when the implanted sample was subsequently annealed by a Q-switched ruby laser. The broad band near 1000 nm in the semi-insulating GaAs samples is due to the presence of the Cr impurity as discussed before.

b. Piezoelectric photoacoustic detection

It has been suggested that the PA measurement is a direct and sensitive probe which gives information about nonradiative processes in semiconductors. These nonradiative processes cannot be studied by using conventional optical absorption or photoluminescence (PL) spectroscopy. Degradation behavior in semiconductor laser diodes (LD) and light emitting diodes (LED) has been extensively studied so far and it has been theorized that the degradation is caused by dislocations and natural defects of the host crystals. However, no definitive conclusion on the degradation mechanism had been advanced.

Wasa *et al.* (1980b) first observed PA spectroscopic signals due to nonradiative states in GaAs and InP by means of a piezoelectric PA technique. They showed that these nonradiative states played an important role in the degradation of light emitting devices. Fig. 9 shows the PA spectra of GaAs which had been held for 6 hours in an O_2 atmosphere at 200°C. The peak A at 0.96 μm appears in the spectra. The knee indicated by E_g corresponds to the band gap of GaAs (1.43 eV) which has also been observed by the microphonic PA detection in Fig. 5(a). When the sample was etched, the signal A disappeared. This shows that the signal A is due to nonradiative states originated from the oxidation of the surface of GaAs and/or the formation of As vacancies.

For a sample of high etch pit density of 10^{15} cm^{-2}, an additional PA signal at 0.89 μm close to the band gap appeared. Wasa, *et al.* (1980b) suggested that this peak was originated from dislocations in GaAs. The PA and photoluminescence (PL) spectra for the chromium and oxygen doped GaAs sample are shown in Fig. 10. There are two peaks at 0.89 (B) and 0.92 μm (C) and a broad but strong band around 1.2 μm (D) in the band gap region of the PA spectra. Since the signals C and D were also observed in other samples of Cr doped GaAs, they suggested that these are due to the Cr impurities in the substrate. PA signals from a Cr^{2+} center at 1.2 μm wavelength have also been observed by

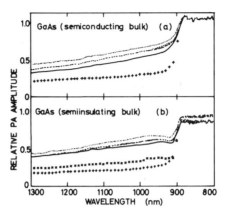

Fig. 8 The PA amplitude versus wavelength of the surface layer of semiconducting (a) and semi-insulating GaAs (b). Surface I(●●●●) is as-implanted, II(- - -) is laser-annealed under 1-bar air, III(——) under 1-bar Ar and IV(——) under 100-bar Ar atmosphere. The plots are for the PA spectra of as grown semiconducting (+) and seminsulating samples (x) of GaAs. [From Morita and Sato (1983)]

Fig. 9 The Piezoelectric PA spectrum of GaAs sample which had been held in a O_2 atmosphere for 6 hours. [From Wasa *et al.* (1980b)]

Fig. 10 The PA and PL spectra of a seminsulating GaAs sample. Each spectrum is indicated by a solid and broken line, respectively. [From Wasa *et al.* (1980b)]

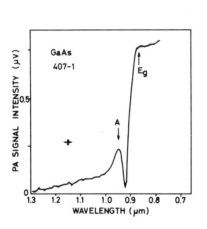

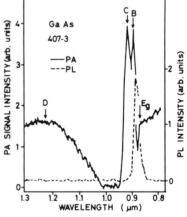

microphonic PA measurements and were discussed in more detail in the preceding subsection.

It should be noted here that states in the gap give a larger PA signal than those above E_g. This is surprising since in an ordinary case the absorption above the fundamental absorption edge is much stronger than that by states in the gap. This indicates that the PA signal does not increase monotonically with the optical absorption coefficient α. These anomalous α dependences of the PA signal were discussed by Ishiguro and Tokumoto (1982). the PA signal amplitude is independent of the sample thickness L and is proportional to α in the low absorption region. It decreases, however, with increasing α when the sample thickness L exceeds $1/\alpha$. Further increase of α so that μ_α ($=1/\alpha$) becomes smaller than the sample thermal diffusion length μ_s makes the amplitude independent of α (photoacoustic saturation). Therefore, care should be taken when interpreting the PA signal amplitude and phase quantitatively.

The PL spectrum of GaAs was also obtained at room temperature by using a He-Ne laser as an excitation source and a Si photodiode as a detector. A peak at 0.89 μm was obtained in the PL spectra of Fig. 10. The energy of the luminescence peak is the same as that of signal B in the PA spectrum. This means that the dislocations responsible for the signal B act as the absorption level of PL and transfer the absorption energy into the crystal as heat.

The PA spectra for InP crystals were also reported by Wasa *et al.* (1980b). A signal at 1.02 μm which may be caused by dislocations of InP was detected. However, no signal corresponding to the energy gap of InP (1.35 eV at room temperature) could be observed.

3. HETEROSTRUCTURES OF III-V SEMICONDUCTORS

Piezoelectric photoacoustic spectroscopy has been applied to investigate the energy band structure of some semiconductor heterostructures (Kubota *et al.*, 1984). Fig. 11 shows that PA spectra of $Ga_{1-x}Al_xAs$ on GaAs wafers. A $Ga_{1-x}Al_xAs$ layer of 1.5 μm thickness was grown on the GaAs substrate of 300 μm thickness by a molecular beam epitaxy (MBE) method. The incident light was on the epitaxial layer and the PZT transducer was attached to the GaAs substrate. A steep rise of the PA signal at 1.42 eV (marked as <u>a</u> in the figure) is due to the direct band to band transition of GaAs. For the samples with epitaxial layer, a decrease around 1.90 eV was observed. This energy is nearly equal to the indirect transition edge of $Ga_{1-x}Al_xAs$. The broken lines were obtained by reducing the excitation light intensity by the ratio indicated in the figure.

Kubota *et al.* (1984) have also measured the PA spectra of $GaAs_xP_{1-x}$ which was grown on a GaP substrate by a liquid phase epitaxy (LPE). The observed PA spectra are shown in Fig. 12, where the excitation light is incident on the epitaxial layer [Fig. 12(b)] or on the GaP substrate [Fig. 12(c)]. For the

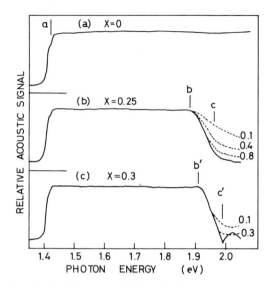

Fig. 11 The piezoelectric PA spectra of $Ga_{1-x}Al_xAs$ on GaAs. The broken lines are obtained by reducing the excitation light intensity by the ratio indicated. [From Kubota *et al.* (1984)]

Fig. 12 The PA spectra of $GaAs_{0.35}P_{0.65}$ p-n junction on GaP. The excitation light was incident on a GaP sample (a); on the GaAsP surface of the p-n junction (b); and on the GaP substrate of the p-n junction (c). [From Kubota *et al.* (1984)]

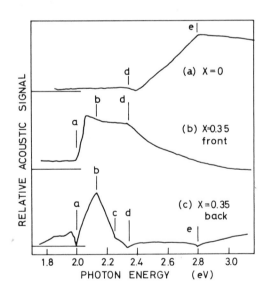

GaP sample with x=0 [Fig. 12(a)], the rise of the signal above ca. 2.35 eV (marked as *d* in the figure) is due to the indirect transition absorption, while the direct transition edge is shown at 2.80 eV (marked as *e*). For the sample with the epitaxial layer, Kubota *et al.* showed that the marks *b* and *c* correspond to the indirect and the direct transition edge of $GaAs_xP_{1-x}$ (x=0.35), respectively. This complex behavior would be due to the complicated bending mode induced by the heat generation. The induced voltage of the PZT detector is determined by the mode pattern of the surface strain. Since the strain would depend strongly on the sample shape, incident light flux, light intensity, incident direction, incident point on the sample surface, chopping frequency, etc., a simple single mode generating the PA signal is not always expected.

In the PA spectra of the heterostructure samples, the PA signal exhibits a substantial decrease at the indirect edge region in the direct transition absorption band (mark *b* in Fig. 11 and mark *d* in Fig. 12). Kubota *et al.* (1984) have considered the conditions under which the decrease of the signal was observed in the PA spectra with the PZT transducer as a detector. Since the PA effect is due to the heat generation by nonradiative energy de-excitation processes of the optically excited carriers in a semiconductor, the induced PA signal would change at excitation photon energies at which the carriers are excited to other energy states having a different type of phonon emission process. Namely, these authors suggested that the decrease of the PA signal amplitude does not necessarily mean the decrease of heat generation, but is an indication that an acoustic wave of a different mode is produced in a PZT detector. To examine this hypothesis in more detail, Kubota and Murai (1984) carried out a space-resolved detection of the excited PA vibrational mode. They concluded that when the indirect transition is added to the direct transition absorption, the PA mode pattern changes to a complicated one which strongly depends on the exciting position, etc. as discussed above. As a consequence, the output voltage of the PZT detector remarkably decreases around the excitation photon energy of the indirect transition edge.

Fig. 13 The microphonic PA spectrum of GaSe at room temperature near the absorption edge. The arrow indicates formation of free excitons. [From Todorovic and Nikolic (1985)]

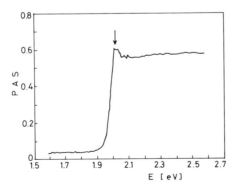

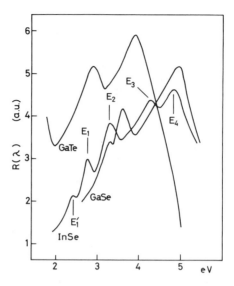

Fig. 14 Reflectivity spectra R(λ) of GaSe, GaTe and InSe at room temperature, obtained by means of saturation PA spectra, where $Q_{PAS}/I_0(\lambda) \sim [1-R(\lambda)]$. The energy positions of the allowed interband transitions are indicated. [From Baldassare and Cingolani (1982)]

Fig. 15 Reflectivity of single crystals of InSe at room temperature (——) and at 4.2 K (- - -). The plots correspond to the reflectivity calculated from the refractive index n from the interference fringes observed in the transmission measurements. [From Piacentini *et al.* (1979)] The notations of the interband transition are changed to compare with Fig. 14.

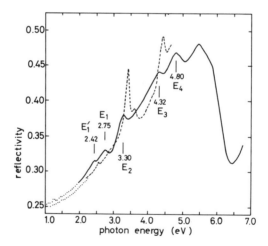

4. LAYERED SEMICONDUCTORS

For the layered semiconductor CdI_2, the PA spectra have been measured in the region of energies near the fundamental absorption edge by the microphonic technique (Baldassare *et al.*, 1984). The samples were grown by the Bridgeman-Stockbarger method with the thickness varying from 20 to 70 μm. For thin film samples, the PA signal is directly proportional to the absorption coefficient in the low absorption region as discussed in section 1. Baldassarre *et al.* have succeeded in obtaining the spectra at low values of the absorption coefficient α from 5 to 100 cm^{-1} in the range between 3.12 and 3.26 eV. They suggested that the PA spectra near the absorption edge showed the indirect nature of the optical gap of CdI_2.

The photoacoustic and absorption spectra of single crystals of the layered semiconductor GaSe have been studied in the neighborhood of the intrinsic absorption edge (Todorovic and Nikolic, 1985). The measurements were carried out at room temperature by a microphonic technique. The sample thickness was 100 μm. The observed spectrum is shown in Fig. 13. A peak appears at 2.02 eV and the PA signal intensity is energy independent above this peak. Comparing these results with optical absorption data, the authors concluded that the peak at 2.02 eV was due to the formation of excitons.

In the PA saturation region, where the PA signal is independent of the absorption coefficient, the only wavelength dependent term of the signal is a term of [1-R(λ)], where R(λ) was the reflectivity at the wavelength λ (see section 1). These conditions are usually satisfied in the energy region above the fundamental absorption edge. Baldassarre and Cingolani (1982) obtained the reflectivity spectra of GaSe, GaTe and InSe by using PA spectra measured by the microphonic technique. Fig. 14 shows their results. The reflectivity was calculated by the equation for the PA signal intensity Q(λ) as

$$Q(\lambda)/I_O(\lambda) = 1-R(\lambda),$$

where $I_O(\lambda)$ is the intensity of the incident light. The spectra show several peaks whose energies are in good agreement with the allowed interband transitions of these materials. The reflectivity spectra obtained by the usual optical reflectance measurements for InSe are shown in Fig. 15 (Piacentini *et al.*, 1979). The peaks at 2.42 (E_1'), 2.75 (E_1), 3.30 (E_2), 4.32 (E_3) and 4.80 eV (E_4) all correspond to the peaks in Fig. 14. Baltassare and Cingolani (1982), upon comparing the PA and reflectivity spectra, concluded that the PA measurements were an effectively useful tool in studying the optical properties of semiconductors in the optically opaque region. More details of the PA spectra for GaSe family compounds will be discussed in Chapter 17.

5. OTHER COMPOUND SEMICONDUCTORS

One great advantage of PA measurements is their ability to detect nonradiative states in semiconductors. A study of the nonradiative states near the fundamental absorption edge in CdS single crystals was carried out by Wasa *et al.* (1980a). The experiments were made at room temperature by means of a piezoelectric detection technique using a sputtered ZnO transducer. Details of CdS photoacoustics are discussed in Chapter 15.

The main interest in ZnS has centered on its luminescent properties. This semiconductor is known as efficient phosphor and has been used, for example, as cathode ray tube or as electroluminescent panels. Photoacoustic spectroscopy offers an excellent means to study powders of this material which cannot be analyzed with other conventional spectroscopic techniques. The absorption spectra of undoped and doped ZnS powders with 0.04% Mn and 0.001% Cu were measured by means of a conventional microphonic PA technique (Simmonds and Eaves, 1981). In addition to the strong near band edge absorption with an onset near 345 nm (3.6 eV), a broad band near 540 nm (2.3 eV) was observed. This broad band has been shown to result from the sum of five Mn^{2+} intracenter transitions. These five energy levels due to the d-d interaction within Mn^{2+} ions were clearly observed in the PA spectra of ZnS:Mn powder phosphor by Morita (1981). A light-induced darkening of ZnS occurs at the surface of both microcrystalline powders and single crystals in the presence of moisture. It is thought to arise from a photolysis effect in which near band gap irradiation produces free carriers which induce a chemical reaction between the ZnS and the water contamination leaving a precipitation of Zn in the surface layer (Simmonds and Eaves, 1981). This degradation changes the color of Mn-doped ZnS from light pastel green to dark gray. These authors have used a simple double-beam arrangement for the PA measurements in which a probe light beam is used to monitor the darkening effect. A sharp increase in the low wavelength PA signal at 600 nm was thus observed with ultraviolet irradiation at a wavelength near 345 nm (3.6 eV).

Photoacoustic measurements of the semiconductor compound of Dy_2S_3 and Nd_2S_3 powders were carried out by the microphonic technique at room temperature (Kurbatov *et al.*, 1982). The rare-earth sulfides represent a theoretically interesting class of semiconducting compounds and applications have already been found in electronics as high resistivity bulk resistors and anti-emission materials. Since the melting points of these materials are high (1500 ~ 1700°C), there are difficulties associated with growing large single crystals. These authors used the spectroscopic capabilities of PAS of powdered samples to obtain absorption spectra of Dy_2S_3 and Nd_2S_3 powders form the PA spectra. Samples were prepared by compacting the powders into pellets using a polymethylmethacrylate as a binder and reflectance spectra were recorded simultaneously. The observed PA spectra were in satisfactory agreement with the absorption spectra of single crystals of these compounds.

The ternary semiconductor compound of $CdIn_2S_4$ is a photosensitive semiconductor and has applications as a good photoconductor in the visible region.

This crystal belongs to a cubic spinel structure and has an indirect energy gap of 2.2 eV at room temperature. PA measurements by the microphonic technique were carried out by Yamashita *et al.* (1982). The PA amplitude spectrum clearly shows an absorption edge. PA phase signals were also measured as a function of the chopping frequency and the sample thickness. This allowed Yamashita *et al.* (1982) to estimate the thermal conductivity and the thermal diffusivity as 1.3×10^{-2} cal/g.cm.K and 2.3×10^{-2} cm^2/s, respectively, by using the RG theory.

V. Conclusions

PA spectroscopy is a very useful tool for monitoring the relaxation of electrons from excited states to the ground state via radiative and/or nonradiative transition pathways. There exist two main techniques to detect photoacoustic signals: either by a conventional microphone or by a piezoelectric transducer. The RG theory and the Jackson and Amer model, have been extensively used to interpret the PA signal amplitudes and the phases from microphonic and piezoelectric experiments, respectively.

The PA spectra of elementary and compound semiconductors have been reviewed in this chapter. In the case of the microphonic technique, the observed spectra can be interpreted well by the RG theory or its modifications. In the linear region of low absorption, the PA signal directly corresponds to the absorption coefficient. Alternatively, in the saturation region of strong absorption, the only wavelength dependent term is the $[1-R(\lambda)]$. Therefore, the peaks in the reflectivity spectra can be calculated from the PA spectra and these agree well with those obtained by the usual reflectance measurements. This demonstrates that PA measurements in the saturation region will be useful for investigating the direct interband transition in semiconductors of which mirror-like surfaces are difficult to prepare.

The Jackson and Amer model can explain the observed piezoelectric PA signal at least qualitatively. In this chapter we discussed important advantages of the piezoelectric technique over the microphonic one. It is the belief of the authors that the PA piezoelectric technique will become a major spectroscopic method in the near future. In the case of semiconductors with a complex structure such as a heterostructure, however, the PA spectra sometimes exhibit extraneous features due to mode pattern of the elastic wave which will be sensed by the piezoelectric transducer. These features are usually not simple to analyze. A more precise theory for the piezoelectric detection technique is necessary to interpret these experimental spectra in order to apply this technique to industrial uses in the future.

VI. References

Baldassarre, L. and Cingolani, A. (1982). Solid State Commun. **44**, 705.

Baldassarre, L., Cingolani, A. and Cornacchia, M. (1984). Solid State Commun. **49**, 373.

Deveaud, B., Lambert, B., Picoli, G. and Martinez, G. (1984). J. Appl. Phys. **55**, 4356.

Eaves, L., Vargas, H. and Williams, P.J. (1981). Appl. Phys. Lett. **38**, 768.

Fathallah, M. and Zouaghi, M. (1985). Solid State Commun. **54**, 317.

Fernelius, N.C. (1980). J. Appl. Phys. **51**, 650.

Fujii, Y., Moritani, A. and Nakai, J. (1981). Jpn. J. Appl. Phys. **20**, 361.

Hata, T., Sato, Y., Nagai, Y. and Hada, T. (1984). Jpn. J. Appl. Phys. Supplement **23-1**, 75.

Herzberg, G. (1950). *"Molecular Spectra and Molecular structure I. Spectra of Diatomic Molecules,"* p. 483, Van Nostrand Reinhold, N.Y.

Ishiguro, T. and Tokumoto, H. (1982). Jpn. J. Appl. Phys. Supplement **21-3**, 11.

Jackson, W. and Amer, N.M. (1980). J. Appl. Phys. **51**, 3343.

Kubota, K., Murai, H. and Nakatsu, H. (1984). J. Appl. Phys. **55**, 1520.

Kubota, K. and Murai, H. (1984). J. Appl. Phys. **56**, 835.

Kurbatov, G.A., Sidorin, K.K. and Chernukha, N.A. (1982). Sov. Phys. Semicond. Engl. Transl. **16**, 1221.

MacFarlane, R.A. and Hess, L.D. (1980). Appl. Phys. Lett. **36**, 137.

Mandelis, A., Teng, Y.C. and Royce, B.S.H. (1979). J. Appl. Phys. **50**, 7138.

Mandelis, A., Siu, E.K.M. and Ho, S. (1984). Appl. Phys. **A33**, 153.

Mandelis, A. and Zver, M.M. (1985). J. Appl. Phys. **57**, 4421.

McDonald, F.A. and Wetsel, G.C. (1978). J. Appl. Phys. **49**, 2313.

Mikoshiba, N., Nakamura, H. and Tsubouchi, K. (1983). *IEEE Ultrasonics Symp. Proc.*, **83CH1947-1**, 685.

Morita, M. (1981). Jpn. J. Appl. Phys. **20**, 295.

Morita, M. and Sato, F. (1983). Jpn. J. Appl. Phys. Supplement **22-3**, 199.

Pankove, J.I. (1971). *"Optical Processes in Semiconductors"*, Prentice-Hall, Englewood Cliffs, N.J.

Patel, C.K.N. and Tam, A.C. (1981). Rev. Mod. Phys. *53*, 517.

Piacentini, M., Doni, E., Girlanda, R., Grasso, V. and Balzarotti, A. (1979). Nuovo Cimento **B54**, 269.

Rosencwaig, A. and Gersho, A. (1976). J. Appl. Phys. **47**, 64.

Rosencwaig, A. (1977). In *"Optoacoustic Spectroscopy and Detection"*, (Yoh-Han Pao, ed.), pp. 193-239, Academic Press, New York.

Rosencwaig, A. (1980). *"Photoacoustics and Photoacoustic Spectroscopy"*, John Wiley & Sons, New York.

Simmonds, P.E. and Eaves, L. (1981). Appl. Phys. Lett **39**, 558.

Todorovic, D.M. and Nikokic, P.M. (1985). Appl. Opt. **24**, 2252.

Tokumoto, H., Tokumoto, M. and Ishiguro, T. (1981). J. Phys. Soc. Jpn. **50**, 602.

Tokumoto, H. and Ishiguro, T. (1983). Jpn. J. Appl. Phys. Supplement **22-3**, 202.

Wasa, K., Tsubouchi, K. and Mikoshiba, N. (1980a). Jpn. J. Appl. Phys. **19**, L475.

Wasa, K., Tsubouchi, K. and Mikoshiba, N. (1980b). Jpn. J. Appl. Phys. **19**, L653.

Yamashita, S., Okushi, H., Matsuda, A., Oheda, H., Hata, N. and Tanaka, K. (1981). Jpn. J. Appl. Phys. **20**, L665.

Yamashita, K., Kasahara, H., Yamamoto, K. and Abe, K. (1982). Jpn. J. Appl. Phys. Supplement **21-3**, 107.

Ziman, J.M. (1972). *"Principles of the Theory of Solids"*. Cambridge Univ. Press, Cambridge, pp. 255.

17

PHOTOACOUSTIC SPECTRA OF SEMICONDUCTORS II: LOW TEMPERATURE SPECTRA

Tetsuo Ikari, Shigeru Shigetomi and Yutaka Koga

Department of Physics
Kurume University
1635 Mii Kurume, Fukuoka, JAPAN

I. Introduction

Room temperature photoacoustic (PA) spectra of semiconductors are reviewed in Chapter 16. The principal advantages of the PA measurements over conventional optical measurements are as follows: We can obtain spectra, similar to the optical absorption spectra, on any type of solids independent of form (i.e. crystals, powders or gels). Furthermore, nonradiative transitions can be used as an important analytical technique for the investigation of the optical properties of semiconductors.

In the case of optical absorption or photoluminescence measurements, much knowledge about the solid's electron and/or phonon system can be gained by decreasing the ambient temperature. A similar situation would also be expected for PA spectroscopy. In this Chapter, we review low temperature PA spectra of semiconductors by using a microphone or a piezoelectric transducer as a detector. In addition, since the amplitude and phase of the PA signal is a function of the thermal parameters of the solid, the PA technique can be very helpful in the investigation of phase transitions in solids. A growing number of examples in this field of PA spectroscopy exists in the literature. We discuss here the temperature variation of the PA signal near the phase transition temperature.

Experimental techniques to obtain a low temperature PA signal are discussed in section II for both the microphonic and piezoelectric configuration. In that section we show that the piezoelectric technique has many advantages over microphonic detection for low temperature measurements.

The low temperature PA spectra of III-VI layered compound semiconductors GaSe and InSe are discussed in section III. Low temperature microphonic PA spectra of the II-VI semiconductor CdS have been reported by Dioszeghy and Mandelis (1986) and they are discussed in Chapter 15. To the best of our knowledge, these are all the published low temperature semiconductor PA spectra up to the present.

In section IV, the application of the low temperature PA technique to the study of phase transitions is reviewed with a proposed model to interpret the amplitude and the phase of the PA signal at the phase transition temperature. The usefulness of the PA technique is also discussed.

II. Experimental

Several reports on apparatus capable of measuring temperature variations of photoacoustic spectra have appeared in the literature. Murphy and Aamodt (1977) have obtained PA signals at the liquid nitrogen temperature. As a PA cell, they used a vacuum-insulated double-windowed cylinder to provide optical access to the working volume of the cell with a good thermal insulation. The sound pressure generated by optical absorption in the cell was transmitted to the room temperature commercial microphone via a needle like stainless-tube with a diameter of 0.1". The entire base of the cell was cooled by direct immersion in liquid nitrogen. Their cell represented a first attempt to construct a low temperature PA spectroscopic cell by the microphonic technique. PA measurements at liquid helium temperature have also been obtained by using a bolometer (Robin and Kuebler, 1977).

Variable temperature PA measurements using a conventional PA cell with a gas have been carried out from 5 to 300 K (Pichon *et al.*, 1979). In these experiments a polypropylene electret with a gold coated surface was used as the low temperature microphone. In order to check the behavior of the whole assembly, a PA signal from mangenese metal was monitored from 5 to 100 K under Argon ion laser illumination. The temperature dependence of the PA signal amplitude could be well explained by the equivalent piston model of the Rosencwaig and Gersho (RG) model (1976). This fact indicated that the response of the experimental device of these authors was reliable when performing PA experiments at low temperatures.

A simple device to measure PA spectra from 77 to 300 K was also designed by Boucher and Leblanc (1981). The sample and microphone cavities were separated, following a similar design principle to that reported by Murphy and Aamodt (1977). In this case, however, the PA cell and microphone were directly linked with shorter distance by a cylindrical high density polyethylene thermal insulator. The presence of two cavities gave a Helmholtz resonator character to the PA cell, resulting in a greater sensitivity at the resonant frequency. These authors further investigated in detail the temperature variation of the resonant frequency. Samples could be cooled down to 77 K without affecting the microphone performance. The first application of this technique by Boucher and Leblanc (1981) was to study an internal conversion process in native visual pigments.

Advantages of the piezoelectric detection technique for the PA spectra have been reviewed in Chapter 16. Briefly, they are as follows: First, the piezoelectric transducer has a wide frequency response range from a few Hz to MHz. Second, it can be operated in a wide range of temperatures and pressures. Third, the sample-transducer configuration is quite compact. These advantages have been exploited in applications to low temperature measurements of PA spectra (Ikari *et al.*, 1984). The configuration in the low temperature cryostat is shown in Fig. 1.

Fig. 1 Schematic configuration of the sample and transducer in the low temperature cryostat. [From Ikari *et al.* (1984)]

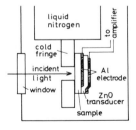

The block diagram of the experimental configuration was in principle, the same as that shown in Fig. 3 of Chapter 16. The PA spectra were normalized by the power spectra of the incident light intensity measured by using a thermopile as a detector. Ikari *et al.* (1984) employed a sputtered ZnO film transducer as the detector of the PA signal. Metal Al film (4000 Å thick) was vacuum evaporated on the cleaved surface as a lower electrode. ZnO film of thickness from 1500 to 4000 Å was then rf sputtered onto the metal electrode and Al was again vacuum evaporated as an upper electrode. The resistivity of the ZnO film was on the order of 10^6 to 10^7 Ω-cm at room temperature and slightly increased with decreasing temperatures toward the liquid nitrogen temperature. The electrical properties of the sputtered ZnO film strongly depended on the history of fabrication and Ikari *et al.* (1984) showed that the high resistivity ZnO film was not adequate for low temperature measurements.

The PZT transducer is generally a more convenient detector for PA measurements than the ZnO sputtered film. But in our experience, the ZnO transducer gives a larger PA signal amplitude with higher signal-to-noise ratio than PZT. The extremely high electrical resistance of the PZT detector may result in a diminished PA signal contribution. In these experiments the sample was mounted on the cold head in the cryostat and silicon grease or thin adhesive tape was applied between the cold finger and the sample to obtain a good thermal contact and a good isolation from cryostat vibrations. The area of the holes of the cold finger through which the incident light illuminated the sample surface was made smaller than the area of the lower Al electrode. This prevented direct unnecessary illumination of the ZnO transducer by the incident light in the transparent wavelength region.

Although there are several advantages to the piezoelectric detection technique of the low temperature PA signal compared to microphonic detection, one must be aware of its deficiencies: The generated surface strain which produces the PA signal strongly depends on the sample shape, chopping frequency, the illuminating point on the sample surface, etc. The amplitude and phase of the PA signal sometimes exhibit a tricky behavior (see for example section 5 in Chapter 16). Care should thus be taken when interpreting the experimental results quantitatively.

III. Low Temperature Photoacoustic Spectra for the Layered Semiconductors GaSe and InSe

The optical properties of layered III-VI compound GaSe have been extensively investigated. The crystal structure is characterized by a strong covalent bonding within the layers and a weak van der Waals bonding between them. This results in a strong anisotropy of the mechanical properties of this compound. On the other hand, less anisotropy has been observed in the electronic band structure (Capozzi and Minafra, 1981). Of further interest is the fact that GaSe has direct and indirect band gaps lying very close in energy to each other.

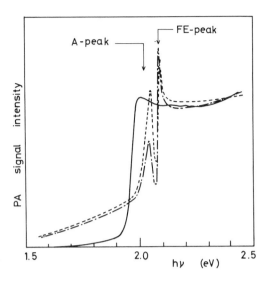

Fig. 2 The PA spectra of p-GaSe at 99 (---), 113 (-●-) and 295 K (–) [From Ikari, Shigetomi, Koga, Shigetomi, Nishimura and Suzuki, 1986]

The observed PA spectra of p-GaSe at 99, 113 and 295 K are shown in Fig. 2 (Ikari, Shigetomi, Koga, Shigetomi, Nishimura and Suzuki, 1986). The thickness of the sample was 350 μm and the sputtered ZnO film transducer was used as a detector. The sample-transducer configuration was the same as that in Fig. 1. For the spectra at 295 K, the PA signal gradually increases with incident photon energy up to about 1.9 eV and then it increases rapidly to exhibit a broad maximum around 2.0 eV. The energy of the maximum is in good agreement with that of direct exciton formation (Todorovic and Nicolic, 1985). The signal is almost constant above this maximum up to 2.5 eV. When the temperature of the sample decreases, the broad maximum becomes a narrow peak and the photon energy at which the peak occurs increases with decreasing temperature. For the spectrum at 99 K, two distinct peaks appear: One is at 2.098 eV (FE peak) and the other is at 2.059 eV (A peak). The energy of the peak at 2.059 eV agrees well with that of the n=1 exciton photoluminescence peak (Voitchovsky and Mercier 1974) and the exciton absorption peak (Antonioli, Bianchi, Emiliani, Podini and Franzosi, 1979). The origin of the peak at 2.098 eV was explained as due to free exciton annihilation by those authors. The temperature dependences of the peak energy for both peaks are shown in Fig. 3. The peak energy of another sample is also shown (white dots) to further clarify the temperature dependence. The plots form a straight line with a slope of 0.51 meV/K and this value agrees with the temperature variation of the band gap. The energy difference of the free exciton and the A peak is estimated to be 40 meV. With the value of 19.5 meV for the ionization energy (Rydberg energy) of the exciton ground state (Voitchovsky and Mercier, 1974), the activation energy of the A level to the direct conduction band was evaluated to be 60 meV. The

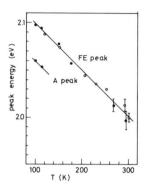

Fig. 3 The temperature dependence of the peak energy of the FE and A peaks [From Ikari, Shigetomi, Koga, Shigetomi, Nishimura and Suzuki, 1986]

activation energy of the A peak was estimated as 36 meV from the temperature dependence of the PA signal intensity.

GaSe is known as a semiconductor which has two conduction band minima with a small energy difference between them (25 meV (Capozzi and Minafra, 1981)). A schematic band diagram is shown in Fig. 4. DCB, ICB and DFE stand for direct conduction band, indirect conduction band and direct free exciton level, respectively. The A level which causes the A peak in the PA spectra is shown in the figure located 40 meV below the direct free exciton level (DFE). Since the energy difference between DCB and ICB is 25 meV, the energy difference of the A level and the indirect conduction band (ICB) was estimated to be 35 meV. This agrees very well with 36 meV obtained from the temperature dependence of the intensity of the A peak. This may indicate that the photoexcited carriers in the A level are thermally released to the ICB and would lead to a decrease of the intensity of the A peak in the PA spectrum when the temperature increases. Ikari *et al.* thus concluded that the A peak in the PA

Fig. 4 A schematic energy diagram of GaSe where DCB is the direct conduction band, ICB is the indirect conduction band and DFE is the direct free exciton level. The A level is the donor which causes the A peak in the PA spectra. [From Ikari, Shigetomi, Koga, Shigetomi, Nishimura and Suzuki, 1986]

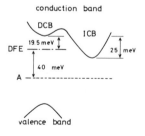

spectrum is caused by the transition of the photoexcited carriers in the donor level to the valence band. Since no radiative transition corresponds to the donor level in the photoluminescence spectrum, the A level was therefore considered to be a nonradiative center.

As discussed in section 2 of Chapter 16, the relation between the PA signal and absorption coefficient α of the sample is not simple in the case of the piezoelectric PA detection technique. The signal also depends on the transducer configuration and care should be taken when discussing a quantitative description of PA spectra. However, low temperature PA spectra have been shown to be useful in investigating the electronic structure of semiconductors.

Low temperature PA measurements were carried out for InSe single crystals by Ikari *et al.* (1984). InSe is one of the GaSe family compounds. Fig. 5 shows typical PA spectra at 116 and 300 K. The sample thickness was 470 μm. The peaks at 1.316 eV (116 K) and 1.250 eV (300 K), were well explained by the n=1 free exciton annihilation. The observed temperature dependence of the peak energy gave the temperature coefficient of 3.6×10^{-4} eV/K. This was in good agreement with that obtained by optical absorption measurements. These authors have noted that the spectrum of the optical absorption coefficient α was well reproduced by the PA spectrum. This means that the PA signal is proportional to α in the entire wavelength region of the measured spectrum. The Jackson and Amer model predicts that, in this so-called proportional region, the inverse of α must be sufficiently large compared with the physical length of the sample. Therefore, the Jackson and Amer model could not be applicable to the present case without any modifications.

Fig. 5 PA spectra of an InSe single crystal at 300 (——) and 116 K (- - - -). [From Ikari *et al.* (1984)]

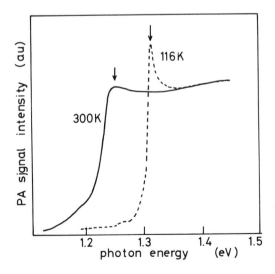

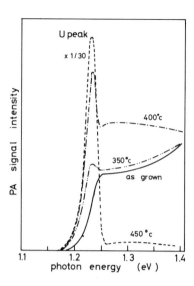

Fig. 6 Annealing behavior of the PA spectra for undoped n-InSe measured at 300 K. The annealing temperatures are indicated on each curve in the figure. [From Shigetomi *et al.* (1985)]

A study of the annealing behavior of the piezoelectric PA spectra of undoped n-type InSe was also reported by Shigetomi *et al.* (1985). In that work, isochronal annealing was carried out to characterize a donor level. The PA spectra of the as-grown and annealed samples at room temperature are shown in Fig. 6. A strong peak at 1.235 eV was observed when the sample was annealed above 350°C. These authors suggested that this peak was due to the nonradiative donor center. The intensity of the peak increased significantly with increasing the annealing temperature and thus it was suggested that the annealing induced an increase of the donor concentration at a donor level of 23 meV below the conduction band. The increase of donor concentration by annealing was also consistent with the isochronal annealing effect of the electron transport properties reported by Shigetomi *et al.* (1984).

Photoluminescence (PL) measurements have traditionally been an important and necessary tool in the study of bandgap (donor, acceptor or defect) states in semiconductors. In PL measurements only radiative transitions may be observed. Since radiative and nonradiative centers usually coexist in semiconductors, the study of nonradiative centers becomes important for a precise understanding of these states in the gap. This is the reason for which PA spectra have attracted attention recently. However, one must keep in mind that the PA spectrum is an excitation spectrum which is related to the total amount of heat generated by many nonradiative processes. Care must therefore be taken in the quantitative analysis of PA spectra of semiconductors.

IV. Application of Low Temperature Measurements to the Determination of Phase Transitions

1. MICROPHONIC DETECTION

Besides the usefulness of the photoacoustic effect for the spectroscopic study of samples with which conventional optical spectroscopy cannot be used, the sensitivity of the PA technique to detect variations of thermal properties can be very helpful to the investigation of phase transitions in semiconductors. There are several reports on the photoacoustic detection of phase transitions of solids in the literature, although, to our knowledge, none is concerned with semiconductors. We feel, however, that familiarity with the potential and the kinds of information that can be derived with the PA probe of phase transitions could be useful to semiconductor workers and justify integrating this knowledge in this Chapter dealing with low temperature PA phenomena.

The first report on a first-order phase transition by a PA technique was published by Florian *et al.* (1978). These authors observed the amplitude and the phase of the PA signal as a function of temperature for the liquid-solid transition of gallium metal and of water and the structural phase transition of K_2SnCl_6 and gave a qualitative explanation of their results.

A quantitative description of the PA effect near the phase transition temperature was proposed by Korpiun and Tilgner (1978). The change of the PA signal as a function of temperature was calculated for an optically and thermally thick sample in the case of a reversible first order phase transition. The RG theory provides a useful description of the PA signal as a function of temperature, however, considerable deviations occur in the temperature range near a phase transition. Korpiun and Tilger, modified the RG model to take into account the latent heat involved in the first order phase transition. They

Fig. 7 Schematic diagram of the arrangement of sample and cell and variation of the stationary temperature in specimen and gas. T_0 is the ambient and T_c is the phase transition temperature. [After Korpiun and Tilger (1980)]

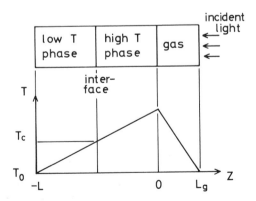

assumed that two different thermal states are present in the sample during the phase transition (Fig. 7). These are separated by a plane interface which is always kept at the transition temperature T_c. The thermal parameters are considered to be different for the low ($T<T_c$) and high ($T>T_c$) temperature phases. The latent heat was taken into account via a boundary condition for the heat flux. Light was assumed to be absorbed within a layer at Z=0 which was very thin compared to the sample dimensions. This is a realistic model for metals and other strongly absorbing materials. In this case, the light absorbed within the layer acts as a plane heat source at the surface. The variation of the temperature at the surface Z=0 results in a change in the position of the interface. Therefore, a periodic variation of the intensity of the light absorbed at the surface leads to an oscillation of the interface which is governed by the absorption and consumption of latent heat in the sample. This periodic storage of heat by the sample should influence the temperature itself as well as the phase angle between the incident chopping light and the oscillation of the temperature at the surface. The former is detectable as the magnitude of the signal. The oscillation amplitude of the interface between the two phases was assumed to be small with respect to the thermal diffusion length.

Under these conditions, Korpiun and Tilger (1980) predicted a decrease of the PA amplitude close to the phase transition temperature T_c. The minimum of the PA amplitude is located at a sample temperature approximately δ_o below T_c, where δ_o is given by

$$\delta_o = (1-R)I_O LL_g/(k_l L_g + k_g L),$$

where k_l is the thermal conductivity of the low temperature phase and R is the sample reflectivity. The geometric parameters are defined in Fig. 7. The PA signal starts to decrease at an ambient temperature δ below T_c, where δ can be approximated by

$$\delta = (1-R)(I_O/k_l)\times(1+\mu_l/\sqrt{2}L),$$

where μ_l is the thermal diffusion length of the low temperature phase (Bechthold *et al.*, 1980). The material is completely transformed to the high temperature phase when the ambient temperature reaches T_m. Korpiun and Tilger predicted that decrease of the PA amplitude would occur simultaneously with an increase of the PA phase. Furthermore, similar changes in signal should occur near the phase transition temperature in warming up as those observed in cooling down experiments.

In order to extend the field of PA spectroscopy towards thermal measurements, Pichon *et al.* (1979) examined thermally thick dielectric samples of $CrCl_3$ and MnF_2 in the neighborhood of a magnetic phase transition. In the case of a thermally thick sample whose thermal diffusion length is smaller than the optical absorption length, the PA signal (RG theory) is found to be proportional to the inverse of the specific heat. The specific heat anomaly as a function of temperature of MnF_2 at the Neel temperature of 68 K was clearly observed in the experiments by Pichon *et al.* (1979) and the observed anomaly was found to be comparable to that observed by conventional calorimetric measurements.

It appears that temperature varying PA measurements above room temperature may be rather easier than those below room temperature. Bechthold *et al.* (1980) have investigated the phase transitions in metal-halogen interstitial alloys by means of the temperature dependent PA technique. A special configuration was employed in their measurements: The microphone was placed at a distance of 9 cm from the sample and the hot region was slightly lifted with respect to the microphone to protect the detector from the high temperature. Measurements were carried out for $TaH_{0.5}$, $NbH_{0.8}$ and $VH_{0.517}$ up to the temperature of 500 K. The first order phase transitions of Ta and Nb alloy at 335 and 386.8 K, respectively, were well resolved. These authors observed a minimum in the PA amplitude, accompanied by a steep rise and succeeding decrease of the PA phase near the phase transition temperature. These results could be well explained by the theoretical predictions of Korpiun and Tilger (1980). The second order phase transition at 445 K was also detected for $VH_{0.517}$ by the PA technique.

Microphonic detection of a phase transition by PA measurements above room temperature was carried out for $NaNO_2$, $CsNO_3$, NH_4NH_3, $BaTiO_3$, CoO, Cu_2HgI_4, VO_2 and V_2O_3 by Somasundaram *et al.* (1986). The phase transitions were well resolved and the authors showed that the amplitude of the PA signal at the phase transition underwent large changes, however, their results did not follow the predictions based on the RG theory especially in the case of the second order phase transition. Recently, photopyroelectric detection has been used to monitor first- and second-order phase transitions in solids (Mandelis *et al.*, 1985). This technique appears promising in calculating the specific heat of materials, as compared to microphonic PAS, which can yield practical information on the *product* of thermal conductivity *and* specific heat of solid samples (Sigueira *et al.*, 1980).

2. PIEZOELECTRIC DETECTION

An alternative method to detect a phase transition by a PA technique is by making use of a piezoelectric transducer. Although there are many advantages to this technique, the appearance of several elastic modes of the transducer sometimes causes trouble to the interpretation of the experimental results. A theoretical model which accounts for the PA amplitude and phase generated at first and second order phase transitions by using a piezoelectric detection was proposed by Etxebarria *et al.* (1984). They extended the Korpiun and Tilger model (1980) to include the case of highly absorbing and thermally thick samples. These authors considered a linear temperature gradient along the solid and two thermal states in the sample during the phase transition as in the model by Korpiun and Tilger (1980). The schematic diagram of the geometry and temperature distribution was essentially the same as that in Fig. 7. In the present case, however, a sufficiently thin piezoelectric transducer was attached as the backing material at Z=-L. The temperature distribution was given in both the sample and gas by the solution of the general heat conduction equations without any sources. The surface heating due to the incident absorbed light was introduced as a discontinuity of the thermal flux at the sample-gas interface Z=0. This alternating heat source generated by the incident light modulation flows into the sample through conduction. A discontinuity of the thermal flow at the interface between the low and high temperature phases was further introduced as a result of phase transition. δ - and Λ - shaped functions were used as theoretical input curves of the specific heat for the first - and the second - order phase transitions, respectively, and the PA signal amplitude and phase were calculated.

The PA signal was detected as a voltage between the transducer electrodes. This voltage depended on two terms: the first term was related to the radial thermal expansion of the sample while the second term appeared due to the sample buckling because of the temperature difference between the sample surfaces. Etxebarria *et al.* (1984) calculated the PA signal amplitude and phase as functions of temperature starting from the explicit expressions for the observed voltage given by Jackson and Amer (1980). The theoretically predicted amplitude and phase variation near first and second order phase transitions are drawn in Figs. 8(a) and (b), respectively. The abscissa is $1-(\delta/\delta_o)$, where $\delta = T_c - T$ and $\delta_o = I_o L / 2k_l$. I_0 is the incident light intensity, k_l is the sample thermal conductivity and L is the sample thickness. For an actual measurement, the term $[1-R(\lambda)]$ caused by the sample reflectance must be taken into account by multiplication with I_0. T_c and T is the phase transition and ambient temperature, respectively. Typical values were further introduced to obtain a simple illustrative example for the PA amplitude and phase. These theoretical results showed that anomalies in the PA signal appear in a very narrow temperature interval ($\sim 10^{-2} K$ for the values chosen) close to the phase transition temperature. If higher-intensity light source is used, then this critical temperature region can be enlarged. However, experimentally observed anomalies in both amplitude and phase usually extended over a wider temperature range than predicted. The discrepancies between the theoretical predictions and the experimental results can be explained from both the experimental conditions and the model construction. One example is the temperature dependence of the elastic constants of the sample, which was ignored in the theoretical model. Such a dependence may be important in the PA signal generation due to the strong sample-transducer coupling.

Direct experimental data of the PA signal as a function of temperature for the first order phase transition of $CsCuCl_3$ is shown in Fig. 9. A continuous decrease of the PA amplitude which becomes more drastic near T_c of 423 K is observed in a relatively wide temperature range. At T_c, a sudden increase occurs, high enough to leave the signal baseline slightly above its room temperature value. The PA amplitude results obtained can be seen to be in reasonable agreement with the theoretical predictions in Fig. 8(a).

Etxebarria *et al.* (1985) further applied the piezoelectric detection technique of PA measurements to the investigation of the ferro-paraelastic phase transition at 89°C of the layered compound $(CH_3CH_2NH_3)_2CuCl_4$. From the

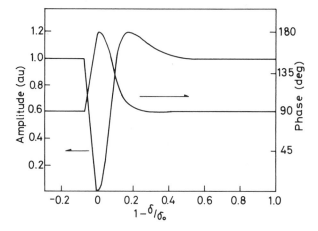

Fig. 8(a) Theoretical predictions for the PA amplitude and phase near a first order phase transition. The PA signals are sensed by a piezoelectric detector. [From Etxebarria *et al.* (1984)]

PA amplitude measurements, they observed an unusual behavior as shown in Fig. 10. The observed spectra could not be predicted from the second order phase transition theory [Fig. 8(b)]. The apparently anomalous result could only be interpreted by taking into account the PA signal dependence on the expansion coefficient, which shows an abrupt change at the phase transition temperature. Therefore, these workers concluded that the dependence of the PA signal on the thermal expansion coefficient becomes important when the piezoelectric detection method is applied.

Fig. (8b) Theoretical predictions for the PA amplitude and phase near a second-order phase transition. The PA signals are sensed by a piezoelectric detector. [From Etxebarria *et al.* (1984)]

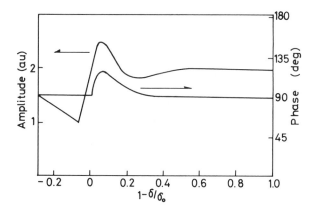

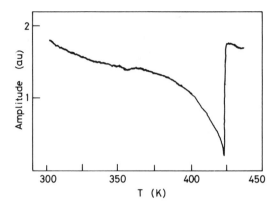

Fig. 9 Temperature dependence of the PA signal for $CsCuCl_3$ by He–Ne laser excitation. The phase transition temperature T_c is 423 K. [From Etxebarria *et al.* (1984)]

V. Conclusions

The low temperature PA spectra of GaSe and InSe have been discussed in this chapter. There are advantages to using PAS as an analytic spectroscopic

Fig. 10 PA amplitude as a function of temperature for $(CH_3CH_2NH_3)_2CuCl_4$ using a He-Ne laser as the excitation source. The continuous curve represents the experimental results and dots show the fitting from the specific heat and the expansion coefficient values. [From Etxebarria *et al.* (1985)]

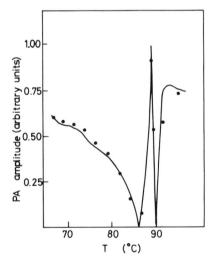

technique to gain knowledge of low temperature semiconductor spectra. The broad band at room temperature associated with the free exciton peak became narrower at lower temperatures. Thermal broadening due to exciton-phonon scattering reduces at low T, which allows one to obtain exciton peak energies with high resolution and thus determine not only the ground state (n=1) but also excited states (n=2,3...) of the direct exciton series. The singlet and triplet ground states may be observed in the high resolution spectra.

In the case of GaSe and InSe, new PA peaks appeared in the low temperature spectra below the fundamental absorption edge. If thermal quenching of these impurity or defect levels is fast, it follows that they can be observed only at low temperature measurements. With a higher spectral resolution, extremely useful information may be extracted for extrinsic energy states in semiconductors.

Reflectivity measurements from the PA spectra in the saturation region, where the PA signal is proportional to $[1-R(\lambda)]$, are also possible at low temperatures. Therefore, one can study the temperature variation of interband transitions. This is especially useful for samples without a mirror-like surface which would be necessary for carring out conventional reflectance spectroscopy.

The last part of this Chapter has been concerned with the investigation of phase transitions by the PA technique and the potential use of this method for semiconductor measurements. We feel that the PA technique will be more commonly used in the future since it has proven to be very sensitive to small changes of the thermal properties of materials at the phase transition temperature. Etxebarria's model of the piezoelectric detection method is promising, as it has shown qualitative agreement with the observed PA signals. However, there remain several discrepancies between experiment and theory. Additional theoretical work is, therefore, necessary for quantitative PA phase transition studies.

Another interesting potential use of low temperature PAS is in the spectroscopic investigation of a phase transition above and below the phase transition temperature. The changes in the electronic structure of semiconductors due to the phase transition should be readily observable by comparing these PA spectra.

VI. References

Antonioli G., Bianchi D., Emiliani U., Podini P. and Franzosi P. (1979). Nuovo Cimento **B54**, 211.

Bechthold P.C., Campagna M. and Schober T. (1980) Solid State Commun. **36**, 225.

Boucher F. and Leblanc R.M. (1981). Can. J. Spectrosc. **26**, 190.

Capozzi V. and Minafra A. (1981). J. Phys. C: Solid State Phys. **14**, 4335.

Dioszeghy T. and Mandelis A. (1986) J. Phys. Chem. Solids, **47**, 1115.

Etxebarria J., Uriarte S., Fernandez J., Tello M.J. and Gomez-Cuevas A. (1984). J. Phys. C: Solid State Phys. **17**, 6601.

Etxebarria J., Fernandez J., Arriandiaga M.A. and Tello M.J. (1985). J. Phys. C: Solid State Phys. **18**, L13.

Florian R., Pelzl J., Rosenberg M., Vargas H. and Wernhardt R. (1978). Phys. Stat. Sol. (a)**48**, K35.

Ikari T., Shigetomi S., Koga Y. and Shigetomi S. (1984). J. Phys. C: Solid State Phys. **17**, L969.

Ikari T., Shigetomi S., Koga Y. and Shigetomi S. (1986). Rev. Sci. Instr. **57**, 17.

Ikari T., Shigetomi S., Koga Y., Shigetomi S., Nishimura H. and Suzuki H. (1986). J. Phys. C: Solid State Phys. **19**, 2633.

Jackson W. and Amer N.M. (1980). J. Appl. Phys. **51**, 3343.

Korpiun R. and Tilgner R. (1980). J. Appl. Phys. **51**, 6115.

Mandelis A., Care F., Chan K.K. and Miranda L.C.M. (1985). Appl. Phys. A**38**, 117.

Murphy J.C. and Aamodt L.C. (1977). J. Appl. Phys. **48**, 3502.

Pichon C., Le Liboux M., Fournier D. and Boccara A.C. (1979). Appl. Phys. Lett. **35**, 435.

Robin M.B. and Kuebler N.A. (1977). J. Chem. Phys. **66**, 169.

Rosencwaig A. and Gersho A. (1976). J. Appl. Phys. **47**, 64.

Shigetomi S., Ikari T., Koga Y. and Shigetomi S. (1984). Phys. Stat. Sol. (a)**86**, K69.

Shigetomi S., Ikari T., Koga Y. and Shigetomi S. (1985). Phys. Stat. Sol. (a)**90**, K61.

Sigueira M.A.A., Ghizoni C.C., Vargas J.I., Menezes E.A., Vargas H. and Miranda L.C.M. (1980). J. Appl. Phys. **51**, 1403.

Somasundaram T., Ganguly P,. and Rao C.N.R. (1986). J. Phys. C: Solid State Phys. **19**, 2137.

Todorovic D.M. and Nicolic P.M. (1985). Appl. Opt. **24**, 2252.

Voitchovsky J.P. and Mercier A. (1974). Nuovo Cimento B**22**, 273.

PHOTOTHERMAL SPECTROSCOPIES OF AMORPHOUS SEMICONDUCTOR FILMS

Keiji Tanaka

Department of Applied Physics
Faculty of Engineering
Hokkaido University
Sapporo 060, JAPAN

I. Introduction

A great deal of work has been expended on amorphous semiconductors in recent years. The field is characterized with both fundamental and technological interests.

From a fundamental viewpoint, an amorphous semiconductor is regarded as a kind of random system in which atoms are connected by covalent bonds. The samples can, in many cases, be prepared easily, and thus these are appropriate to investigate the properties peculiar to disordered systems, such as the anomalies in specific heats at low temperatures (Phillips, 1981). Amorphous semiconductors are also suitable for studying the nature of randomness, since crystals having the same compositions as amorphous specimens are easily available. Electrons in semiconductors can be excited optically or thermally, and the dynamics of carriers excited in the random potential fields is of considerable fundamental interest. Here, the electronic carriers may play fascinating roles in non-periodic potentials.

Technologically, amorphous semiconductors are becoming widely used as electronic and optical materials (Hamakawa, 1982). Important examples are amorphous hydrogenated Si (a-Si:H) employed as solar batteries and a-Se layers as xerographic photoreceptors. The high optical absorption and high dark resistivity, respectively, of these materials are favorable for such applications. Although these devices might be inferior to those made from crystalline semiconductors in reliability, the easier preparation of large-area semiconducting films is an indispensable advantage.

When investigating amorphous semiconductors, we sometimes need the experimental data of optical absorption edges in thin film forms. For amorphous tetrahedral semiconductors such as a-Si:H, the samples can be prepared only as thin films with thicknesses of $1 \sim 10$ μm. For chalcogenide specimens, e.g. a-Se, we can prepare both deposited films and bulk glasses. As will be described in section III, however, properties of amorphous semiconductors depend on preparation methods and on treatments after the preparation, therefore, accordingly studies of the thin films are required. Among the various properties information on the electronic structure is of primary importance, so that evaluation of the absorption coefficients around the optical absorption edge in thin amorphous semiconductors is necessary.

Conventional optical spectroscopy is confronted with a problem when the optical absorption edges of thin films are studied: In general, the absorption coefficient α of a specimen having thickness d is calculated from the transmittance J by using the equation

$$J = \exp(-\alpha d), \tag{1}$$

where the reflection at the sample surface is neglected for simplicity. In amorphous semiconductors, α changes drastically from *ca.* 10^{-1} cm^{-1} to *ca.* 10^{5} cm^{-1} at around the optical absorption edge (see, Fig. 9). If the thickness of an

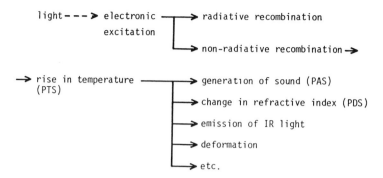

Fig. 1 Conversion of radiation energy absorbed in a semiconductor. The photon energy is assumed to be comparable to the energy gap.

amorphous film is 1 μm, then J = 0.99 when $\alpha = 100$ cm^{-1}. Therefore, investigation of the absorption edges down to $\alpha = 100$ cm^{-1} requires the measurement of the transmittance with an accuracy better than 1%. This accuracy is difficult to obtain, owing to fluctuations of light intensity and so forth. Further, in low absorption regions spectral fringes due to the multiple interference in thin specimens appear, making the evaluation of α more complicated.

If an absorbing material is irradiated with light, most of the radiation energy is transferred into heat, which causes a rise in temperature of the sample (Fig. 1). We can directly detect the temperature rise and evaluate its absorption coefficient by the method called (in a restricted terminology) photo-thermal spectroscopy (PTS). The principle is essentially the same as that of photo-calorimetry, which is employed to investigate the absorption coefficient of optical fibers at a single laser wavelength (Skolnik, 1975). In PTS, however, measurements of spectral dependences are emphasized. The rise in temperature induces several effects in surrounding media, which can be detected acoustically or optically. The relevant thermal wave methods are termed as microphone photoacoustic spectroscopy (PAS) (Rosencwaig, 1980) or photothermal deflection spectroscopy (PDS) (Jackson *et al.*, 1981).

In this Chapter, we review the applications of these novel spectroscopic methods to studies on amorphous semiconductors. In section II, we introduce the thermal spectroscopic methods capable of measuring small absorptions in amorphous films, and compare the characteristics with those of non-thermal techniques. In section III, some unique features of amorphous semiconductors are briefly summarized, and the results obtained with the thermal methods are discussed. In the final section, perspectives for future studies on the thermal spectroscopies are described.

II. Spectroscopic Methods

1.   THEORETICAL BACKGROUND FOR THERMAL METHODS

We derive an expression for the change $\Delta T(t)$ in temperature of a given medium when illuminated with chopped light, $I(t) = I \exp(i\omega t)$, where ω is the chopping angular frequency. In the following the sample is assumed to be thermally thin, i.e. $d \ll l$ where d is the thickness and l is the thermal diffusion length (see, Table I) defined as

$$l \equiv (2k/\omega\rho c)^{\frac{1}{2}}, \tag{2}$$

where k, ρ and c are, respectively, the thermal conductivity, the density and the specific heat. Since $d \ll l$, $\Delta T(t)$ in the specimen becomes independent of depth from the illuminated surface, so that a simple theoretical analysis can be applied.

Table I

Densities (ρ), specific heats (c), thermal conductivities (k) and thermal diffusion lengths (l) of several materials, evaluated when the chopping frequency is 10 Hz.

	$\rho(\text{g/cm}^3)$	$c(\text{J/g.K})$	$k(\text{W/cm.}K)$	$l(\mu\text{m})$
a–Se	4.29	0.325	5.07×10^{-3}	110
a–As_2S_3	3.20	0.66	4.40×10^{-3}	81
Si	2.33	0.70	1.42	1600
SiO_2	2.27	0.84	1.4×10^{-2}	150
CCl_4	1.59	0.87	1.03×10^{-3}	49
PVDF	1.78	1.4	1.3×10^{-3}	41
$LiTaO_3$	7.45	0.42	4.1×10^{-2}	200
air	1.29×10^{-3}	5.25	2.4×10^{-4}	790

If the material having a heat capacity Q is heated by monochromatic radiation, $\Delta T(t)$ can be approximated by

$$Qd\Delta T(t)/dt = W(t) - G(t)\Delta T(t), \tag{3}$$

where $W(t)$ is the absorbed power, and $G(t)$ represents the heat conduction loss defined as

$$G(t) = k(1/\Delta T(t)) \int \nabla T(t) \, dS, \tag{4}$$

in which the integration is carried out over the sample surface. Energy dissipation by radiation from the heated area is neglected here. Since the input radiation is chopped at frequency ω, $W(t)$, $G(t)$ and $\Delta T(t)$ are written so that $W(t) = W \exp(i\omega t)$. Therefore, from Eq. (3) we obtain,

$$\Delta T = W/(G+i\omega Q). \tag{5}$$

In conventional experimental situations, $\omega \gg 1 \text{ sec}^{-1} > G/Q$, and accordingly Eq. (5) is approximated by

$$|\Delta T| = W/\omega Q. \tag{6}$$

W is related to the absorption coefficient α of the film having the thickness d *via* the equation

$$W = I\{1-\exp(-\alpha d)\}, \tag{7}$$

where reflections as well as the multiple interference are neglected in order to keep the mathematics simple. Combining Eqs. (6) and (7), we obtain a proportionality

$$|\Delta T| \propto 1-\exp(-\alpha d), \tag{8}$$

shown in Fig. 2, or an equality

$$|\Delta T|/|\Delta T_s| = 1-\exp(-\alpha d), \tag{9}$$

where ΔT_s denotes the temperature change from saturation when $\alpha d \gg 1$. Eq. (9) shows that by measuring the ratio $|\Delta T|/|\Delta T_s|$ or other related quantities (see Fig. 1), we can evaluate α. This is the principle of thermal spectroscopic methods.

Fig. 2 Dependence of $|\Delta T|$ on α (see Eq. (8)) for a film of 1 μm thickness.

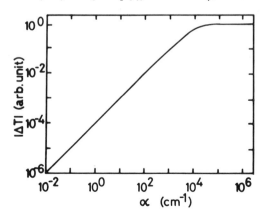

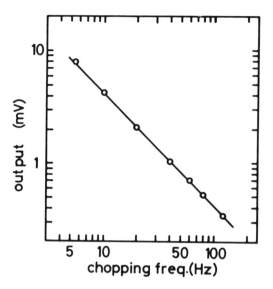

Fig. 3 Dependence of the output corresponding to $|\Delta T|$ on the chopping frequency of a PTS system (see Eq. (6)). The sample is an a–As_2S_3 film of 0.61 μm thickness and the temperature sensor is pyroelectric $LiTaO_3$ of 50 μm thickness.

The above analysis reveals some points worth mentioning. Eq. (6) shows that the thermal signal is inversely proportional to the heat capacity Q of the material inspected. Thus, the thermal wave methods are particularly suitable for thin films having low Q values. Eq. (6) also shows that $|\Delta T| \propto \omega^{-1}$. This frequency dependence is observed with the PTS method (see the next section) as shown in Fig. 3. The proportionality is modified in some cases, e.g. when thin samples are attached to thick substrates and the assemblies are effectively thicker than the thermal diffusion lengths. Mostly, however, $|\Delta T|$ monotonically decreases with increasing ω (Rosencwaig, 1980; Jackson et al., 1981). As is seen in Fig. 2, Eq. (8) gives $|\Delta T| \propto \alpha$ when $\alpha d \ll 1$. This implies that the thermal wave methods are convenient for measuring small absorptions. Conversely, they are not appropriate for high absorption coefficient measurements, since the signal is saturated when $\alpha \gg d^{-1}$.

Evaluation of α by using Eq. (9) may yield errors of ~ 20%, since the refractive index of the material is assumed to be unity there. To improve the accuracy, we must know the spectral dependence of refractive indices of the sample and surrounding media, and evaluate the reflections and multiple interference of light by using more rigorous equations, such as Eq. (2) reported in the paper by Tanaka et al. (1983). In addition, precise estimation of Q and G in Eq. (5) is actually not feasible. The thermal spectroscopic results are, therefore, calibrated with the data at $\alpha \approx 10^4$ cm^{-1} obtained with the transmission method.

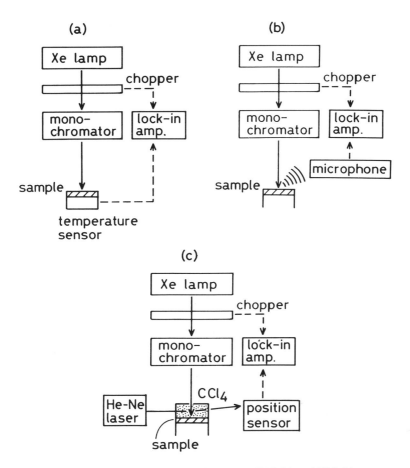

Fig. 4 Schematic illustrations of PTS (a), microphone PAS (b), and PDS (c) systems. Solid and dashed lines denote optical paths and routes of the electric signal.

2.   THERMAL SPECTROSCOPIC METHODS

As far as thermal spectroscopic methods are concerned, the rise in the temperature of various related quantities (Fig. 1) is measured for the evaluation of absorption coefficients. We briefly review several thermal methods and their features in this section.

The thermal spectroscopic methods have common advantages over conventional transmission methods. The first is that high accuracy is not required for measuring small absorptions. This arises from the fact that when the absorption of a material is zero, the thermal output becomes also zero. Regarding the example described below Eq. (1), the magnitude corresponding to the absorption of 1% can be measured with thermal methods. High sensitivity is needed but high accuracy is not essential, therefore the situation is much more favorable than the conventional methods. The second advantage is that the signal can be enhanced by increasing the light intensity incident upon the sample, since the absorbed radiation energy is proportional to the light intensity. This fact, however, should not be overestimated here, since properties of amorphous semiconductors are often modified with intense illumination (see, Figs. 14 and 17). Third, the thermal methods are mostly insensitive to light scattering. Consequently, the absorption, not the attenuation, can be *directly* measured. Fourth, since the thermal methods require no photoelectric devices detecting the light intensity, in principle, they can be used to inspect the spectral dependence over wide photon-energy regions, where efficient photodectors may not be available.

PTS is the most direct method of obtaining the spectral dependence of the optical absorption coefficient. Fig. 4a illustrates a typical system. A light beam from a Xe lamp is chopped mechanically, passed through a monochromator, and illuminates a thin film under investigation. The sample is periodically heated by absorbing the optical energy, and the rise in temperature is detected by a temperature sensor in conjunction with a lock-in amplifier. This method has become practical for investigating amorphous films only in recent years (Tanaka *et al.*, 1983 and 1986).

The appropriate selection of the temperature sensor is an important point of PTS systems. A fundamental requirement for the sensor is, as is implied from Eq. (6), a small thermal capacity, since we must include its heat capacity into Q. In addition, optical properties of the sensor are important. In a wavelength region where an investigated film is nearly transparent, radiation energy may be absorbed by the temperature sensor, which generates residual signals. The background signal can be cancelled when a pair of thermocouples is utilized differentially as the temperature sensor (Tanaka *et al.*, 1983). Another method for reducing the signal is to use temperature sensors transparent in the wavelengths of interest. For this purpose, pyroelectric materials such as polyvinylidene difluoride (PVDF) films sandwiched in between CuI electrode films (Tanaka *et al.*, 1986) and thin $LiTaO_3$ crystals coated with CuI or indium-tin oxide (ITO) electrodes can be used. Alternatively, one may employ a temperature sensor, at the surfaces of which the light is completely reflected. At present, however, we have not succeeded in finding detectors having such a property.

Distinctive characteristics of PTS are summarized as follows: The most important one is its simple and direct detection scheme. Neither critical alignment of optical components nor rigid optical tables are needed. The system is free from acoustic and vibrational noises. In contrast, as described above, stringent requirements are sought for the temperature sensor, since this devices should be in thermal contact with the sample. Among a few sensors investigated so far, $LiTaO_3$/ITO seems most appropriate to study amorphous films. It has been employed for investigating optical absorptions in a temperature range between 20 K and 891 K (Glass and Line, 1976).

PAS is widely employed in the investigation of amorphous films. Its principle is shown in Fig. 4b. A chopped monochromatic beam from a Xe lamp irradiates a specimen, which generates acoustic waves having the same frequency as that of the chopper, typically $10 \sim 50$ Hz. The acoustic waves are detected by a microphone, and the electrical output is measured synchronously. The acoustic amplitude is proportional to $|\Delta T|$ in Eq. (5), so that absorption coefficients can be evaluated. PAS is applicable to absorption measurements at cryogenic temperatures (Rosencwaig, 1980), although the system becomes fairly complex. One of the demerits of the system is that it is substantially affected by acoustic noise such as shock waves generated when doors of experimental rooms are closed. PAS systems are now commercially available.

Fig. 4c schematically shows at typical PDS system of the transverse type, originally developed by Jackson *et al.* (1981). A sample, e.g. an a-Si:H film deposited on an oxide glass is illuminated with monochromatic chopped light from a Xe lamp. The specimen is heated by absorbing the radiation energy W. The temperature rise in the film induces a spatially graded change in the refractive index of liquid CCl_4. A light beam from a He-Ne laser probing the index gradient is deflected (mirage effect). The magnitude S of the deflection, measured by a position-sensitive photodetector, can be expressed as (Amer and Jackson, 1984; and Chapter 10)

$$S \propto W \exp(-z_0/l) dn/dT, \tag{10}$$

where z_0 is the separation between the sample surface and the probe beam, l and dn/dT are the thermal diffusion length and the temperature coefficient of the refractive index of CCl_4, respectively. Here, the thermal diffusion length of the sample is tacitly assumed to be smaller than the diameter of the focussed spot of the chopped light.

A main advantage of PDS lies in its high sensitivity capable of measuring the absorption coefficient down to less than 1 cm^{-1} in films of 1 μm-thickness. Thus, the method is especially convenient for the study of the small absorption originating from midgap states in semiconductor films. The sensitivity is augmented partly by the great temperature coefficient of liquid CCl_4, 5×10^{-4}/K. If we use air instead of CCl_4, the sensitivity will be restricted to $\geq 10^2$ cm^{-1}. Thus, PDS is valuable only at around room temperature where the liquid can be used. It may be also advantageous to this method that the position-sensitive detector can be remotely located. In contrast, since the probe beam must graze the sample surface, i.e. $z_0 < l$, critical alignment of the optical system is required. Stabilization of the probe beam is necessary, and air turbulence along the optical path and particulates in the deflecting medium should be carefully removed. PDS systems should be constructed on a vibration-isolated table.

There are other thermal methods which may be applied for detecting the temperature rise in thin films. One such method is photo-emittance spectroscopy (Skolnik, 1975), in which the blackbody radiation from a material heated by monochromatic light is detected by an infra-red sensor. The method seems, however, to be less sensitive than the techniques described above. In other methods, the strain induced by thermal expansions in a heated solid is detected optically (Tanaka and Ohtsuka, 1977; Dersh and Amer, 1985) or acoustically (Rosencwaig, 1980). The acoustic method is called piezoelectric PAS, since the

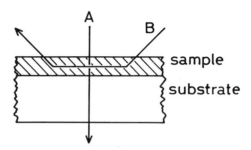

Fig. 5 The principle of the guided wave spectroscopy. A and B indicate the optical paths of plane and guided waves.

strain is monitored by piezoelectric transducers such as PZT. This method is useful only when the sample is thicker than the piezoelectric material. If otherwise, the strain is not efficiently conducted to the transducer, or expansive and compressive strains in the transducer cancel the electrical output. Therefore, piezoelectric PAS may not be applied to thin amorphous films.

3.   NON-THERMAL SPECTROSCOPIC METHODS

Several non-thermal methods that can produce similar results given by the thermal methods have been and are being developed. In this section, we briefly survey these techniques.

Conventional transmission methods can be improved. Ritter and Weiser (1986) recently demonstrated that the absorptance of thin films normalized by their transmission is practically free of interference fringes. The absorption coefficient of the film between 10^4 cm^{-1} and 10^{-1} cm^{-1} may be obtained from this ratio in a simple way.

Guided wave spectroscopy (Olivier, 1984) is useful in the investigation of small absorptions in thin films. The principle is very simple as shown in Fig. 5. In the figure, A denotes the optical path employed in conventional transmission measurements. In contrast, the absorption of the film can be probed with the optical guided wave (B in Fig. 5), which can be propagated when the refractive index of the film is higher than that of the surrounding media, mostly substrates and air. Since the path of the guided wave can be lengthened to ~ 1 cm, absorption coefficients around 1 cm^{-1} can be measured. In addition, by using guided waves of TE and TM modes, we can investigate dichroism. This technique has been applied to the evaluation of a-Si:H films. It has, however, at least two practical problems: The first is that guided waves are substantially scattered if inhomogeneities are present at the surfaces of the investigated films. Thus, this spectroscopy gives the coefficient of attenuation, *not* absorption. The second is that, in order to propagate guided waves, severe optical alignment is needed, and spectral investigation is not easy. Plane optical waves can be converted into guided waves by prism couplers which are located closely to the film, i.e. with the air-gap separation of ~ $\lambda/4$, where λ denotes the wavelength. The conversion efficiency critically depends on the separation, and accordingly on the wavelength of light when spectral investigation is carried out with a constant separation. Thus, the efficiency should be calibrated by some other techniques. Several ideas have been proposed for this purpose (Sasaki *et al.*, 1980; Won *et al.*, 1980; Gal *et al.*, 1985).

A review of photocurrent spectroscopy is given by Crandall (1984). Photocurrent J results from two serial phenomena: generation of photocarriers and their transport to electrodes. Thus, it can be expressed as

$$J \propto \eta \tau \mu W, \qquad (11)$$

where η denotes the quantum efficiency for carrier excitation, τ the lifetime and μ the mobility of the carriers. W is the absorbed light energy approximated by Eq. (7), so that the absorption coefficient α can be evaluated from measurements of J. This serves as a simple method for the inspection of small absorptions. The photocurrent signal is enhanced when generated by guided optical waves. By using this technique, Olivier and Bouchut (1981) were able to measure the absorption coefficient of a-Si:H films down to $\sim 10^{-3}$ cm^{-1}. The problem lies, as is easily speculated, in the assumption that $\eta \tau \mu$ is independent of the photon energy. Amer and Jackson (1984) demonstrated through a comparative study of PDS and photocurrent spectroscopy that this factor varies more than three orders of magnitude in some cases. Thus, the validity of the correspondence between J and α should be carefully inspected for each study.

Fig. 6 Relations among several definitions indicating random systems.

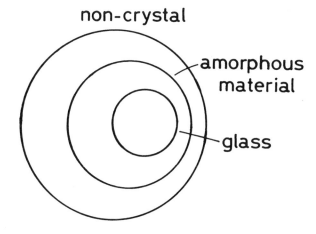

III. Applications to Physics of Amorphous Semiconductors

1.   GENERAL REMARKS

It may be appropriate first to clarify the relations among some definitions used to specify random systems. Fig. 6 illustrates Mott's definition (Mott and Davis, 1979). According to that definition, a "non-crystal" denotes a random condensed matter, which includes most of liquids and some solids having non-periodic atomic structurs. "Amorphous" is a general term representing non-crystalline solids. We, then, refer to an amorphous material prepared through quenching the melt as a "glass", well-known examples of which are the oxide glasses used as windowpanes. Therefore, conventional a-Si:H films, which are deposited from gas phases by glow-discharge or other techniques, are not grouped into glassy semiconductors. Phillips (1979) and others (Elliott, 1984) have used slightly different definitions.

Glassy semiconductors normally exhibit the glass-transition phenomenon, and, in contrast, amorphous tetrahedral films undergo crystallization when heated in furnaces (Elliott, 1984). Both phenomena manifest that the internal energies of these materials are higher than those of the crystals. That is, these materials are thermodynamically in non-equilibrium states. Note that the liquids having random atomic structures are in equilibrium states, and that diamonds having the perfect periodic structure are not thermally equilibriated under 1 atm at room temperature.

We may summarize, therefore, that three essential properties of amorphous semiconductors are i) structural randomness, ii) non-equilibriation or meta-stability in a thermodynamic sense, and iii) existence of bandgaps of 0.1 ~ 3 eV. These conditions are fulfilled in several kinds of materials, in which chalcogenides and tetrahedrals are typical substances.

When studying amorphous semiconductors, we must recognize a situation largely different from that familiar with crystalline physics. Structures of single crystals can, in principle, be determined through diffraction experiments, and this knowledge is a firm basis or investigations of various properties. We, however, have no detailed structural information of amorphous semiconductors. Thus, one of the purposes of studying their electronic or optical properties is to get some idea on these microscopic structures. The next section gives concise description of structures of amorphous semiconductors.

2.   STRUCTURAL PROPERTIES

Structures of amorphous semiconductors can be classified into two components. One is on the structure of normal bonds having densities of $\sim 10^{23}$ cm^{-3}. The other concerns the topology of defects such as dangling bonds, valence alteration pairs, local structures around dopants and homopolar bonds in stoichiometric materials (Elliott, 1984). Although the density of these defects is at most 10^{21} cm^{-3}, the energy levels may be located in the band gap. Therefore,

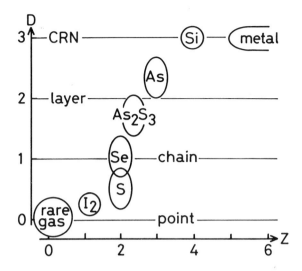

Fig. 7 Classification of amorphous materials in terms of the average coordination-number Z and the network dimensionality D. CRN means the continuous-random-network structures. For detalis, see the text.

electronic properties of amorphous semiconductors are substantially affected by defects.

In order to grasp the normal-bonding structures of amorphous materials, we can adopt two quantities shown in Fig. 7 (Zallen, 1983). One is the average coordination-number of covalent bonds per atom, here denoted as Z, which represents the short-range order of local atomic units (Phillips, 1979). If the composition of a sample is expressed as $Si(Ge)_x As_y S(Se)_{1-x-y}$ then

$$Z = 4x+3y+2(1-x-y), \tag{12}$$

since coordination numbers of Si(Ge), As and S(Se) in materials of interest are, respectively, 4, 3 and 2. As shown in Fig. 7, $Z = 2$ for Se, 2.4 for As_2S_3 and 4 for Si. Further, $Z = 1$ for I_2 and $Z = 0$ for rare gases, although these are not grouped into semiconductors, of course. The coordination number of an atom, whether this is included in crystals or in glasses, can be uniquely defined. This empirical observation is known as the Ioffe-Regel rule (Mott and Davis, 1979).

The other quantity is the network dimensionality D, defined as the number of dimensions in which the covalent amorphous clusters are embeded (Zallen, 1983). this gives information on medium-range structural orders. For instance, D=1 for a-Se, since it is considered to be composed with flexible chain molecules, which are held together with weak intermolecular forces of the van de Waals type. Most strictly, D is assumed to be fractional, since the chain molecules are neither infinite (D=1) nor very short (D≈0) in length. However, we have no reliable and direct means identifying the medium-range orders in random systems, and assignments of D are sometimes controversial. It is mentioned here that D of amorphous materials depends on preparation methods, e.g. D≈2 for glassy As_2S_3 and D≈0 for As_2S_3 films evaporated onto cooled substrates (Daniel and Leadbetter, 1980). The dependence of structures on preparation methods is attributable to an inherent property of amorphous materials, namely their meta-stability (Fritzsche, 1980).

Thermal properties of amorphous materials may be connected with D. For instance, materials having greater D exhibit higher thermal conductivities. Table I compiles several thermal properties of materials relevant to thermal spectroscopic methods.

3. ELECTRONIC AND OPTICAL PROPERTIES

The electronic structures of chalcogenide and tetrahedral materials are different from each other. Fig. 8 exemplifies the difference with Se and Si. In Se, the outer-most electron configuration is $4s^2p^4$. The energy separation between the s and p states is ~ 10 eV, so that the s state is negligible in the following argument. When Se atoms are bonded with the two-fold coordination, the p state splits into three levels, the bonding σ state, the antibonding $σ^*$ state and the lone-pair electron state. The energy levels of these states are broadened in Se solids, in which the conduction band arises from the $σ^*$ state and the upper part of the valence band is composed by the lone-pair electron state. In Si, as is well known, the outer-most electron configuration of $3s^2p^2$ in the isolated atoms

Fig. 8 Electronic structures of Se and Si. The atomic structures are shown on the left. The middle column shows the spatial densities of bonding (o) and lone-pair (LP, hatched) electrons. Dotted lobes indicate bonding electrons of neighboring atoms. The right illustrations show energy levels of the outer electrons in the atoms and the solids.

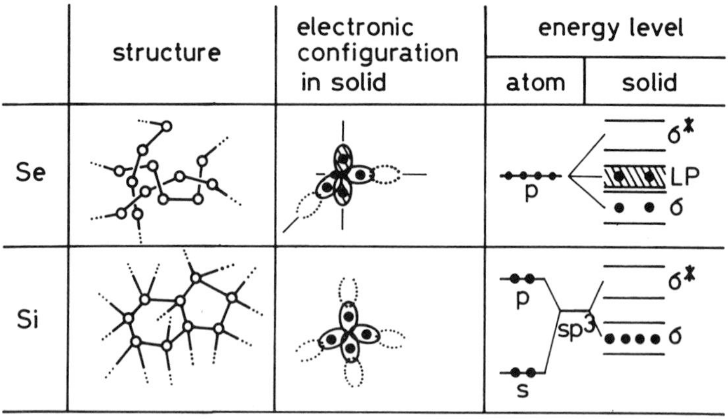

is hybridized into the sp^3 orbital which splits into the bonding and antibonding states. These states correspond to the valence and conduction bands in the solid.

In tetrahedral amorphous semiconductors, extrinsic conduction had not been obtained for a long time. The reason is now understood as follows. In amorphous networks of tetrahedral semiconductors much strain is included, because of the high coordination number (Phillips, 1979). The strain cannot be relaxed in real materials, and as a result many dangling bonds, typically 10^{20} cm^{-3}, are contained. It is believed that wavefunctions of electrons in the dangling bonds resemble that of the atomic sp^3 state, since the bonds are isolated from surrounding atoms. Accordingly, as is implied from Fig. 8, the energy levels of the defects are present in the middle of the band gap (Bar-Yam and Joannopoulos, 1986). These mid-gap states of $\sim 10^{20}$ cm^{-3} pin the energy position of the Fermi level, so that control of the conduction type is impossible.

Spear and LeComber (1975) discovered that introduction of H atoms into the network is effective for terminating the dangling bonds. Bonding and antibonding states of Si-H bonds are not located in the energy gap (Adler, 1984), and the density of the mid-gap states can be reduced. They also succeeded in preparing n-type and p-type a-Si:H films by doping P and B atoms when depositing the films.

Chalcogenide glasses exhibit some interesting photoeffects (Elliott, 1984), since the structures are flexible and the lone-pair electrons are liable to be excited by illumination of visible light. As will be discussed in Section 5, the photoeffects accompany modifications of optical properties.

Fig. 9 The optical absorption edge of a–As$_2$S$_3$ evaluated from transmission measurements. Absorption coefficients α above and below $\sim 10^3$ cm^{-1} are studied by using evaporated films and bulk glasses.

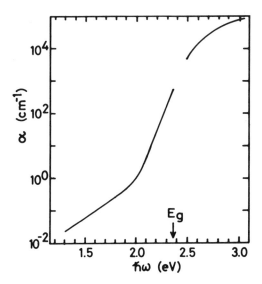

With regard to the nature of the defects, there remain several open problems. Since the density of the defects is at most two orders of magnitude less than that of the normal bonds and lattice structures are non-periodic, direct structural determination of the defects in amorphous semiconductors seems very difficult. We can only speculate the structure from information obtained from electronic studies, e.g. optical absorption, photoluminescence and ESR experiments. The results are discussed in light of theoretical models.

Mott and Davis (1979) and Kastner and Fritzsche (1978) proposed the ideas of charged dangling bonds and valence alteration pairs in chalcogenide materials. They argued that defects such as negatively-charged one-fold-coordinated chalcogen atoms were included in the glasses. Direct experimental evidence of the defects has not been obtained, nonetheless many experimental results such as absence of ESR signals in As-chalcogenide glasses, can be understood with the assumption of the defects in a coherent picture. It is considered that these defects contribute to the mid-gap states. Similar concepts are currently applied to tetrahedral materials (Adler, 1984).

Fig. 9 shows a typical result of the optical absorption edges in amorphous semiconductors (Kosek and Tauc, 1970; Wood and Tauc, 1972). The characteristics, obtained by measuring the transmittance of bulk and thin As_2S_3 samples, are divided into three parts. These have been the subject of much research and controversy with respect to their origins.

In the region where $\alpha \gtrsim 10^4$ cm^{-1}, the spectral behavior follows

$$\alpha \hbar \omega \propto (\hbar \omega - E_g)^n, \tag{13}$$

where $\hbar \omega$ denotes the photon energy and $1 \leq n \leq 3$. This behavior extends over photon energies of ~ 0.5 eV. It is common to define the optical energy gap E_g, sometimes called the Tauc gap, by using this relationship. What E_g means, however, has not been confirmed. The following shows a simple argument for this spectral dependence.

Under the assumption that the wavevector is not a good quantum number in disordered systems, an expression for α can be given from Fermi's golden rule:

$$\alpha \hbar \omega \propto \int D_v(E) \, D_c(E + \hbar \omega) \, |P|^2 \, dE, \tag{14}$$

where the integral is carried out over all filled states, normally the valence band, and D_v and D_c represent the densities of the initial and empty states. We assume that the transition matrix element P is independent of E, and that $D_v(E) \propto (E_v - E)^i$ and $D_c(E) \propto (E - E_c)^j$, where E_v and E_c denote the energies of the top of the valence band and the bottom of the conduction band, respectively. Then Eq. (14) is reduced to

$$\alpha \hbar \omega \propto (\hbar \omega - E_g)^{i+j+1}, \tag{15}$$

where $E_g = E_c - E_v$. For conventional three-dimensional materials, $i=j=\frac{1}{2}$, resulting in $n=2$, which is observed in the majority of amorphous semiconductors. In contrast, $n=1$ in a-Se. This observation was interpreted by Mott and Davis (1979) as originating from the low-dimensional structure ($Z=1$), in which $i=j=0$ might be assumed. A similar result has been discovered in amorphous As-S films of S-rich compositions (Tanaka, 1980).

In the region between $\sim 10^1$ cm^{-1} and $\sim 10^4$ cm^{-1}, α obeys a simple relation,

$$\alpha \propto \exp(\hbar\omega/E_0), \tag{16}$$

which is termed the Urbach tail. Such a spectral dependence is often observed in ionic crystals, where E_0 is a function of temperature. In amorphous semiconductors, E_0 hardly depends on temperature.

The Urbach tail in amorphous semiconductors is considered to originate from electronic transitions in random potential fields (Mott and Davis, 1979), so that E_0 can be used as a measure of the fluctuations. It is reported that E_0 varies linearly from ~ 50 meV to ~ 100 meV as Z increases from 2 to 4. This can be interpreted as an enhancement of internal strains in high Z materials.

It is assumed that the density-of-states of the tails below the conduction band and above the valence band has an exponential dependence:

$$D_i(E) \propto \exp(E/E_0^i), \ i=c \text{ and } v, \tag{17}$$

Fig. 10 The energy-configuration (E-q) diagram for electron-lattice coupling systems exhibiting exponential tail in optical absorption edges.

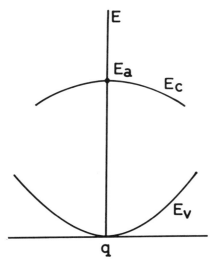

where E_0^i are defined for the conduction-band tail ($E_0^c > 0$) and the valence-band tail ($E_o^v < 0$). Then, E_0 in Eq. (16) can be deconvoluted into E_0^c and E_0^v by using Eq. (14) under several assumptions (Tanaka and Yamasaki, 1982; Amer and Jackson, 1984).

It should be mentioned that in the deconvolution analysis the rigidity of lattice structures is implicitly assumed. The assumption, however, is not necessarily justified (Cohen *et al.*, 1983), particularly in chalcogenide systems. As was described previously, one inherent property of the chalcogenides lies in its small Z, which implies that the lattices are substantially flexible. Thus, the atomic structures are probably distorted when electrons are excited. We show in the following that such systems may exhibit the exponential dependence of the absorption coefficient.

Optical properties of the strong electron-lattice coupling systems can be discussed by using configuration-coordinate diagrams. An example of such diagrams (Kolobov and Konstantinov, 1979) is shown in Fig. 10, in which the horizontal axis represents the configuration q of certain atoms and the vertical axis denotes the energy summed up for electronic and lattice systems. Upper and lower lines denote the states associated with the conduction and valence bands. The absorption coefficient α can be approximated by

$$\alpha \propto \int \exp(-E_v/k_B T)\delta(E_v - E_c + \hbar\omega)\, dq, \qquad (18)$$

where $k_B T$ is the thermal energy, the exponential term represents the population of the lower state, and $\delta(x)$ indicates the Dirac delta function. If $E_v = q^2$ and

Fig. 11 Optical absorption edges of undoped and P-doped a-Si:H evaluated by using PDS. The film is glow-discharged with the substrate temperature of 230°C and the rf deposition power of 2 W. In P-doping, the gas-phase concentration of PH$_3$ is 10^{-2}.

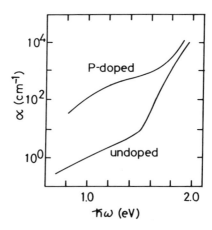

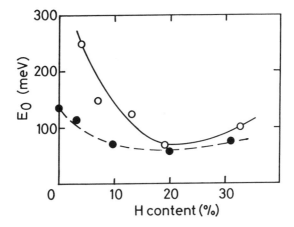

Fig. 12 Dependence of the Urbach slope parameter E_0 on the H content in a-Si:H films prepared by the ion-beam deposition (open circles) and in a-Ge:H films deposited by reactive sputtering (solid circles).

$E_c=E_a-\gamma q^2$, in which E_a and γ are constants, we can derive an exponential dependence from Eq. (18):

$$\alpha \propto \exp\{(\hbar\omega-E_a)/(1+\gamma)k_B T\}. \tag{19}$$

Below $\sim 10^2 cm^{-1}$, a higher absorption than that expected from an extrapolated Urback tail is observed. Its origins can be attributed to impurities (Tauc, 1975), charged or neutral defects, or extrinsic dopants (Mott and Davis, 1979). Investigation of the absorption profile is vitally important, since it gives information on states in the energy gap. Cody (1984) shows that, under several assumptions for electronic transitions from defect states to the conduction band, the low-energy absorption can be expressed by

$$\alpha/\hbar\omega=A(\hbar\omega-E_d)^{1/2}. \tag{20}$$

In this equation, A and E_d are constants relating to the density of the defects and their energy level. The constant A can be determined through fitting the dependence to experimental curves, so that the density is evaluated. Amer and Jackson (1984) theoretically derive an alternative expression.

4.   TETRAHEDRALS

Main results obtained by PAS and PDS studies give information on the Urbach tail and gap states in a-Si:H and a-Si alloy films. The samples are prepared through several deposition methods under various conditions, and are either doped or irradiated*.

Fig. 11 (Amer and Jackson, 1984) shows a typical PDS result of undoped a-Si:H films prepared by the glow-discharge technique. We see the quadratic dependence (Eq. (13)), the Urbach tail (Eq. (16)) and the mid-gap absorption (Eq. (20)). Extensive work suggests that the slope of the Urbach tail is not influenced appreciably with changes in deposition parameters, if preparation method and H content are fixed (see, e.g. Cody, 1984). When the H content is less than 20%, the slope becomes smaller with decreasing the H content as shown in Fig. 12 (Ceasar et al., 1984; Cody, 1984). Yonezawa and Cohen (1981) demonstrated theoretically that the decrease in the slope is induced by increasing strain in amorphous networks resulting from the reduction of the H content. A slight increase in E_0 above the 20% H-range is attributable to mixing of monohydride and dihydride coordinations.

In contrast, the residual absorption below the Urbach tail is substantially influenced by deposition procedures and doping. Fig. 11 also shows a result of P-doped a-Si:H films. We see that by P doping the Urbach tail remains unaffected and the absorption below ~ 1.6 eV increases. A similar result is confirmed by PAS studies (Yamasaki et al., 1981).

Fig. 13 A typical diagram showing density-of-states for undoped and P-doped $(PH_3/SiH_4 = 10^{-2})$ a–Si:H films. E_c is the conduction-band edge, E_v the valence-band edge and E_F the Fermi energies.

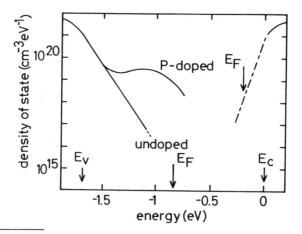

* Detailed results recently obtained for tetrahedral amorphous semiconductors can be found in the Journal of Non-Crystalline Solids (1985), Vol. 77 & 78, *(Proceedings of 11th International Conference on Amorphous and Liquid Semiconductors).*

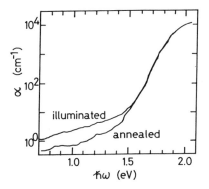

Fig. 14 Effect of illumination on the absorption spectrum of undoped a-Si:H. Illumination creates the mid-gap absorption which can be annealed away by heating the films above 150°C.

By using Eqs. (14) and (17) under several assumptions, data such as those shown in Fig. 11 can be converted to density-of-states spectra. Fig. 13 illustrates the example reported by Tanaka and Yamasaki (1982). Several characteristics are confirmed in the spectra: i) The tail above the valence band is expressed by Eq. (17) with $E_0^v \approx -60$ meV. The conduction-band tail is much sharper, $E_0^c \leq 40$ meV. ii) Even in the undoped film, a hump may exist at 0.6 ~ 0.8 eV above the valence-band edge (Amer and Jackson, 1984). A possible interpretation of the hump is based on the electronic transitions from the states accompanying Si dangling bonds to the conduction band. It is considered that H atoms contained in the Si matrix cannot completely terminate Si dangling bonds, and defects of ~10^{15} cm^{-3} still remain. iii) P doping produces the mid-gap states of ~10^{19} cm^{-3} eV^{-1}, the magnitude of which does not necessarily correlate with the dopant density. It is estimated that only ~ 10% of doped P atoms are effective in moving the position of the Fermi level, and most of the remainder contributes to defects having 3-fold coordinations with the p^3 orbital (Adler, 1984). This doping inefficiency is contrasted with that in crystalline semiconductors. The p-type dopant B in a-Si:H can lower the position of the Fermi level in principle, however, the resulting spectral feature seems to be more complicated (Adler, 1984).

In P-doped a-Si:H, Tajima *et al.* (1986) recently discovered with PAS studies at room and low temperatures that the mid-gap absorption is affected by the strong electron-lattice coupling. Bar-Yam and Joannopoulos (1986) theoretically predicted this coupling. These observations are expected to modify the considerations based on Eq. (20), which neglects the lattice distortion.

Fig. 14 shows an effect of illumination to an undoped a-Si:H film. It is known that by intense photoillumination the photoconductivity response of glow-discharged a-Si:H is degraded, the so-called Staebler-Wronski effect (Staebler and Wronski, 1977). Amer *et al.* (1983) demonstrated by using PDS that the gap-state absorption increases with illumination. This increase is

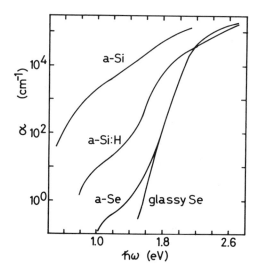

Fig. 15 Optical absorption edges of a-Si, a-Si:H, and a-Se films and bulk Se glasses.

annealed away by heating the films to 175°C. They tentatively attributed the effect to the Si dangling bonds which are created by photoinduced breakdown of Si-Si bonds. There is also a possibility that impurities are responsible for the effect.

Optical absorption edges of a-SiC:H and a-SiGe:H films have been studied by PDS and PAS in recent years (Skumanich *et al.*, 1985; Boulitrop *et al.*, 1985; Asano *et al.*, 1986). These alloys exhibit a wide range of energy gaps from 1 to 3 eV, which may produce versatile applications of these materials. However, the optical studies reveal that new kinds of defects seem to exist in these alloys. It is plausible that in glassy alloys the compositional disorder becomes another factor influencing electronic properties.

5.   CHALCOGENIDES

The first application of PAS to chalcogenide thin films, evaporated a-Se, was reported by Ceasar *et al.* (1984). For chalcogenide materials, bulk glasses quenched from melts are available so that comparison of characteristics between glassy samples and deposited films is of interest. Fig. 15 shows the result of these authors together with that from glassy Se obtained with conventional transmission measurements by Siemsen and Fenton (1967). We see that the two results agree well when $\alpha \gtrsim 10^1$ cm^{-1}, and that an excess absorption at around 1.4 eV is detected in the films. Whether this excess absorption is intrinsic to evaporated Se or not should be investigated further.

It is interesting to note in Fig. 15 that the Urbach parameter for a-Se, $E_0 \approx 60$ meV, is comparable to that of highly passivated a-Si:H, *ca.* 70 meV.

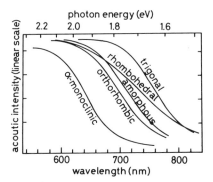

Fig. 16 Photoacoustic spectra of various forms of Se.

Weinberger *et al.* (1984) discovered by using PDS that E_0 in polyacetylene is also 70 meV. This agreement may be coincidental, or it may imply an essential feature of random systems such that an ultimate degree of randomness built into amorphous semiconductors is unique. As is described in Section III-3, an empirical proportionality between the average coordination-number Z and E_0 has been known. The coordination number for H atoms in a-Si:H can be assumed to be unity, so that it is calculated that Z=3.4 in a-Si:H containing 20 atomic % H. Nonetheless, E_0 of the a-Si:H film is similar to that of a-Se (Z=2). Thus, the empirical relationship breaks down here.

The optical absorption edges of a-Se and various forms of Se crystals have been studied by Nagata *et al.* (1985). Fig. 16 shows their results. Since most of the Se crystals are obtained as minute flakes, conventional transmission measurements are difficult to be applied, therefore PAS was employed instead. It is seen in Fig. 16 that the various Se crystals differ in the position of the absorption edge by ~ 0.3 eV. These workers also find that the locations of the absorption edges in the Se crystals correlate with the macroscopic densities, varying between 4.40 g/cm³ for α-monoclinic Se and 4.81 g/cm³ for trigonal Se. The increase in the density can be related to the shrinkage in the intermolecular distance, which broadens with width of the valence band (Fig. 8), resulting in a decrease of the band gap. They argue, therefore, that in Se the bandgap decreases with increasing intermolecular interaction. As is known, however, the density of a-Se is the smallest, ~ 4.29 g/cm³, although the optical absorption edge is located at a position similar to that of orthorhombic Se. This fact indicates that randomness is another factor determining the location of the absorption edge. If we assume that the energy gap of a-Se is governed by microscopic density fluctuations, the result of Nagata *et al.* (1985) implies a fluctuation of ~ 10%. This value seems to be reasonable.

One of the strengths of PAS, namely its applicability to the investigation of samples of various shapes, is also demonstrated in the following studies. Bhatnagar *et al.* (1985) investigated the optical energy gap of a-Se powders as a function of annealing temperature. On annealing, the gap decreased linearly

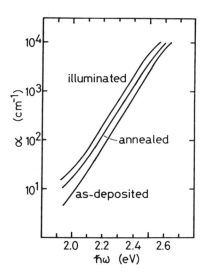

Fig. 17 Optical absorption edges for several states of As$_2$S$_3$ films of ~ 1 μm thickness evaluated by using PTS.

toward that of crystalline Se. Tamura *et al.* (1986) studied PAS signals from iso-lated Se chains contained in channels of mordenite crystals in a temperature range between 20 K and 300 K. They found that an additional absorption around the optical absorption edge was induced by photoillumination. The induced absorption which was observed only at low temperatures may be con-nected with the photoinduced mid-gap absorption originally discovered by Bishop *et al.* (1975). Further thermal wave studies have been valuable in shed-ding some light on the mechanism responsible for creating photoinduced defects. Harris and Tenhover (1986) recently demonstrated that PAS investiga-tion is capable of measuring the optical absorption edges of chemically unstable glasses such as Si-Se and Si-S systems. The absorption spectra could be evaluated by using samples with irregular shape.

Fig. 17 shows the optical absorption edges of As$_2$S$_3$ films investigated with PTS at room temperature by the present author[*]. As is seen, the Urbach tail is more gradual than that of the bulk materials (Fig. 9). This characteristic is attributable to homopolar bonds present in stoichiometric compounds (Street *et al.* 1978; Tanaka *et al.* 1985). The figure shows that the optical absorption edge of as-evaporated films shifts to lower photon energies by illumination or by annealing at the glass-transition temperature, ~180°C. The change between the illuminated and the annealed state can be repeated reversibly, the phenomenon known as the reversible photodarkening (Elliott, 1984). The result shown in Fig. 17 reveals that the photoinduced red-shift in the absorption edges extends down to ≥ 10 cm^{-1}.

---

[*]   The details will be publised elsewhere.

A comparison of the photoeffects in $a-As_2S_3$ and a-Si:H films is intriguing. In the former the optical absorption edge is red-shifted (Fig. 17), and in contrast a-Si:H exhibits the photoinduced mid-gap absorption (Fig. 14). As is noted above, a similar mid-gap absorption is induced in chalcogenide materials when exposed to light at low temperatures (Bishop *et al.*, 1975; Tamura *et al.*, 1986). These mid-gap absorptions can be attributed to creation or structural transformations of defects (Mott and Davis, 1979). On the other hand, the red-shift accompanying the reversible photodarkening is thought of to originate from enhanced randomness of normal bonding structures, specifically from variations of the intermolecular distance by $\sim 10\%$ (Tanaka, 1986). The photodarkening is, accordingly, intimately connected to the low coordination numbers of chalcogenide glasses.

IV. Summary and Future Perspectives

In this Chapter it has been demonstrated that the thermal wave spectroscopic methods are powerful for evaluating small absorptions, $\alpha d \gtrsim 10^{-4}$, in thin solid films. It has also been shown that PAS is especially useful for spectroscopic measurements of media with various forms. When they are applied to amorphous semiconductors, these methods give valuable information on band tails and mid-gap absorptions. This application, however, has appeared only in recent years, and much work remains to be done.

There seems to be no study available that closely compares the results obtained for a single material via several thermal methods. PAS and PDS signals are supposed to be sensitive to texture of samples. Thus, surface roughness and inhomogeneity in films including compositional variation with depth from surfaces may substantially influence the data. In contrast, PTS probes the temperature rise at the bottom sides, and the surface effects may be detected differently from other methods. A careful comparative study is needed in order to obtain more detailed and reliable results and to clarify distinct characteristics among the various thermal wave methods.

The sensitivity attained so far is nearly satisfactory at present, and future progress will be concentrated on versatile applications. One of these is to measure the dependence of the optical absorption on temperatures ranging between cryogenic temperatures to ~ 700 K. Measurements at low temperatures yield the relation between radiative and non-radiative recombinations, in which only the latter part can be detected by the thermal spectroscopic methods (Jackson and Nemanich, 1983). High-temperature studies of small optical absorptions will reveal annealing kinetics of defects and effects induced at around the glass transition and crystallization temperatures.

The thermal wave methods will be widely employed to evaluate the photoinduced absorption. In this respect, it may be worthwhile to note that in PAS and pyroelectric PTS only the response induced with chopped irradiation can be detected. In other words, these are insensitive to CW illumination. (Chopping illumination is not essential in PDS. We utilize it to enhance the sensitivity *via* synchronous detection.) Therefore, intense CW light which can induce the additional absorption may be directed to a sample during a thermal spectroscopic measurement. The results for the absorption induced under illumination are valuable to the investigation of defects, the nature of which is the subject of considerable debate.

V. References

Adler, D. (1984). In *"Semiconductors and Semimetals"* Vol. **21A** (J.I. Pankove, ed.), Academic Press, New York, pp. 291-318.

Amer, N.M., Skumanich, A., and Jackson, W.B. (1983). Physica **117B & 118B**, 897.

Amer, N.M. and Jackson, W.B. (1984). In *"Semiconductors and Semimetals"* Vol. **21B** (J.I. Pankove, ed.), Academic Press, New York, pp. 83-112.

Asano, A., Ichimura, T., Sakai, H. and Uchida, Y. (1986). Jpn. J. Appl. Phys. **25**, L388.

Bar-Yam, Y. and Joannopoulos, J.D. (1986). Phys. Rev. Lett. **56**, 2203.

Bhatnagar, A.K., Venugopal R.K., and Srivastava, V. (1985). J. Phys. D **18**, L149.

Bishop, S.G., Strom, U. and Taylor, P.C. (1975). Phys. Rev. Lett. **34**, 1346.

Boulitrop, F., Bullot, J., Gautier, M., Schmidt, M.P. and Catherine, Y. (1985). Solid State Commun. **54**, 107.

Ceasar, G.P., Abkowitz, M. and Lin, J.W-P. (1984). Phys. Rev. B **29**, 2353.

Cody, G.D. (1984). In *"Semiconductors and Semimetals"* Vol. **21B** (J.I. Pankove, ed.), Academic Press, New York, pp. 11-82.

Cohen, M.H., Economou, E.N. and Soukoulis, C.M. (1983). Phys. Rev. Lett. **51**, 1202.

Crandall, R.S. (1984). In *"Semiconductors and Semimetals"* Vol. **21B** (J.I. Pankove, ed.), Academic Press, New York, pp. 245-295.

Daniel, M.F. and Leadbetter, A.J. (1980). J. Non-Cryst. Solids **41**, 127.

Dersh, H., and Amer, N.M. (1985). J. Non-Cryst. Solids **77 & 78**, 615.

Elliott, S.R. (1984). *"Physics of Amorphous Materials"*. Longman, London.

Fritzsche, H. (1980). In *"Fundamental Physics of Amorphous Semiconductors"* (F. Yonezawa, ed.) Springer Verlag, Berlin, pp. 1-13.

Gal, M., Ranganathan, R. and Taylor, P.C. (1985). J. Non-Cryst. Solids **77 & 78**, 543.

Glass, A.M. and Lines, M.E. (1976). Phys. Rev. B **13**, 180.

Hamakawa, H. (ed.) (1982). *"Amorphous Semiconductors"*. Ohm, Tokyo and North-Holland, Amsterdam.

Harris, J.H., and Tenhover, M.A. (1986). J. Non-Cryst. Solids **83**, 272.

Jackson, W.B., Amer, N.M., Boccara, A.C. and Fournier, D. (1981). Appl. Opt. **20**, 1333.

Jackson, W.B. and Nemanich, R.G. (1983). J. Non-Cryst. Solids **59 & 60**, 353.

Kastner, M. and Fritzsche, M. (1978). Philos. Mag. **B 37**, 199.

Kolobov, A.V. and Konstantinov, O.V. (1979). Philos. Mag. **B 40**, 475.

Kosek, F., and Tauc, J. (1970). Czech. J. Phys. **B 20**, 94.

Mott, N.F. and Davis, E.A. (1979). *"Electronic Processes in Non-Crystalline Materials"*, 2nd ed., Clarendon Press, Oxford.

Nagata, K., Miyamoto, Y., Nishimura, H., Suzuki, H. and Yamasaki, S. (1985). Jpn. J. Appl. Phys. **24**, L858.

Olivier, M. (1984). In *"New Directions in Guided Wave and Coherent Optics II"* (D.B. Ostrowsky and E. Spitz, eds.). Martinus Nijhoff Publishers, The Hague, pp. 639-657.

Olivier, M., and Bouchut, P. (1981). J. Phys. Suppl. **C4**, 305.

Phillips, J.C. (1979). J. Non-Cryst. Solids **34**, 152.

Phillips, W.A. (ed.) (1981). *"Amorphous Solids, Low-Temperature Properties"*. Springer-Verlag, Berlin.

Ritter, D. and Weiser, K. (1986). Opt. Commun. **57**, 336.

Rosencwaig, A. (1980). *"Photoacoustics and Photoacoustic Spectroscopy"*, John Wiley & Sons, New York.

Sasaki, K., Takahashi, H., Kudo, Y. and Suzuki, N. (1980). Appl. Opt. **19**, 3018.

Siemsen, K.J. and Fenton, E.W. (1967). Phys. Rev. **161**, 632.

Skolnik, L.H. (1975). In *"Optical Properties of Highly Transparent Solids"* (S.S. Mitra and B. Bendow, eds.), Plenum, New York, pp. 405-433.

Skumanich, A., Frova, A. and Amer, N.M. (1985). Solid State Commun. **54**, 597.

Spear, W.E. and LeComber, P.G. (1975). Solid State Commun. **17**, 1193.

Staebler, D.L. and Wronski, C.R. (1977). Appl. Phys. Lett. **31**, 292.

Street, R.A., Nemanich, R.J. and Connell, G.A.N. (1978). Phys. Rev. **B 18**, 6915.

Tajima, M., Okushi, H., Yamasaki, S. and Tanaka, K. (1986). Phys. Rev. **B 33**, 8522.

Tamura, K., Hosokawa, S., Endo, H., Yamasaki, S. and Oyanagi, H. (1986). J. Phys. Soc. Jpn. **55**, 528.

Tanaka, K. (1980). Thin Solid Films **66**, 271.

Tanaka, K. (1986). Jpn. J. Appl. Phys. **25**, 779.

Tanaka, K. and Ohtsuka, Y. (1977). J. Opt. (Paris) **8**, 27.

Tanaka, K. and Yamasaki, S. (1982). Solar Energy Mater. **8**, 277.

Tanaka, K., Satoh, R. and Odajima, A. (1983). Jpn. J. Appl. Phys. **22**, L592.

Tanaka, K., Gohda, S. and Odajima, A. (1985). Solid State Commun. **56**, 899.

Tanaka, K., Sindoh, K. and Odajima, A. (1986). Rept. Progr. Polym. Jpn. (in press).

Tauc, J. (1975). In *"Optical Properties of Highly Transparent Solids"* (S.S. Mitra and B. Bendow, eds.), Plenum, New York, pp. 245-260.

Weinberger, B.R., Roxlo, C.B., Etemad, S., Baker, G.L. and Orenstein, J. (1984). Phys. Rev. Lett. **53**, 86.

Won, Y.H., Jaussand, P.C. and Chartier, G.H. (1980). Appl. Phys. Lett. **37**, 269.

Wood, D.L. and Tauc, J. (1972). Phys. Rev. **B 5**, 3144.

Yamasaki, S., Okushi, H., Matsuda, A., Oheda, H., Hata, N. and Tanaka, K. (1981). Jpn. J. Appl. Phys. **20**, L665.

Yonezawa, F. and Cohen, M.H. (1981). In *"Fundamental Physics of Amorphous Semiconductors"* (F. Yonezawa ed.), Springer-Verlag, Berlin, pp. 119-144.

Zallen, R. (1983). *"The Physics of Amorphous Solids"*, John Wiley & Sons, New York.

Guide to Abbreviations

PA(S)	:	Photoacoustic (Spectroscopy)
PZT	:	Lead Zirconate Titanate (Piezoelectric)
PAM	:	Photoacoustic Microscopy
PBD	:	Photothermal Beam Deflection
PBDM	:	PBD Microscopy
OBD	:	Optical Beam Deflection
FFT	:	Fast Fourier Transform
NDE	:	Nondestructive Evaluation
PMOS	:	P-channel Metal-Oxide-Semiconductor
MOSFET	:	Metal-Oxide-Semiconductor Field Effect Transistor
PTR	:	Photo-Thermal Radiometry
PL(S)	:	Photoluminescence (Spectroscopy)
EPD	:	Etch-pit Density
$P^2E(S)$	:	Photopyroelectric (Spectroscopy)
PVDF	:	Polyvinylidene Difluoride (also PVF_2)
PD(S)	:	Photothermal Deflection (Spectroscopy)
PT(S)	:	Photothermal (Spectroscopy)
PEC	:	Photoelectrochemical
PC(S)	:	Photocurrent (Spectroscopy)

Index

471